Environment
and
Plant Ecology

Environment
and
Plant Ecology

JOHN R. ETHERINGTON

Lecturer in Botany,
University College of South Wales, Cardiff

with a contributed chapter by

W. ARMSTRONG

Lecturer in Botany,
University of Hull

JOHN WILEY & SONS

London · New York · Sydney · Toronto

Library of Congress Cataloging in Publication Data:

Etherington, J. R.
Environment and plant ecology

1. Botany—Ecology. I. Title

QK901.E85   1974   581.5   74-3725

ISBN 0 471 24615 8 (cloth)

ISBN 0 471 99737 4 (Pbk.)

Text set in 11/12 pt. Photon Baskerville, printed by photolithography
and bound in Great Britain at The Pitman Press, Bath

Preface

The terrestrial plant, rooted on one spot and originating from seed which may be distributed almost at random, is at risk to a whole range of environmental hazards which a mobile organism can escape. From the moment that germination begins, the developing plant must cope with soil chemical and physical conditions, climatic and microclimatic effects, pathogens and herbivores. The germinating seed carries a specific blueprint of genetic data which dictates its response to a complex of environmental pressures. If the genetic scope for phenotypic plastic response is exceeded by one, or more, of these pressures, then germination or establishment will fail. Another seed of the same species, but carrying a different set of gene recombinants, may survive, but, if it cannot, it may be that the environmental conditions are entirely outside the genetic capability of the species.

Within the first few days after germination, even though the arrival of the seed was a random event, a degree of ecological pattern is generated. One species survives, another is lost; one seed dies but its genetic brother, differing in a few genes, persists, perhaps as the progenitor of a new ecotypic stock. With time, the microclimate oscillates and alters, the plants grow and change in their morphology and physiology and, as the seedlings occupy more space, so competition begins. Competitive interaction causes not only visible change but also distorts 'normal' physiological responses, sometimes exceeding their limits. By the time the first seeds have developed to adult plants, an interacting web of environmental and biological limitations has either dictated the survival of individuals or at least profoundly influenced their phenotypic development.

At any time, the ecosystem is a legacy of previous events: its component species, their morphology and physiology, the distribution of soil organic matter and inorganic nutrient elements, all of these factors and others preserve evidence of past development. As an ecosystem grows and stabilizes the wastage rate is enormous and, for every plant which survives to reproduce, perhaps many thousands are lost in failure to germinate, attack by herbivore or pathogen, killing by drought or frost, competitive exclusion or a multitude of other environmental effects. Out of this carnage emerges a group of individuals which have passed the genetic sieve of intense selection and whose offspring will, in turn, inherit combinations of attributes which best fit them to adapt to the localized environment and to withstand the competitive pressures of their neighbours. This complex species-environment interaction, coupled either with sharp geological or topographic discontinuity or with large-scale climatic gradients, produces the ecological associations so easily recognized by the field worker, in the first case as discrete communities and, in the second, as some form of continuum.

This book attempts to examine some of the details of this intricate jigsaw of plant, animal, microorganism and environmental interactions as it manifests itself in plant physiological ecology. The problem is to dissect the network of energy fluxes, elemental cycles and control systems and to present them in the light of the individual species' behaviour and competitive interaction. In a decade which has seen the increasing application of techniques for analysing continuous multiply interacting variables, combined with the introduction of systems analysis to ecological modelling, it might be thought unwise to attempt this dissection. It is the author's view, however, that the ultimate in ecosystem modelling is to describe not only general functional relationships but also the population dynamics of the species concerned.

The systems approach, at the present time, is admirably suited to the modelling of generalities such as whole ecosystem energy flow, water transport or nutrient cycling, but it is much less useful in describing species behaviour within the system. It is very easy to insert hypothetically feasible values in a general model of ecosystem nutrient cycling and to compare the computed results with the real world. It is much less easy to make the same approach at the individual species level without the backing of physiological–ecological experimentation. Unlike the whole–system situation, the requisite values for individual species modelling cannot be inferred from the general. Herein lies the need for a comparative experimental study of physiological ecology in the field and in controlled environmental conditions.

With the exception of the first chapter, this book is intended to provide a background of information concerning the environment and plant response which the undergraduate will not otherwise easily find in a single book. It has been fairly liberally laced with references which open the doors to more specialist textbooks and to some more important or interesting original papers. The first chapter is rather different, being a thumbnail sketch of the present situation in ecology. It also provides more general reference to those parts of plant ecology which are not discussed in detail in this text.

I would like to thank all of those colleagues and students who, by discussion and listening, have brought me to a fuller understanding of ecology. In particular, I cannot overstate the debt which I owe to teaching as an aid in marshalling ideas. The best learning aids are insatiable reading, note-taking and a captive audience! To countless authors, a word of thanks for the discovery that so many different viewpoints and interpretations exist, for me the greatest pleasure of literature research. I must also express my gratitude to Professor A. J. Rutter, from whom I first realized that physiological ecology would be more rewarding than a conventional 'lab. bench' study of physiology or biochemistry. As a consequence I enjoy a paid hobby; an immensely satisfying experience.

Inevitably the text must include errors and omissions for which I must be responsible. I shall be grateful for any comment or discussion from readers concerning future amendments.

University College of South Wales J. R. ETHERINGTON
Cardiff
May 1974

Contents

Chapter 1

The aims and development of plant ecology

The words *oikos* (house) and *logos* (study of) were first put together by Reiter in 1865 but Haekle, in 1866, defined ecology in its modern sense: 'the body of knowledge concerning the economy of nature—the investigation of the total relations of the animal to its inorganic and organic environment.' (Kormondy, 1965). Long before these words were written, throughout the course of history and prehistory, man had been aware of the mutual bond between himself and the organisms which gave him sustenance. The hunting life, slowly changing and evolving to primitive agriculture, bred an awareness of the changing seasons, the need of plants and animals for water, the sudden catastrophe of fire, flood and drought, the inevitable regrowth which followed them. Man, at the mercy of the elements, endowed them and the organisms of his environment with a mystical or religious significance which led to their conservation in the deepest sense of the word. Man in small groups, man with no possessions or resources, could not afford the vandalisms of community or individual which civilizations were to bring.

From these unspoken beginnings, the subject developed, and a body of knowledge has accumulated, particularly during the last century, which has revealed the web of immense complexity which is our ecosystem. Man as a species has broken the rules; has learned to postpone the inevitability of death and to avert the decimation at birth which is the lot of other animals. Human thought, more powerful than any living force so far liberated on this planet, has led to a source of destruction through population pressure which must be our source of ultimate concern. The advance of industrialized societies and their political–economic structure now comes into conflict with responsibility for the future, for population control and conservation of irreplaceable resources. Not surprisingly, ecology has been described as the 'subversive science' (Sears, 1964). Nicholson (1970) suggests that the delayed development of the subject was not entirely fortuitous, resources being devoted to the physical and chemical sciences which are 'less productive of thoughts dangerous to the "system" '. The paramount need of our time is to solve the problems of human ecology within a humanitarian framework in which thought, for the first time, must replace biological necessity.

HISTORICAL

Clements (1916, 1928) cites Petrus de Crescentius in 1305 as the first writer to appreciate the existence of plant competition and King, in 1685, as the first to describe the concept of succession. De Crescentius was aware, in forestry practice, of the 'choking' effect of competitive plants and, nearly 400 years later, in his account of the Irish boglands King described the succession from open water to a peaty terrestrial surface by plant colonization and accumulation.

The concept of succession, the progressive development of differing types of plant cover, found growing support in the 1800s and in 1891 Warming produced his classic description of the successional relationships in Danish dune systems. The sand dune formation provides an ideal setting for such work, presenting, as it does, soil surfaces of known chronology. Under Warming's influence Cowles (1899) published a series of papers describing the dunes of Lake Michigan; these, with Warming's work and the later publications of Clements were the seeds of much 20th-century ecological thought.

In 1905 Clements's text on research methods in ecology appeared and helped to establish the tradition of measurement, pioneering the use of the quadrat and the adoption of instrumental techniques for defining the habitat. A few years later Cowles (1911), in discussing vegetational cycles, realized the threefold importance of climate, physiography and biota in ecology and, in Britain, Tansley (1911) for the first time used the successional concept in describing a great vegetation (*Types of British Vegetation*). In its stressing of succession it is almost unrivalled and grew into the 1939 classic *The British Isles and their Vegetation*.

By 1916 and 1920 Clements was able to produce the two works, *Plant Succession* and *Plant Indicators,* which contain an enormous body of information and serve still to stimulate and amaze the most sophisticated of readers. A little later the trilogy was completed with *Plant Competition* (Clements, Weaver and Hanson, 1929). These works alone must place Clements in an unrivalled position as a pioneer of modern ecology.

During this same period plant physiology was also growing in stature, for example the textbooks of Pfeffer (1880) and Sachs (1887) portend much of the accomplishment of the next century. The relation between plant structure, function and distribution, the basis of modern ecology, was also well documented by the beginning of the present century. In 1884 Haberlandt published *Physiological Plant Anatomy* and Warming (1896) and

Schimper (1898) described the physiological foundation of plant geography. These three publications, like Pfeffer's and Sachs's textbooks, disclose countless ideas at each reading: it is a little humbling to find how much fundamental botanical thought stems from the great men of these early years.

During the same period, soil science grew from the concepts of Dokuchayev in the 1870s. Soils were conceived, for the first time, to be independent and unique entities resulting from the combination of climate, living matter, parent material, relief and time. By 1900 Dokuchayev had arrived at a 'final' classification of the soils of Russia. In Britain, Hall (1903) produced one of the first surveys of soil physical and chemical conditions as they relate to plant growth, but his concepts of classification were rudimentary by comparison with Dokuchayev's. This is probably a reflection of the limited climatic range in Britain, resulting in soil types which are more strongly differentiated in relation to parent material than to climate and contrasting with the Russian situation within which the genetic concepts of pedology evolved.

The relationship of climate and vegetation to soil formation was an essential part of Coffey's (1912) classification of North American soils and by 1932, in Britain, Robinson had written *Soils, their Origin and Classification* in which he fully realized that neither climate nor geology alone could be a sufficient basis for a classification of soils or an explanation for their genesis. Joffe (1936) produced a text on pedology which, in essence, contained most of the still current ideas on the interaction of climate, geology, topography and biosphere. From that time onward the majority of advances in pedology have been those of detail rather than formulations of new concepts.

The simultaneous advance of ecology, physiology, agricultural science and pedology during the first half of the 20th century resulted in many diverse attempts to classify the complex systems of organism and environment and to analyse the functions and interrelationships of single species within the ecosystem. With the advent of more sophisticated measuring devices and increasing availability of computer facilities there is growing interest in whole-system analysis: it seems likely that ecology during the remainder of this century will see, at one extreme, the detailed analysis of physiological and genetic characteristics which fit individual species for life in circumscribed habitats and, at the other, attempts to produce mathematical models of whole ecosystems utilizing quantitative information drawn from field studies and controlled environment experiments.

THE CURRENT SITUATION

The synthesis of ideas which has emerged and developed from these early works has been much influenced by the concepts and techniques of individual workers or research groups. In some cases investigations have continued for 50 years or more in a particular institution and very often modes of research have been disseminated to other centres by the emigration of individual workers. As a result, the subject has advanced on different fronts which may be considered in three main divisions. (i) The classificatory front, stimulated perhaps by the early plant geographers and latterly coming under the influence of growing statistical theory. (ii) The physiological–ecological front stemming from the early influence of Cowles and Clements in the U.S.A. and from Warming, Schimper, Haberlandt and many others in Europe. (iii) The ecosystem approach in which all components, physical, chemical or biological, are treated in their total interdependent complexity. Some early ecosystem analyses were made, for example Lindeman's (1942) classic paper on ecosystem energy flow and trophic structure, but here only one factor of the system was analysed. Attempts to analyse more extensive relationships in ecosystems have awaited easy access to computing facilities and the development of the multivariate procedures necessary to define the interactions of the various parameters. Full ecosystems analysis is still in its infancy and is likely to take its place in the forefront of ecological advance during the next few decades. Growing population pressure and the need for decisions concerning the management and conservation of natural resources have been the greatest driving forces in the introduction of systems analysis to ecology, the techniques having proved their value to controlled decision-making in the industrial–economic context.

CLASSIFICATION

During the early part of this century Clements's (1916) concepts of succession and habitat relationships were brought together in a philosophy of vegetation study which centred on the assumption that the plant community behaved as an organismic whole: 'The developmental study of vegetation necessarily rests upon the assumption that the unit or climax formation is an organic entity. As an organism the formation arises, grows, matures and dies'.

This view was in direct opposition to that of Gleason (1926), who considered that the apposition of species within the community was a random occurrence and rejected Clements's organismic hypothesis in favour of the individualistic concept: 'The vegetation of an area is merely the resultant of two factors, the fluctuating and fortuitous immigration of plants and the equally fluctuating and variable environment'.

This divergence of opinion led to a polarization of thought which, in a sense, has been perpetuated to the present day. On the one hand the Clementsian organism is reflected in all of those attempts to establish hierarchical classifications of plant associations. These attempts imply the existence of discrete communities which have a degree of internal organization and abut on other communities with sharp delimitation or, at the most, shallow ecotones or 'zones of tension'. On the other hand, the Gleason concept implies no internal organization within the plant community and suggests that composition will change as a continuous reflection of changing environmental factors. Sharp discontinuities will occur only either when geology or topographic factors impose sudden changes of environment or as a consequence of invasion of one community by another, an immature situation.

Tansley's (1911; 1939) descriptions of the British vegetation and his modified concept of the 'pseudo-organism' bear the mark of the Clementsian approach as perhaps do the later statistical attempts to establish hierarchical classifications. The association analysis technique (Williams and Lambert, 1959), for example, used a matrix of species–pair correlations established for occurrence in quadrats. These were tested against a random distribution hypothesis using the χ-squared test which thus quantifies the degree of interaction between species. High values are found for strongly associated species and for species which rarely appear together. The highest value of χ-squared, summed for each species irrespective of positive or negative association, is used to establish a primary split into one group with, and one group without, species x: this species is considered to be a key factor for the classificatory splitting as it shows the greatest degree of interaction with all other plants in the sample quadrats. The process is repeated to split and subsplit these groups, giving a branching hierarchy of subgroups defined by presence or absence of the key species. The basic philosophy of this type of classification closely resembles the Clements–Tansley system in which associations were split according to their dominants taken in relation to habitat and seral stage.

At the opposite pole, Gleason's work and the rise of the individualistic concept led, in N. America, to a questioning of the reality of the plant association and the validity of hierarchical classifications. The concept of continuous variation, with habitat factors and time, stimulated the first description of continuum analysis (Curtis and McIntosh, 1951) but by 1957 the inadequacy of a linear continuum had been realized in the adoption of a pluridimensional approach, or ordination (Bray and Curtis, 1957). Lambert and Dale (1964) pointed out that this early ordination was preceded by Goodall's (1954) technique of factor analysis which was, in fact, a principal component ordination.

The last decade has seen the detailed consideration and testing of a number of ordination techniques: early ordinations used various coefficients of stand similarity to extract axes of variability and some attempt was made to interpret these axes as reflections of environmental variables. Orloci (1966) described an ordination as a summarization of the information content of a matrix by the projection of the points into a space which has less dimensions than the original. The Bray and Curtis simple ordination establishes axes of variability by reference to the most dissimilar stands and, as a result, a single misplaced stand can cause serious distortion. Principal component analysis overcomes this problem by establishing ordination axes which reflect maximum variation between all stands in the data cluster. As a result it is less seriously affected by the presence of a single aberrant stand.

During the early years of quantitative ecology in the U.S.A. and Britain similar developments took place in Europe. Whittaker (1962), in a detailed review of the profuse literature of classification, makes a separation into Southern and Northern European and Russian classificatory traditions which are contrasted with those of Britain and America in Table 1.1. The strongest features differentiating the European approach from the Anglo–American are the adoption of fidelity or constancy rather than dominance as association criteria, and the collection of stand data into association tables from which abstract definitions of the associations are constructed. There has, in the past, been considerable criticism of the subjective nature of the data collection of the Braun-Blanquet school of phytosociology, but for rapid description of the vegetation of a large formation or extensive continental areas it is extremely efficient. The techniques are described by Braun-Blanquet (1932) and reviewed at length by Poore (1955a, b, c; 1956). General descriptions of the whole range of classificatory techniques may be found in Greig-Smith (1964), Kershaw (1973) and Pielou (1969).

PHYSIOLOGICAL ECOLOGY

Clements (1920) wrote: 'Every plant is a measure of the conditions under which it grows. To this extent it is a measure of soil and climate'. From this concept he developed his very detailed appreciation of plants as indicators of a host of environmental variables. In surveying the history of the subject he noted that it could have made little headway until the foundations of plant physiology were established but by the beginning of the 20th century it had entered a rapid growth phase which continues to the present day.

Limiting factors
Liebig (1840) first drew scientific attention to the subject of mineral nutrition and formulated what has subsequently been called the 'Law of the Minimum': 'by the deficiency or absence of one necessary constituent, all the others being present, the soil is rendered barren for all those crops to the life of which that one constituent is indispensable'.

Blackman (1905) crystallized this notion in quantitative form by approaching all physiological factors, photosynthesis in particular, with the generalized van't Hoff chemical rule that for every 10°C rise in temperature the rate of reaction is about doubled or trebled. He found that this was not

Table 1.1 The geography of ecological classification

U.S.A.	Britain	S. Europe	N. Europe	Russia
Early development of a cause—effect philosophy and adoption of Clementsian notions. Units Formation—an organic entity, based on climatic climax, which grows, matures and dies. Subdivided into: Associations—defined by dominant species. Subdivided into: Consociations—defined by single dominants. Subdivided into: Societies—defined by subordinate species. Accepted until about mid 1930s followed by considerable experiment with quantitative methods. Development of continuum and ordination methods in late 1940s and to present day.	Most extensively influenced by Tansley. Formation definition similar to Clements' and subdivisions into association, consociation and society but these include all mature major communities—not just the climatic climax communities of the Clements technique. Definition by dominants. 1950s onward—experimentation with phytosociological methods leading to the description of the Scottish vegetation by McVean and Ratcliffe (1962) and Burnett (1964). Also widespread experimentation with various forms of statistical approach e.g. association analysis and ordination in the 1960s.	Mainstream of phytosociological thought from schools of Zurich and Montpellier — based on rich Alpine and Mediterranean vegetation. Early adoption of characteristic species approach. (Species of low amplitude confined to one or two associations.) Characteristic described by fidelity. Attempts to define community units on a floristic basis, establishing units comparable to genera and species in taxonomy. Association tables for many stands are brought together to define abstract associations by characteristic species.	Extensively influenced by Uppsala school of phytosociolqgy. Use of small quadrats and definition of normal vegetational units by stratal structure and concept of constancy. (Species present in 90%† of stands.) Formations classified by physiognomy without reference to species. Associations uniform in both physiognomy and stratified species structure. Leads to definition of numerous small associations which are regarded as more or less real and as having fairly sharp boundaries. Close attention is paid to minimal area for stand sampling as constancy is a function of sample size up to minimal area.	Difficult to generalize but development has followed a pattern of relating vegetation to environmental gradients and definitions of biocoenoses—complexes of organisms in critical relationship to environment.

†The tabulated data affords only a brief summary of the various approaches to vegetational classification. Details of technique may be found in Braun-Blanquet (1932), Grieg-Smith (1964), Kershaw (1973) and Pielou (1969). An extensive and critical discussion of the philosophy of classification (Whittaker, 1962) was the source of this summary.

true for organisms as the rate approaches a maximum and then either remains constant or falls off again with increasing temperature. To explain the relationship he propounded the 'Law of Limiting Factors': 'When a process is conditioned as to its rapidity by a number of separate factors, the rate of the process is limited by the rate of the "slowest" factor'. He illustrated this relationship by considering the limiting effect of CO_2 availability on photosynthesis and plotted hypothetical curves for different light intensities and CO_2 concentrations (Figure 1.1). The sharp inflections shown by these curves do not occur in reality, a more characteristic form being that of the pecked empirical curves: the transition zone of changing gradient of the curve being related to the exponential nature of the limiting

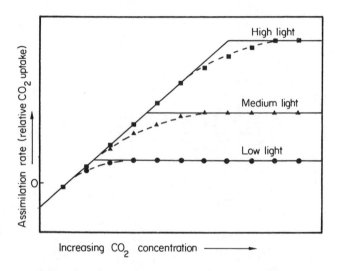

Figure 1.1 The relationship of photosynthetic carbon assimilation to ambient carbon dioxide concentration. The unbroken lines represent the hypothetical Blackman relationship and the pecked lines are based on experimental results. Rising carbon dioxide concentration increases the photosynthetic rate until the light intensity becomes limiting: at this point the curve inflects and further increase of carbon dioxide concentration has no effect unless the light intensity is raised.

process. Blackman's consideration of high temperature, supraoptimal conditions, showed that the law of limiting factors is valid not only for minimal supply conditions but also in circumstances where physiological tolerance of, or capacity for, a high level of some external factor is exceeded (light, temperature, nutrients at toxic levels, toxins etc.).

The concept serves to delimit, for each genotype-phenotype, a response-range to the environmental complex which may be described as its physiological tolerance. If the environmental factors are plotted as multidimensional axes then, for each species, there will be a circumscribed hypervolume of tolerance in which the plant can survive. The concept has been used by Hutchinson (1965) to define the term *niche* but in the ecological sense the effects of competition and predation must be considered, hence a plant may be confined to a niche with a less extensive hypervolume than that defined by its physiological range and its niche boundaries may even be displaced from the physiologically satisfactory range. As Major (1969) has written: '. . . plant physiology does not deal very much with ecosystems and it deals with them less daily. Plant competition is an unmentioned physiological fact'. Figure 1.2 illustrates, in simplified form, the concept of a three-dimensional physiological and ecological tolerance volume; a good ordination of environmental

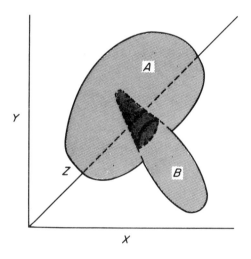

Figure 1.2 Tolerance volumes. The volumes *A* and *B* represent physiological tolerance volumes of two species. The portion of *B* which is included in the volume of *A* (dotted) shows the loss, by competitive exclusion, in the ecological relationship between the two species. *X*, *Y* and *Z* are three environmental variables: in the multivariate system physiological and ecological tolerances may be represented by *n*-dimensional hypervolumes.

parameters might result in the same type of representation. Multidimensional representation which is needed to cope with the immense range of environmental variables takes the concept beyond visualization into the realm of multifactorial analysis.

The limiting factor approach has led to the development of traditions of autecology in which the behaviour of a single species, or a small number of species taken comparatively, is assessed in physiological detail in the hope that relationships to environmental gradients may be defined.

Analysis of growth

A great deal of work has been done in the autecological–physiological field during the last two decades and a great bulk of information has accumulated. By its very nature the work requires specialists in many fields and, consequently, the literature is fragmentary, single species having been studied in relation to a few factors. The literature is filled with accounts of a single species' response, for example, to light, to nutrient levels: to the whole host of environmental variables. The approach is further overspecialized as most of its foundations lie in the work of the early agricultural physiologists, the majority of the techniques having been developed in this field.

The early work of Blackman (1919) on the analysis of plant growth stimulated a whole series of detailed and critical plant physiological studies which have contributed greatly to our knowledge and thought in this field. Blackman described growth in terms of the 'Efficiency Index' which he derived from Kelvin's Law of compound interest. In concept it drew attention to the fact that plants gain weight in proportion to their existing weight capital and thus show exponential weight increase during periods of unlimited growth. West, Briggs and Kidd (1920) extended the concept and provided the classic basis for the growth analytical approach. Overall growth was defined by the Relative Growth Rate, a product of Leaf Area Ratio and Unit Leaf Rate (subsequently termed Net Assimilation Rate).

F. G. Gregory, who worked with Blackman during the development of the efficiency index concept, was subsequently responsible, with many coworkers, for the pursuit of a range of investigations of plant nutrition, photoperiodism and thermoperiodism, often using growth analytical techniques. The bulk of this work was done before controlled environments became available and for this reason they pioneered the use of multifactorial experiments and realized the importance of massive replication of treatments. The introduction of statistical methods played a prominent part in the evolution of the work and was particularly influenced by Fisher (1925) at Rothamsted Experimental Station, where much of the field work was carried out. Regression and multiple regression provided a powerful tool for identifying correlations in multifactorial work. At the same time, analysis of variance and the blocking of experimental designs were introduced. From this time onward the use of such techniques has become commonplace in all aspects of biology and medicine and is still spreading into other spheres of science and industry.

Concurrently with the utilization of growth analytical techniques a great deal of definitive work was done on the measurement of photosynthesis, respiration and biochemical composition of plants in relation to varied environments, laying an extensive foundation for subsequent developments in whole-plant physiology and physiological ecology. A series of autecological investigations was initiated within this intellectual climate, exemplified by the work of Blackman and Rutter (1959: summary of four papers) on the factors influencing the growth and development of *Endymion non-scripta*. The authors stressed the importance of linking general ecological observation to precise field studies and multifactorial experiments.

Developing from the growth analytical approach, studies have been made of the behaviour of the whole crop. This has been much assisted by Watson's (1947) concept of the 'Leaf Area Index': the ratio of total leaf area to ground area (abbreviated LAI). As Watson noted, the measure which is relevant to plant dry matter accumulation is the integral of LAI with time, since this takes into account both leaf area and its persistence. The concept has been studied extensively, particularly in relation to light penetration and competition for light (Donald, 1963). Further complications are caused by variation in canopy architecture, the phyllotaxis and angle of leaf insertion affecting the optimal LAI at different light intensities (Loomis, Williams and Duncan, 1967). One current breeding programme for higher crop efficiency centres around the LAI/angle relationship, high leaf angle permitting better light penetration and higher leaf area carrying capacity (Army and Greer, 1967).

Controlled environments

During the late 1930s and after the war, approaches to whole-plant physiology and physiological ecology were revolutionized by the development of controlled environment facilities which liberated the experimenter from the vagaries of the weather and enabled him to replicate in time as well as space. The best known early work, the development of the Pasadena Earhart Plant Research Laboratory, has been described in some detail (Went, 1957). By systematic experimentation the Pasadena Labs. have established 'physiological profiles' for a number of species chosen either for their commercial value or their inherent suitability for research on such problems as photoperiodism. The needs of research also initiated a search for 'botanical *Drosophilas*', plants of small size and rapid life cycle which permit the greatest return from the smallest space investment.

Went argued that batteries of controlled environment units facilitate much more than the mere physiological 'fingerprinting' of plants. They also provide circumstances in which ecology can become a precise experimental science and the worker no longer has to guess about the influence of uncontrolled environmental fluctuation. More recently Lang (1963) and Evans (1963) have discussed the role of 'phytotrons' in research and the difficulty of extrapolating results to the field. Two of the greatest difficulties which arise are the simulation of the spatial diversity of habitat microclimates and the physical problems of providing a satisfactory long- and shortwave energy environment. Profiles of wind speed, light, temperature, water and CO_2 develop in the field and plants are subject to varying levels of longwave input and output, causing heat-loading effects which do not arise in fluorescent-lit cabinets. Some attempts have been made to simulate the real environment energy balance but it is not easily achieved, either in spectral quality of the radiant input or in disposal of surplus heat within the environmental chamber.

One of the first publications from the Pasadena phytotron was an account of the temperature and photoperiod requirements of a range of Californian annuals (Lewis and Went, 1945) followed by Clausen, Keck and Hiesey's (1948) study of ecotypic variation in geographical races of *Achillea millefolium*. These factorial experiments with differing day lengths and temperature regimes established the importance of night temperature to the growth and survival of some plants while the study of *A. millefolium* permitted the isolation of ecotypes which could be precisely defined in terms of day length and diurnal temperature regime. Such experimental work provides physiological tolerance profiles which can be measured in no other way and is essential for the understanding of plant behaviour in the context of the ecosystem. The problem is that, for efficient research, the units are necessarily large and costly. Lang (1963) has suggested that there is a minimum size for such units below which efficiency is lost and noted that the Pasadena Laboratories have 50 controlled rooms and greenhouses. Clapham (1956), in a review of the current status of ecology, made the same point and suggested that a small number of such units should be established at various centres rather than dispersing large numbers of small and inefficient units more widely.

Physiological ecology in the field

Careful experimentation in the field and in controlled environments permits the specification of the minimum demands which a species will make on its habitat during its life-cycle. If conditions for germination are satisfied the seedling then proceeds to the establishment phase with a differing set of physiological requirements. As it grows and develops a more complex canopy, it encounters competition for light; below the surface the roots begin to compete with those of other individuals. The physiological tolerance range then becomes distorted by the influence of other plants and possibly by the selective effects of predation and disease.

All of these characteristics of the adult plant association take the investigator from the laboratory and growth chamber into the field, and require the detailed recording (often continuous recording) of many climatic and soil factors coupled with detailed observations of plant behaviour. The below-ground environment is most complex and difficult to monitor continuously though a few characteristics such as soil water potential and temperature are measurable in this way. Chemical and physical attributes are generally accessible only by spot sampling and destructive processing. None of the techniques of soil analysis are sufficiently precise, as yet, to specify the conditions at the absorbing root surfaces. Above ground the situation is less complex and, in consequence, micrometeorology has a fairly long history of satisfactory measurement, compared with attempts to measure short-term changes in the soil.

Geiger's classic *The Climate Near the Ground* was first published in German in 1927 and even at this early date showed that a great deal of knowledge had accumulated concerning the surface energy balance of the earth and the stratification of temperature, wind speed, water vapour and CO_2 above the surface. The remarkable increase in the availability of recording instruments, particularly the potentiometric recorder, since the war has redoubled efforts in this field. Penman and Long's (1960) essay on microclimate in a shallow crop canopy during three consecutive summers is an excellent example of the use of continuous recording methods and also drew attention to the great horizontal discrepancies in air parameters which make for difficulty in defining profiles. Lemon (1965) reviewed the use of micrometeorological data for calculating vertical transfer of mass, momentum and energy, thus providing the means of calculating the ecosystem CO_2 budget of assimilation/respiration and transpired water loss.

Despite the inability to record soil parameters continuously many workers have attempted to solve the problem by massive replication of spot samples in space an time. It has sometimes also been assumed that limited sampling serves to define some characteristics when low diffusion rates within the soil matrix prevent rapid changes. Such approximations may be valid for a large soil volume but must be suspect in considering the soil adjacent to the roots. The earliest work of this type was directed at plant species of obviously contrasted habitats, for example, the development of the calcicole/calcifuge problem, investigation of unusual soil–plant associations such as the serpentine flora and such obvious differences as those associated with soil wetness. Such investigations as these have utilized a whole armoury of chemical analytical, physicochemical, experimental ecological and pedological methods which have developed during the past century.

Pearsall (1964), in describing the development of British ecology, ranks the investigation of species–habitat relationships as second in chronology to classificatory techniques. He suggested that the use of correlation analysis between habitat and vegetation is the necessary precursor to experimentation. The most fruitful work of this type has embodied comparative studies of more than one species or ecotype so that differences in field distribution may be correlated with habitat factors and subsequently investigated with suitable experimental techniques.

Grime (1965a), in an excellent review of the value of the comparative experiment, suggests that it must necessarily contribute more to our knowledge of causal mechanisms than the simple establishment of spatial and temporal correlations which are subject to so many difficulties. The number of environmental variables which can be measured is limited; some are not yet precisely measurable (e.g. mineral nutrient availability); at community boundaries the changes of many variables are correlated with changes of vegetation but few are the cause of the change; many environmental variations are imposed by the plants rather than the reciprocal relationship; time lags in plant response may mask many environmental correlations and, finally, seedling and adult tolerances differ from each other. To confuse the field situation further the catastrophic event, though of rare occurrence, may be definitive in limiting the distribution of a particular species.

By contrast, the use of information acquired in the field to design comparative experiments permits the variation of one or more factors, while the background environment is held constant: each factor which is suspected of causality in the ecological response may be investigated and relative susceptibilities of various species investigated. A good early example of this type of study is seen in the work of Mooney and Billings (1961), with various arctic and alpine populations of *Oxyria digyna*. This plant has a circumpolar low-elevation distribution in the Northern hemisphere and a high-elevation distribution through the mountain ranges of N. America and Eurasia. The work was undertaken to locate the ecological attributes permitting this wide distribution: whether the explanation lies in extreme ecological plasticity or whether the various populations constitute a range of ecoclines or ecological races? Field measurement and controlled environment studies overwhelmingly supported the latter explanation, there being marked latitudinal variations in photoperiodicity, flower production, rhizome growth, chlorophyll content, respiration rate, temperature dependence of photosynthesis, light saturation level, perennating bud formation and night temperature tolerance. This is an excellent case of a study based upon general field observation which, by field physiological measurement and comparative experiment, has unearthed a wide range of differences in metabolic potential which are partially responsible for the differences between the component populations.

THE ECOSYSTEM CONCEPT

Tansley (1935) first used the term ecosystem, in the awareness that, for full understanding to be achieved, ecology must include all aspects of the interactions of organisms and environment. Sadly, few individuals have possessed the intellectual equipment, or had the training, to cope with the animal/plant interface and until recent years the subject has been underfinanced so that multidisciplinary teams could rarely be assembled. Consequently, the full development of this conceptual approach has been much delayed and is, in many ways, still in its infancy.

There have been a few notable exceptions to this generalization, for example the extensive studies of ecosystem energy flow made in recent years owe much to the stimulus given by Lindeman (1942) in formulating the trophic–dynamic concept of ecosystem function. His treatment of ecosystem production and efficiency in the form of an energy budget at the various trophic levels (primary production, primary, secondary and tertiary consumption) laid the foundations for an integrated approach which is reflected in the Odum brothers' classic papers on coral reef (Odum and Odum, 1955) and freshwater spring (Odum, 1957) trophic structure and productivity. During this same period Odum (1953; 1959) presented the student, for the first time, with a textbook based upon the ecosystem approach which brought wide attention to bear upon the problems of energy flow, biogeochemical cycling and ecological regulation.

These approaches have, however, been fragmentary: attempts at integrated analyses of partial systems functions, very often undertaken as an aid to, and mode of, teaching ecology. The great transformation has come about in the last one or two decades during which the holistic approach to systems, originating in military and industrial operational research, has been transferred to ecology. Watt (1966) wrote 'Systems analysis is a body of techniques and theories for analyzing—complex problems viewed as systems of interlocking cause and effect pathways'. The term 'ecosystem' is merely a contraction of 'ecological system' and how should such a system be studied if not by systems analysis and computer? (Spurr, 1969).

The fundamental tenet of the systems approach is that an extremely complex process can be most easily dissected into a large number of very simple unit components rather than a smaller number of relatively complex units (Watt, 1966). Secondly, complex processes in which all variables have changed with time can be dealt with most easily by recurrence formulae which express the current status as a function of the state at time zero. The third necessary concept is that of optimizing processes, which relates to mathematical theories of maximization and minimization of functions.

The synthesis of these ideas leads to ideas concerning optimal choice of alternative strategies at each of a sequence of times. Multistage decision processes of this sort have two computational features: they have a high dimensionality and they have to be solved by repetitive procedures. Here we have the key to the long-delayed birth of the ecosystem idea: massive computation is required and it has not been feasible to solve the problems until low-cost, high-speed computational facilities became widely available.

Further difficulties have arisen in environmental instrumentation for data acquisition purposes: the analysis of processes in terms of small component units obviously demands numerous replicated sampling points, posing problems of cost and data logging. Solution of these instrumentation problems is a current preoccupation amongst ecologists, a wealth of publications having appeared in the last few years (Eckardt, 1965; Bradley and Denmead, 1967; Eckardt, 1968; Wadsworth, 1968).

Cooper (1969) outlines the successive steps needed for a systems approach to ecology.

(a) Establish the objectives of the system—for example, a photosynthetic system has, as one objective, the 'aim' of maximizing radiant energy fixation.

(b) Determine the relationships between the variables of the system and the identified objectives. This process is a mathematical modelling of the system and is based upon expression of the component functions and interrelationships of the system in computer language. The values assigned to the functions may be determined from actual measurements or from 'informed guesses' if data is lacking. At this stage the simulated behaviour of the ecosystem may be checked on the computer by comparing the responses of both the real world system and the simulation to experimentally imposed constraints. Errors in the simulation may then be reduced by refining the model, the end product being an abstract entity which provides a simplified imitation of the behaviour of those parts of the system which interest the analyst.

(c) Further steps may be taken to determine input–output (cost–benefit) ratios of the system for any number of alternative pathway choices. The simulation may be used predictively to assess the future behaviour of an ecosystem (its natural evolution or its response to imposed conditions), providing an entirely new tool for the ecologist, who has never before been able to cope intuitively with the consequences of multiple changes in interacting factors. As the subject develops it is likely to be of immeasurable value in all fields of natural resource management.

Modelling may be empirical or predictive: in the former case statistical tools such as multiple regression, factor and principal component analysis are brought to bear on data collected in the field. All factors which are thought pertinent to the system are measured, and changes in the system are recorded with time. The mathematical screening will eliminate many of the variables as irrelevant and the others may be built-in as components of the model. No preknowledge of any empirical or logical relationship between the factors is required but measurements need to be made with reasonable precision. In such models predictions may be made with considerable accuracy but, because of difficulties in causal interpretation of correlation analyses, the omission of a necessary factor may, under some circumstances, cause major deviations from the behaviour of the real world system. Provided that such errors are detected in the model testing stage they may be eliminated if the missing factor can be measured and included in the new model.

Predictive models, by contrast, are constructed to locate those factors which should be measured in the real world, thus eliminating much of the laborious screening involved in the empirical approach. The first step in this kind of analysis is the construction of a block diagram representing the major components, storages and flows within the system (Figure 1.3). The enforced discipline of this mental process is in itself very valuable, requiring the precise identification of previously vague concepts in terms of subsystem blocks, storage sites, flowpaths and feedback. Once the block diagram is complete its components are assigned hypothetically feasible values based on general rather than specific measurements in the field. The model may then be run on the computer and the outcome compared with the real world system, in order to locate discrepancies. At this stage the analyst must correct for omission of components in the block diagram and adjust values of incorrectly assessed components. This process is continued until the model gives acceptable agreement with reality, at which stage it may be used to discover the sensitivity of the system to specific inputs, thus defining limits for its normal functioning and highlighting those aspects about which more information is needed. Much orthodox thought in science has required detailed knowledge of all aspects of a system before progress in understanding may be made. Preliminary computer experiments with sensitivity analyses have, however, shown that simple systems are much more sensitive to changes in the organization structure (flow pathways) than to numerical changes in the values of the individual components (Cooper, 1969).

Predictive models are likely to be very valuable in ecology and biology generally but have not yet reached the degree of precision which can be achieved with factor analysis. Van Dyne (1969) suggested that this type of approach is so important that a specific training effort should be made in undergraduate programmes to increase the output of students with the requisite academic background for utilizing and advancing the systems approach to ecology. It is possible that this is the only route through which man may solve the imminent problems of his own impact on that largest and most complex of ecosystems: the biosphere. The present state of ecosystems analysis is reviewed and summarized by Watt (1966) and Van Dyne (1969).

THE DESIGN OF EXPERIMENTS, DATA COLLECTION AND ANALYSIS

The origins of ecology in its rather loose, descriptive phase, coupled with the difficulties of precisely measuring many environmental variables, have led unjustly to its being considered a 'soft' science: contrasting with the

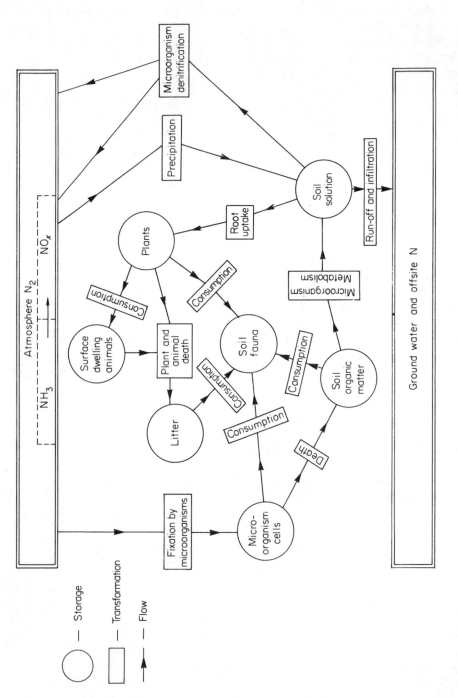

Figure 1.3 Block diagram for a systems analysis of the processes involved in ecological nitrogen cycling. The lines represent flow paths, the boxes are transformations that regulate flow quantities and the circles are storage reservoirs. The two double-walled boxes represent storages outside the boundary of the ecosystem.

traditionally precise or 'hard' disciplines of physics, chemistry, biochemistry and molecular biology. This attitude has been a consequence of the excessively complex multifactorial nature of ecological interrelationships: for example, despite all attempts to control environmental factors during ecological experiments the genetic variability between replicated individuals may impose much greater error variances than would be either expected or tolerable in physical or chemical experimentation. It is, in fact, no coincidence that the development of statistical techniques for coping with sample variability originated to a large extent in biological research and has contributed so much to knowledge of experimental design and sampling.

The investigation of a phenomenon demands a rigorous approach based upon a repetitive sequence of operations.

(a) Observe and specify the nature of the problem.

(b) Establish a hypothesis, or preferably more than one, which will explain the observations. Multiplication of hypotheses prevents the worker from developing a 'product loyalty' which may interfere with the objective approach.

(c) Design experiments which, as far as possible, test the validity of one hypothesis and exclude the others.

(d) Perform the experiments sufficiently rigorously to obtain 'clean' results.

(e) Eliminate hypotheses which are not supported and, if necessary, establish new or substitute hypotheses.

(f) Repeat the sequence from (b) to (e) until a single hypothesis remains.

(g) If necessary restate the problem and establish new hypotheses to carry explanation to a deeper level, for example, an ecological problem may be reduced to the physiological, or the physiological to the biochemical level.

Adoption of this approach in ecology is not always easy, for example the specification of the problem becomes difficult when based upon collected field data, as correlation cannot always be assumed to imply causal relationships. Many environmental characteristics are correlated together and most ecological systems are so complex that multiple correlation is the rule rather than the exception. Consequently, there may be difficulty in specifying (i) the details of the problem and, (ii) the hypotheses necessary to explain it. Finally, the design of experiments is not always easy as uncontrolled environmental changes may become confounded with experimental variates. Comparative experiments with a range of plant species, preferably of similar size and morphology, help but do not entirely eliminate this problem. The use of controlled environments further improves the situation by reducing unpredictable effects. Provided that fairly large-scale replication is used and the experiments are designed with suitably randomized layouts, statistical analysis subsequently gives the worker a mathematical estimate of the trust which may be placed in the data.

The first stage of problem formulation involves qualitative and quantitative field study of plant distribution and its relationship to environmental variables. The design of a suitable sampling programme is of crucial importance to the legitimacy of subsequent statistical analysis. Sampling of plants for quantitative presence or for more sophisticated measurements such as weight determination, organ measurement or chemical analysis, must follow the procedures of replication and randomization outlined in various standard texts (Fisher, 1935–66; Snedecor, 1956; Greig-Smith, 1964). The location of sampling points for monitoring microclimatic and soil variables must likewise be governed by considerations of randomization and sample size but the problem is further complicated by the vertical stratification of many parameter values. Sampling and measurement of environmental variables is extensively reviewed in Platt and Griffiths (1964—general), Black (1965—soil); Eckardt (1965—ecophysiology); Bradley and Denmead (1967—general); UNESCO (1968—agroclimatology) and Wadsworth (1968—general). The handling of plant material for weight determination, calorific content, nutrient content and water status is detailed in Jackson (1958—nutrients); Eckardt (1968—dry matter production and energy fixation); Barrs (1968—water status).

Field data of this type may be analysed by a whole range of statistical techniques based upon correlation or analysis of variance. There is a danger arising from the fact that many environmental factors are themselves cross-correlated, leading to difficulties in causal interpretation. However, the experimental stages of the investigation should be designed to eliminate any such misinterpretations of field data.

Interpretation of field data may lead to the establishment of a number of hypotheses concerning the cause of ecological phenomena. These are amenable to experimental testing using controlled field experiment, glasshouse or, more precisely, controlled environmental investigation. The simplest approach is the variation of a single factor while all others are held constant: if the variate is applied at different levels of intensity, correlation, contingency or variance analysis may be used to establish the degree of trust which may be placed in the results. A more sophisticated and economical approach is that of multivariate experimentation, in which several factors are varied simultaneously in a factorial design. Fully orthogonal experiments, which contain all possible permutations and combinations of the variated, are the simplest to perform and analyse but various incomplete and confounded designs may be used for economy of experimental effort. Either multiple correlation or analysis of variance techniques is needed to cope with these more complex experimental designs.

Experimental design in both simple and multifactorial experimentation demands replication in order to provide comparisons which separate treatment responses from the random variation of individual organisms. Increase in replicate numbers improves the 'cleanliness' of experimental means by reducing standard error; it is, however, a response of diminishing returns as the increased precision is proportional to the square root of the replicate number. Some compromise must be made between additional experimental labour and increasing the accuracy of estimation. Treatment replicates require randomization according to recognized principles providing either totally randomized designs or suitable randomized block layouts: without this precaution spatial variations become confounded with treatment effects so that it is no longer possible to assess the outcome of an experiment. A great deal of literature exists concerning these problems, both of design and analysis, examples being Fisher (1935–66), Snedecor (1957), Mather (1943–65; 1967) and Lewis and Taylor (1967). The latter text includes discussion of many of the simpler problems of data collection, analysis and presentation. A firm grasp of experimental design principles is necessary before a student undertakes any expermental work.

SUMMARY

The contents of this introductory chapter are intended to set the scene: to indicate to the student the state of play in ecology and to provide source references for those parts of the subject which are not further discussed in this text. The remainder of the book is devoted to the outlining of those aspects of environment and plant physiological response in which the foundations of ecology lie and to consideration of the interactions of organisms which permit integrated function of ecosystems. As a textbook of the early seventies it may be criticized for reducing the subject to component units rather than discussing it as a systems network. The authors' justification is twofold: clarity, for the student, is possibly better achieved by dissection rather than agglomeration and, secondly, the linear format of a book is not well suited to a multiple pathway approach at this level.

Chapter 2

Energy exchange and productivity

It has been traditional for ecologists to describe an environment in terms of its associated organisms and such factors as soil conditions, temperature, sunlight, rainfall, wind and humidity. The meteorological factors are a reflection of the pattern and utilization of solar radiation which, on a global scale, drives the atmospheric circulation and its associated cycles, and, at a microenvironmental level, provides the main long-term control which is rarely overruled by such catastrophic (but nevertheless determinant) events as extreme drought, fire, flood or frost.

Terrestrial organisms live in a thin boundary layer between the solid fabric of the earth's crust and the inhospitable environment of interplanetary space. This boundary layer, the earth's atmosphere, is strongly affected by the net flux of radiant energy between sun, earth and extraterrestrial space and it is with the multitude of energy environment effects in the boundary layer that this chapter is concerned.

TRANSFER OF ENERGY

All objects above the temperature of absolute zero emit energy by radiation and, in turn, may absorb radiant energy. The amount of energy radiated by a body is a function of the fourth power of the absolute temperature of its surface (Stefan's Law):

$$R = \sigma T^4$$

where R is the energy radiated (W m^{-2})
   σ is the Stefan–Boltzmann constant ($5\cdot57 \times 10^{-8}$ W m^{-2} T^{-4})
   T is the absolute temperature ($^{\circ}$K).

The radiation emitted by a body will have a wavelength distribution which is also related to its surface temperature as wave propagation is a function of molecular oscillation which increases in frequency with temperature rise. The equation linking wavelength to frequency is:

$$\lambda = c/v$$

where λ is the wavelength (cm);
    c is the velocity of light (cm sec^{-1});
    v is the frequency (cycles sec^{-1}).

Thus increase of temperature increases the frequency of propagation and a hot body will have a shorter wavelength emission than a cold body.

The peak of the wavelength distribution (λ_{max}) is related to temperature by the Wein displacement law::

$$\lambda_{max} = 2864/T$$

where λ_{max} is expressed in μm. The distribution of radiant energy around this peak is so asymmetric that only 25% is of shorter wavelength and 75% is of longer wavelength. Some examples of λ_{max} are: for the sun's surface (5793°K), c. 0·05 μm (blue-green); for the tungsten filament of an electric lamp (c. 3000°K), 0·96 μm (shortwave infrared) and for the earth's surface mean temperature (14°C = 287°K), c. 10μm. The latter value is well into the longwave infrared and it is in this spectral region that terrestrial organisms exchange radiant energy with their environment.

The quantity and quality of radiation is thus a function of the surface temperature of a body, assuming that it is a perfect radiator of energy. A perfect radiator or 'black body' has an emissivity factor of one and shows the theoretical Stefan–Boltzman energy emission at all wavelengths. An object, placed in a stream of radiant energy, may either absorb all of it or absorb some and transmit and/or reflect the remainder. It may also absorb differentially in various parts of the spectrum. Good emitters are equally good absorbers and, consequently, a perfect black body will totally absorb all incident radiation.

Within any closed system the balance of energy exchange will then be a function of the relative surface temperatures and the absorbance and emittance characteristics of all component parts of the system. It will, furthermore, depend on any other modes of energy transfer and energy-dependent changes of state. For this reason energy balance studies demand, in addition to radiation measurement, the measurement of conduction, convection and changes of state such as the evaporation of water or the melting of ice involve latent heat exchange.

THE EARTH'S RADIATION BALANCE

Radiant energy from the sun reaches the upper atmosphere with a flux density of 1360 W m^{-2} (2·0 cal cm^{-2} min^{-1}) integrated over the whole of the solar spectrum; this value is referred to as the solar constant. Passage through the atmosphere to the earth's surface depletes the incoming radiation of energy at discrete wavelengths by absorption and throughout the spectrum by scattering (Figure 2.1). Oxygen, ozone and nitrogen absorb at wavelengths less than 0·3 μm and eliminate most of the ultraviolet, while water and ozone absorb a considerable part of the infrared. Terrestrial organisms of a normal type would not be able to survive direct exposure to

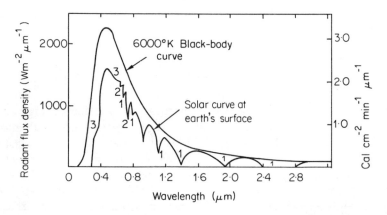

Figure 2.1 The radiant flux density from a 6000°K black-body at solar distance compared with the solar curve at the earth's surface. The absorption bands due to water vapour (1), oxygen (2) and ozone (3) are shown. After D. M. Gates, *Energy Exchange in the Biosphere,* Harper and Row, New York, 1962, Figure 3.

sunlight if its ultraviolet content were not reduced in this way; furthermore, the attenuation of the infrared reduces the maximum heat loading of exposed plants to a level which is just tolerable.

At the earth's surface the direct solar flux is reduced to something over 50% of the extraterrestrial flux, but no overall figures can be given for income as it will range from a very low value under full cloud conditions to a very high value when direct sunlight is reinforced by reflected cloudlight. It is more satisfactory to look at mean daily totals of incoming radiation as the controlling factor in transpiration, the daily total in the 0·4–0·7 μm band as a photosynthetic source and short-term integrated values of the total flux when heat loading is considered.

Figure 2.2 shows the midday and midnight radiation balance of a vegetated surface. The downward flux of solar shortwave (0·3–3·0 μm) radiation rises from zero before dawn to a maximum at midday and returns to zero shortly after sunset. The exchange of longwave (3·0–100 μm) radiation between earth, atmosphere and space continues more or less unchanged during night and day. During the whole 24-hour period there

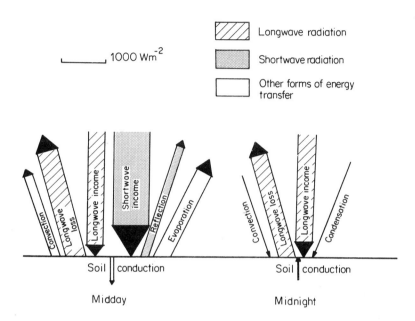

Figure 2.2 The day and night energy balance of a vegetated surface. Reproduced with permission from R. Geiger, *The Climate Near the Ground,* translated from the German 4th Edition (1961), Harvard U.P. Cambridge, Mass., 1965.

is a net black-body loss by longwave radiation from the earth which is almost balanced by the incoming black-body radiation from the atmosphere. Geiger (1965) presents an annual budget for a site in temperate Europe which illustrates these points.

		MJ m^{-2} year^{-1}	cal cm^{-2} year^{-1}
Shortwave income:	Direct	(+1429)	+34,153
	Scattered	(+1814)	+43,444
Longwave income (atmospheric)		(+10,053)	+240,533
Longwave loss from earth's surface		(−11,236)	−268,837
Reflection loss of shortwave income		(−602)	−14,376
Net radiation balance		(+1459)	+34,926

Eighty-six per cent of this net balance was utilized in evaporation of water and the remaining 14% in atmospheric heating; conduction to and from the soil cancelled out over the year. The utilization of the greater part of the net flux in evaporation is the reason for the very strong correlation between solar radiation and plant water loss (p. 179).

The fraction of the solar income used in photosynthesis is very low by comparison, perhaps reaching 2–5% in very efficient ecosystems but only a fraction of 1% on a worldwide basis. The consequences, for the biosphere, of the photosynthetic process are, however, enormous and the influence of the energy balance on plant water loss and leaf temperature is determinant in the efficiency of overall productivity.

LEAF TEMPERATURE

Provided that sufficient information is available concerning the physical characteristics of a leaf and the influence of its physiological status on such characteristics, then its energy balance may be entirely specified by the following equation (Gates, 1968):

$$Q_{abs} = R \pm C \pm LE \pm M$$

where Q_{abs} is the radiant energy absorbed (W m^{-2});
R is the reradiation from the leaf (W m^{-2});
L is the latent heat of evaporation of water (J kg^{-1});
E is the transpiration rate (kg m^{-2} s);
M is metabolically absorbed radiation (W m^{-2}).

R is governed by leaf temperature through the Stefan's Law relationship but the leaf temperature is a complex function of many environmental variables, being governed not only by radiant input, but also directly by air temperature and windspeed, and indirectly by these factors and relative humidity through the influence of evaporative heat loss. Due to the complexity of this relationship, individual cases require separate calculation and, to aid this approach, Gates and Papian (1971) have compiled an atlas of computed energy budgets for a range of environmental conditions. This provides comprehensive information on the control of leaf temperature and transpiration by radiant input, air temperature, relative humidity, airspeed, internal diffusion resistance and leaf dimensions.

Figure 2.3 shows the relationship between radiant energy input and leaf temperature for a number of conditions. Several points of ecological interest emerge from this type of analysis: with low radiant input, high humidity and low wind speed, leaves may be considerably above air temperature. This also occurs if the internal diffusion resistance is high so that transpiration is limited under high radiation conditions. In this case, with an air temperature of 20°C the highest leaf temperature achieved is 43·1°C but with air temperatures of 30 and 40°C, respectively, the leaf temperatures would be 49·4 and 55·7°C. Leaves may thus, in full sunlight, reach temperatures which are physiologically supraoptimal, biochemically damaging or sometimes barely sublethal. The role of transpiration in controlling leaf temperature cannot be doubted; for example, a tenfold increase in transpiration resistance from 1·0 to 10·0 s cm⁻¹, with an air speed

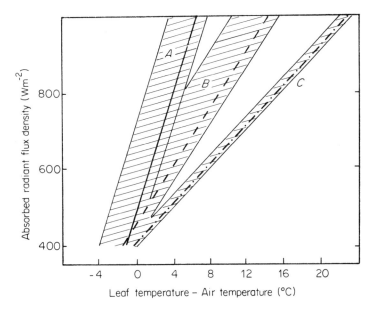

Figure 2.3 The relationship between solar radiant input, air temperature and leaf temperature. The plotted lines are for relative humidity 75% while the hatched bands represent the r.h. range from 0–100%. Intermediate values may be obtained by linear interpolation. All values are for an air temperature of 20°C and leaf dimensions 5 cm (parallel to air flow) by 5 cm (right angles to air flow). A: Airspeed 100 cm sec⁻¹; leaf internal diffusive resistance 1 sec cm⁻¹. B: Airspeed 10 cm sec⁻¹; resistance 1 sec cm⁻¹. C: Airspeed 10 cm sec⁻¹; resistance 10 sec cm⁻¹. Plotted from Gates and Papian (1971).

of 10 cm s⁻¹, a radiant input of 1·0 × 10³ W m⁻² at 75% relative humidity and 20°C, causes a leaf temperature increase of 7·8°C.

  In a low humidity environment with low radiation input, the leaf may be several degrees cooler than the surrounding air, due to evaporative heat loss. This may be a significant factor in leaf survival with an air temperature near freezing point but the balance between leaf temperature and radiant input also shows a strong interaction with wind speed.

Wind speed, in nature, is rarely below 50 cm s^{-1} and the plotted value of 10 cm s^{-1} corresponds approximately to air movement in free convection without any contribution from wind. By reducing the value of the boundary layer resistance (p. 31), increasing wind speed tends to increase transpiration loss in the low wind speed range. However, if the leaf is above air temperature due to radiant heating then increasing wind speed causes cooling and reduces the gradient of leaf to air water vapour pressure. In a humid environment, such an increase of wind speed may actually reduce transpiration. By contrast, if the leaf is in a low humidity, low radiation environment it will be cooler than the surrounding air; increasing wind speed will warm the leaf and the transpiration rate will be increased by both temperature and boundary layer effects.

These effects have been discussed in relation to the single leaf. Within natural leaf canopies the air is often more humid and cooler than the external air, hence wind movement is likely to introduce drier air and increase transpiration loss. Occasionally, if the canopy is strongly warmed by sunlight during the midday period, increasing wind speed may reduce transpiration. If the external air is either humid or cold then increasing wind speed may have little influence on water loss.

The measurement of leaf temperature is, therefore, rather important in field studies of energy balance, water use or photosynthetic productivity. Direct measurement is possible with small thermocouples, thermistors or resistance thermometers either cemented or clipped to the leaf surface but great care is needed to avoid interference with radiation absorption, reradiation or transpiration. Provided that this type of interference is minimized, thermocouples probably provide the most accurate measurements of leaf temperature and are very suitable for localized measurement. Recently, the technique of infrared radiation thermometry has become widely-applied. The thermometer utilizes a radiometer suitably filtered to have a bandpass in the infrared (often 8 to 13 μm) and fitted with a lens, mirror or diaphragm to define the field of view. Modern instruments may be adjusted to compensate for the emissivity of the source and permit direct, contactless measurements. A further sophistication of the technique is i.r. thermography, in which the field of view is scanned to produce a picture in which either optical density or colour are related to surface temperature. The methods of temperature measurement are discussed fully by Perrier (1971).

Gates (1965) noted that photosynthesis is light- and temperature-dependent and that other physiological processes are temperature-dependent only. The ecological significance of varying radiation load and temperature is enormous as they strongly influence productivity and competitive ability. Using data of Thut and Loomis (1944), Gates utilized the energy environment approach to estimate the diurnal variation in photosynthetic rate of *Zea mais*. Figure 2.4 shows the marked correspondence between his estimates and Thut and Loomis' measurements of leaf growth rate. The technique affords a model, sufficiently precise, to forecast the occurrence of a midday slump in photosynthetic rate followed by a slow recovery in the afternoon. Gates suggested that this slump is a consequence of leaf overheating which may directly affect the biochemical mechanism of photosynthesis; may increase respiratory CO_2 output, thus raising the substomatal CO_2 concentration to induce stomatal closure; or, finally, may increase the leaf water deficit which would accentuate the stomatal closure.

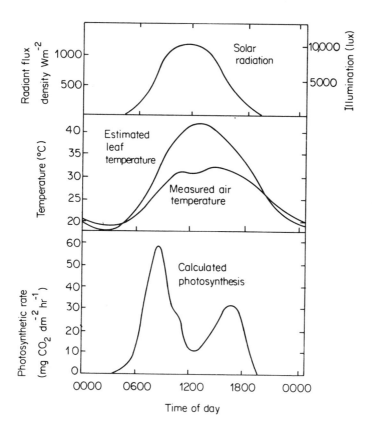

Figure 2.4 Calculated diurnal variation of photosynthetic rate for maize corresponding to diurnal change of solar radiation and air temperature. Reproduced with permission from D. M. Gates, Energy, plants and ecology, *Ecology*, **46**, 1–10, Figure 7 (1965).

MICROCLIMATOLOGY

The complex radiation balance of an ecosystem results in the establishment of localized variations in meteorological conditions which provide the field of study for the micrometeorologist. The subject has developed enormously since the 1950s and now embraces topics ranging from epidemiology in phytopathology, through the manifold uses of microclimatology in the study of plant behaviour in the field, to techniques in the measurement of atmospheric pollution.

The microclimate of an ecosystem may be specified in oversimplified form by a series of profiles, showing variation with height, of radiation, air-temperature, water vapour pressure, wind speed and carbon dioxide concentration. These profiles show a wide diurnal variation which may be seen by comparison of the night and day values in Figure 2.5: the vertical scale of the profiles would be varied according to the height of the vegetation cover.

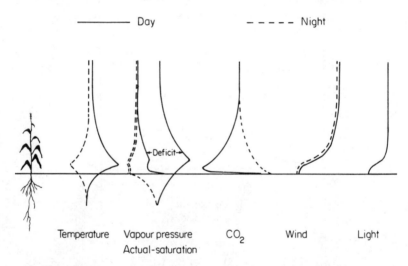

Figure 2.5 Micrometeorological profiles through a vegetation canopy to show the contrast between day and night conditions.

The penetration of radiation into a canopy generally obeys the Bouguer–Lambert law (Loomis, Williams and Duncan, 1967):

$$I = I_0 e^{-KL}$$

where I and I_0 refer to the radiant intensity on a horizontal
surface within and above the canopy respectively;
L is the leaf area index;
K is the extinction coefficient.

The extinction coefficient (K) is a constant for a given species and is related to canopy architecture and leaf chlorophyll content and reflectivity. The value lies between 0.3 and 0.5 for grass-type canopies and approaches 1.0 for leaves which are nearly horizontal. The value shows an inverse relationship to chlorophyll content and reflectivity of the leaves. As a result there is an exponential gradient of reduction of radiant intensity downward into the canopy which may sometimes reach very low values at ground level.

During the day, radiative heating of the canopy causes a convectional transfer of sensible heat so that the air temperature within the upper canopy may be higher than that above or below. At night the relationship is reversed as the canopy air layer is cooled by contact with leaves which are both transpiring slowly and losing heat by radiation.

Wind speed shows no strong diurnal variation but overall wind speeds tend to be higher in the daytime as a consequence of convectional effects. The development of wind speed profiles in the canopy is a problem in steady-state boundary layer flow and is a component factor in establishing profiles of water vapour and CO_2, measurements of which may be used in making estimates of water vapour flux due to transpiration and CO_2 flux due to photosynthesis. Figure 2.5 shows the wind velocity distribution with height in a plant community; the profile above the canopy is logarithmic and, within the canopy, becomes exponential. If the logarithmic curve is extrapolated downward the zero velocity intercept lies at a height of $D + z_0$ in which D is the zero plane displacement and depends on plant height and z_0 is the roughness height, a measure of the surface roughness of the community. The roughness height is effectively the thickness of a laminar sub-layer through which individual elements project, its value being related to the height variation and spacing of the individual elements. If the canopy were rigid it would have a constant value but reported variations of z_0 with wind speed can be accounted for by the variation of surface roughness depending on leaf flutter, branch movement and leaf streamlining. The surface frictional characteristics of the canopy are entirely specified by z_0 and D (Lemon, 1965).

During both night and day, as would be expected, the saturation vapour pressure profile shows a close correspondence with the temperature profile (Figure 2.5). At night the actual vapour pressure almost reaches saturation as the air and the canopy are cooled by radiation and convection. Some water vapour is transferred from the canopy by transpiration though the rate is low. During the day the actual vapour pressure curve departs considerably from the saturation curve, the deficit increasing as the canopy is approached from above; this is a consequence of the convectional air heating at this level. As the canopy is entered the deficit becomes more marked but deeper in the canopy the actual vapour pressure curve inflects and, toward the bottom of the canopy, reapproaches the saturation curve as a consequence of transpiration coupled with the fairly low rate of air movement at this level and the lower temperatures at the base of the plant cover.

The carbon dioxide concentration profile shows a marked diurnal change, depending on photosynthetic depletion during the day and respiratory enrichment at night. The gas is present in dry air at a concentration of c. 300 v.p.m. with an average fluctuating between 270 and 360 v.p.m. (Geiger, 1965). A slow increase has been observed over the last 50 years, consequent on the burning of fossil fuels, but the buffering effect of solubility in ocean water has prevented a very large change.

Because the total concentration is so low, and utilization by photosynthesis in the canopy so inhomogeneous, profiles must be based upon means from fairly long sampling periods. Figure 2.6 shows the variations at any one sampling point to be greater than the differences measured in the mean profiles. This imposes a serious limitation on the technique as a method of measuring gross community photosynthesis and

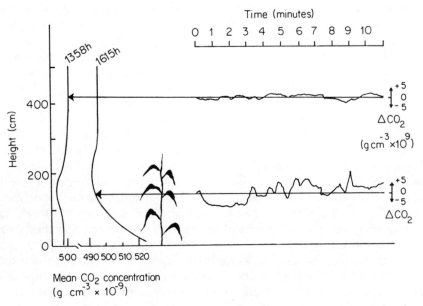

Figure 2.6 Mean CO_2 concentration (C) profiles in a *Zea mais* canopy for the indicated times, and the variation (ΔC) during the ten-minute sampling period for the times and levels shown by the horizontal arrows. The profiles are plotted as means for a ten-minute sample: all samples in the same profile were taken at the same time. The variations from sample means, plotted on the right-hand graphs, are to the same scale as the profiles. These results were collected with a clear sky in a uniform crop canopy under 'steady state' conditions! After E. R. Lemon, *Harvesting the Sun*, Academic Press, New York, 1967, Figure 21, p. 288.

for this reason it can only be used under the most ideal and uniform conditions.

During the day, as the canopy is approached from above, the CO_2 concentration declines, reaching a minimum at a point near mid-canopy level (Figure 2.5). Below this the concentration increases steeply with depth, becoming equal to the external concentration at a level roughly corresponding with the compensation light intensity and reaching a fairly high value at soil level. This profile indicates the photosynthetic depletion of CO_2 in the upper canopy, the equilibrium corresponding to the compensation point lower in the canopy and enrichment by respiratory CO_2 from the lowest shaded leaves and from soil organisms. By contrast the night profile shows a gradual downward increase of CO_2 concentration consequent on respiratory CO_2 evolution. Lemon (1965, 1967) and Tanner (1968) describe techniques by which profile measurements of wind speed, water vapour and CO_2 concentrations may be used for the calculation of the vertical fluxes of water vapour and CO_2 to or from the canopy surface. An example is cited on page 173.

MEASURING TECHNIQUES

Developments in micrometeorology during the last decade have refined the techniques which are available for microenvironmental measurement and it is no longer necessary to modify the more cumbersome instruments of macrometeorology for the purpose. Long (1968) has reviewed the construction and characteristics of instruments for measuring temperature, humidity and air flow, all of which were designed to give continuous records; an essential requirement of a system which is being used to estimate the diffusive fluxes of water vapour and CO_2 which show immediate dependence on the energy balance and considerable temporal fluctuation due to turbulence effects. Long also pointed out that limitations of the instruments make micrometeorology a 'fine weather' subject, and continuous recording permits the choice of suitable periods for analysis.

Long suggested that the platinum resistance thermometer is the best sensing element for temperature measurement and psychrometry as it is robust, sensitive and requires only a low cost recorder. Thermocouples are useful for point measurement and sensing of temperature differences, but for field use careful screening of the leads is required, their sensitivity is low and an expensive potentiometric recorder is needed. For psychrometric determination of humidity the wet-bulb sensor is provided with a muslin sheath and a reservoir to give an uninterrupted water supply.

Above the canopy small Sheppard cup-anemometers may be used to measure wind speed: the instruments are adapted to record by means of a photoelectric or mechanical attachment which gives one impulse per revolution. Within the canopy turbulence reduces the performance of cup-anemometers and Long described a hot-bulb anemometer for use under these circumstances.

Measurement of CO_2 profiles is more difficult as the concentrations are extremely low and the fluctuation due to turbulence is large (Figure 2.6). Air samples have to be drawn from the requisite height, filtered, dried and delivered to a conventional infrared gas analyser. Monteith (1968) gave details of a system with automatic switching which sequentially records CO_2 concentration at six sampling points. Gates (1962) has suggested that the sampling problem could be solved by using direct path infrared CO_2 analysis with i.r. sources and sensors mounted on adjacent towers within the canopy.

Radiation measurement presents more difficulty than photoelectric measurement of 'light' intensity. Szeicz (1968) indicates the extent of the problem by categorizing radiation measurements:

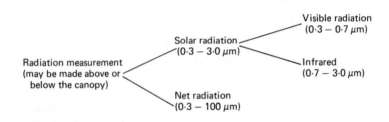

Various sensors may be used, based on one of the following principles: (i) thermoelectric (ii) photoelectric (iii) bimetallic distortion (iv) distillation and (v) photochemical. Of these only the thermoelectric, photoelectric and distillation methods now find wide use. Photoelectric techniques using selenium barrier layer cells, cadmium sulphide photoresistivity cells, silicon solar cells or photo-tubes suffer the disadvantage of strongly biased spectral response, which raises considerable problems of calibration and measurement if energy content of the radiation is to be measured at different times of day or at different levels in the canopy. They are useful for comparative purposes if the spectral composition of the light does not change and have the advantage of a high electrical output for a low level of illumination.

The Bellani pyranometer utilizes the distillation principle, radiant energy impinging upon a blackened copper sphere containing a volatile liquid; the amount of liquid distilling over into a graduated collecting tube is a measure of the integrated energy input during the period of exposure.

Thermoelectric instruments utilize batteries of thermocouples arranged in series to form a thermopile; one set is blackened with optical black paint and another is coated with a white or reflecting material. The two sets are connected in opposition so that the resultant thermal e.m.f., when exposed to radiation, is linearly related to the incoming radiant flux. If the thermopiles are protected by a glass dome the instrument responds to both visible and infrared up to $3 \cdot 0$ μm. To measure the energy input in the visible part of the spectrum ($0 \cdot 3$–$0 \cdot 7$ μm) a second unit is required, fitted with a Schott RG-695 (ex RG-8) filter with a sharp cut-off below $0 \cdot 7$ μm; this instrument measures in the infrared and the visible component is computed by difference.

Net radiation, the balance between upward and downward flux of radiant energy at all wavelengths up to 100 μm, is more difficult to measure as it represents a small difference between two large and opposing fluxes. Very few materials transmit radiation at all wavelengths, consequently naked thermopiles have been used in some instruments, one black and white pair facing downward and another pair upward. Such instruments are very sensitive to forced convection by wind and have to be fitted with a ventilating fan to swamp this effect. Polythene has no strong absorption bands in the visible or infrared wavelengths and has been used to form domes for net radiometers to avoid the need for ventilation. Gates (1962) and Sestak, Catsky and Jarvis (1971) both give extensive accounts of the instruments which may be used in radiometry. As described earlier in this chapter (p. 28), radiometry is also taking its place as a remote temperature sensing technique.

CLIMATE

Long-term climatic records are an essential background to the local measurements so far discussed. The seasonal cycles of temperature, daylength, rainfall, humidity and wind exert a strong control over physiological and reproductive processes which is reflected in ecosystem structure and function. The incidence of short-term oscillations such as droughts, floods, gales and extreme frosts may also be determinate of species distribution or competitive behaviour, sometimes even when their incidence is on a time scale of many years. The patterns of climatic and altitudinal species distribution which characterize the world's ecosystems are basically determined by these interactions and, of course, by their relationship to soil formation.

Climatic effects of this type are excessively complex as they interact differently at various stages in the plant life cycle. At all stages, population phenotypic plasticity is a reflection of genetic constitution, hence the enormous variety of species which are confined to different ecosystems throughout the world.

The following scheme illustrates some of the possible interactions of climatic factors with plant growth at different life cycle stages:

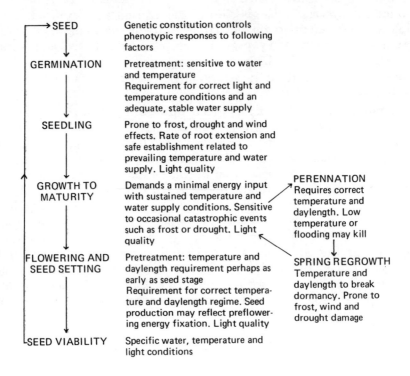

PHOTOSYNTHESIS

The origin of living organisms and their mechanism for replication of heritable characteristics radically changed the character of the primitive earth's surface, setting the stage for a progressive evolution and selection, by survival, of ever more complex organisms. Second only to the appearance of life, in its global impact, must have been the emergence of photosynthetic energy fixation. Our world, as we now know it, is moulded by this process; the very oxygen which we breathe is its product, the primitive reducing atmosphere, through the intervention of chlorophyll, having been replaced by a gaseous mixture containing over 20% of oxygen.

Limiting factors in photosynthesis

Under field conditions photosynthesis may be limited by environmental or by genotypic–phenotypic characteristics.

Environmental variables
1. Energy
   (a) Quantity, quality and duration of 0·4–0·7 μm photosynthetic energy input.
   (b) Quantity and duration of > 0·7μm input influencing heating and water loss effects.
   (c) Other photomorphogenic and photoperiodic effects.
2. CO_2 supply—concentration and leaf ventilation.
3. Air and Leaf temperature.
4. Soil water
   (a) Potential.
   (b) Quantity available.
5. Mineral nutrition.
6. Seasonal cycle.
7. Pathological condition.

Genotypic–phenotypic variables
1. Leaf diffusion resistance
   (a) Stomatal and cuticular.
   (b) Internal.
2. Carbon pathway in anabolism and catabolism (e.g. C-3 versus C-4 fixation pathway; presence or absence of photorespiration).
3. Composition of photosystems.
4. Chloroplast shape, structure and distribution in cells.
5. Lead structure
   (a) Anatomical
   (b) Optical.
6. Plant leaf area.
7. Leaf display
   (a) Angle
   (b) Phyllotaxy, spacing and overlapping (self-shading).
8. Concentration of photosynthesate and translocation rate.
9. Endogenous rhythms (e.g. in stomatal aperture).
10. Developmental stage.
11. Leaf age.
12. Adaptation and pretreatment effects.
13. Factors relating to environmental conditions, e.g. plant water potential, nutrient status, temperature etc.

Various workers have attempted to model this enormously complex interaction but usually by simplification of the internal plant variables to a single photosynthesis vs. energy input characteristic curve. Duncan (1967), for example, discusses computer simulation of photosynthetic energy fixation in relation to nine variables: (1) leaf area; (2) leaf angle; (3) leaf position; (4) leaf transmissivity; (5) leaf reflectivity; (6) the characteristic curve relating photosynthesis to leaf illumination; (7) intensity of sunlight;

(8) intensity of skylight and (9) position of the sun. Assuming all other factors to be optimal, this approach permits adequate prediction of photosynthetic rates under a range of environmental conditions.

During plant growth, *leaf area index* (leaf area/ground area) increases from zero to a maximal value which is characteristic for the species but may not be optimal for the individual. Leaf angle and leaf position interact with leaf area index in governing penetration of light into the canopy (see p. 30); leaves with a steep angle of insertion (grass-like) permit deeper penetration than those held horizontally. Leaf position may also vary radially with phyllotaxy so that varying degrees of projection overlap occur. Higher values of leaf area index are, however, inevitably accompanied by projection overlap and mutual shading of leaves. All of these variables ultimately affect variable (5), the response of the individual leaf to illumination level. Interpretation of canopy response to light intensity is complicated by the penetration of full intensity sun-flecks deep into the canopy; due to leaf movement and the changing position of the sun these may have a very short duration for any one leaf. A further difficulty is that diffuse light deep inside the canopy is spectrally different from daylight, having selectively lost more of the photosynthetically active red and blue wavelengths compared with the central part of the spectrum. The subcanopy light is usually, also, proportionately enriched in the far-red and near infrared part of the spectrum, perhaps with morphogenetic consequences.

Measurements of transmission through canopies suggest that attenuation of radiation follows the Beers law relationship (see p. 30) so that diffuse light intensity falls off exponentially into the canopy. The photosynthetic response of individual leaves at various levels in the canopy is not, however, simply related to the mean diffuse light intensity as the selective filtering of red and blue light by the upper leaves makes the deep canopy, green-rich light less photosynthetically useful. The effect of moving sun-flecks also causes departures from the Beer's law distribution of light but this is probably not very important.

Deep canopy leaves may be morphologically suited to low light intensities (shade leaves) by, for example, reduced mesophyll thickness but a more common complication is the fact that leaves may become adapted to low light intensities, showing a higher photosynthetic and lower respiratory rate for a given intensity. The consequence of these changes is that the lower part of a canopy shows a higher photosynthetic rate than would be predicted from the light-response curve of a single, normal leaf. This seriously complicates any mathematical model approach to canopy photosynthesis based on the single leaf light response. However, the very failing of such models usefully draws attention to those parts of the system which are difficult to simulate.

A good example was the simulation of crop surface photosynthesis attempted by de Wit (1959). Making the assumptions detailed in Figure 2.7

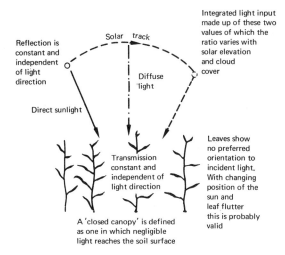

Reflection is constant and independent of light direction

Solar track

Integrated light input made up of these two values of which the ratio varies with solar elevation and cloud cover

Diffuse light

Direct sunlight

Transmission constant and independent of light direction

Leaves show no preferred orientation to incident light. With changing position of the sun and leaf flutter this is probably valid

A 'closed canopy' is defined as one in which negligible light reaches the soil surface

Figure 2.7 Diagrammatic representation of a crop canopy illustrating the various simplifying assumptions required by de Wit's (1959) mathematical model of crop surface photosynthesis.

he was able, for any combination of radiant flux density and solar elevation, to compute the energy input to leaves not already above the light saturation intensity. This value, multiplied by the gradient of the photosynthetic response line (g CH_2O per unit of radiant energy input) is the potential photosynthetic rate of the crop surface.

Integration of these rates for different seasons and periods of the day, after correction for catabolic loss, gave the potential photosynthetic production of a closed crop canopy. Under favourable conditions the model gave satisfactory correspondence with real photosynthetic production, but problems arise with longer term estimates. For Netherlands conditions the model gave values of 36 tonnes ha^{-1} during the summer months (April–September) and 13 t ha^{-1} for the six winter months. This may be compared with actual grass yields of 15 t ha^{-1} during the summer and zero in winter.

The lack of agreement between de Wit's long-term predictions and the real photosynthetic production may be partially explained by the effects of spectral change, sun-flecking and physiological adaptation but the difference between winter and summer values provides the clue to more important effects. During the winter in the Netherlands low temperature and/or climatic wetness almost entirely prevent photosynthetic energy fixation, hence the great departure from the predicted winter values. The major loss in the summer months is attributable to water shortage. De Wit's model was, of course, based on the assumption that all conditions other than radiant input were optimal for photosynthesis and thus did not allow for these departures from the ideal.

Adequate simulation thus demands allowance for all of those environmental factors which may inhibit photosynthesis or, like increasing temperature, cause a relative increase of respiration rate. The model must also compensate for inherent factors such as canopy leafiness, leaf distribution and departure from uniform physiological response such as leaf conditioning to shade conditions.

The interaction of inherent and environmental factors in limiting photosynthetic production are strikingly illustrated in Black's (1963) study of *Trifolium subterraneum* at Adelaide, S. Australia. Figure 2.8 shows the annual course of potential production, predicted from radiation income, compared with actual production measured by sampling. Adelaide has a

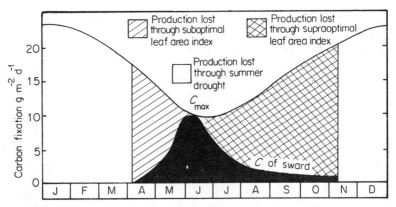

Figure 2.8 Sources of loss of potential production of *Trifolium subterraneum* under southern hemisphere, arid zone conditions. C_{max}: annual potential production. C: actual production. Reproduced with permission from J. N. Black, An analysis of potential production. . . .,*J. App. Ecol.* **1**, 3–18, Figure 9 (1964).

Mediterranean type of climate with a hot, dry summer in which growth is stopped by drought, accounting for the departure from potential production during the period November to March. Growth recommences after rewetting of the soil in April but does not become maximal until June; the reason for this is that leaf area index is suboptimal during April and May so that much light is lost through the canopy to the soil surface. The build-up of leaf area index continues, through an optimum when real and potential production are equal, to a supraoptimal condition when the plants are carrying so much leaf that serious self-shading causes a large reduction of photosynthetic production. In this late season condition the leaves in the lowermost part of the canopy are below their compensation points and act as respiratory sinks.

Except for adaptation effects it is probably reasonable to accept the fact that light saturation curves are similar for most temperate zone crops but this is not likely to be true for many other plants. Decker (1955), for example, has drawn attention to the light saturation curves of sun and shade plants (see Figure 10.4). Shade plants often reach photosynthetic saturation at light intensities as low as one fifth of full sunlight while sun plants have much higher saturation values; indeed in some crop plants such as *Saccharum officinarum* light saturation has not been achieved at twice full sunlight intensity (Zelitch, 1971). Light saturated rates in sun-adapted species may be more than an order of magnitude greater than those of shade species (Bjorkman, 1968). Grime (1965) showed that shade plants often have lower inherent respiratory rates than sun plants, thus conserving resources when they are below compensation point and making more efficient use of photosynthesis when above compensation.

The most significant, recent contribution of photosynthetic physiology to the study of the sun and shade plant situation has been the discovery of plant species which are capable of reducing substomatal CO_2 concentration to near zero compared with most plants which do not reduce the value below about 50–100 v.p.m. By analogy with the light compensation point this value is called the CO_2 *compensation point* as it represents a balance between photosynthetic absorption and respiratory evolution of CO_2. The low CO_2 compensation species include *Saccharum officinarum, Zea mais,* various tropical grasses of the Panicoideae and Chloridoideae–Eragrostoideae, some *Amaranthus* and *Atriplex* species and various algae which have compensation values of less than 10 v.p.m. (this is the CO_2 concentration to which they are able to reduce the atmosphere of an illuminated enclosure at intensities well above the light compensation point). Discussion may be found in Moss (1962), Tregunna and Downton (1967) and Downton and Tregunna (1968).

The common factor linking these low CO_2 compensation species is the possession of the Hatch–Slack CO_2 fixation pathway in which early labelling first appears in C-4 compounds such as oxaloacetate and malate rather than in the C-3 phosphoglycerate of the more usual Calvin pathway (Hatch and Slack, 1966). Hatch–Slack plants also differ from Calvin plants in lacking the light-stimulated respiratory pathway termed photorespiration.

Photorespiration limits maximum photosynthetic rate in Calvin plants as it depends completely on concurrent photosynthesis for its substrates. By liberating CO_2, at the same time as photosynthesis is removing CO_2 from the leaf intercellular space, it is the cause of the high CO_2 compensation point of Calvin plants and its absence permits Hatch–Slack plants to reduce CO_2 to a very low level (2–3 v.p.m. in some species). Photorespiration is also stimulated by normal atmospheric oxygen concentrations, with the consequence that apparent photosynthesis is increased by reducing the atmosphere O_2 level (Warburg effect).

Hatch–Slack plants almost certainly have the very high light saturation values, noted above, because their low CO_2 compensation point establishes a larger gradient of CO_2 concentration between the leaf air space and the external atmosphere. Related to this efficient use of high light intensities it is not just fortuitous that sugarcane (*S. officinarum*) and maize (*Z. mais*) are amongst the highest yielding crop plants.

PRODUCTIVITY

Primary productivity is the integrated effect of ecosystem photosynthesis and represents the fixation of solar energy in organic matter. Gross primary production is the overall photosynthetic production of dry matter before correction for respiration, and net primary production is gross production minus plant respiratory loss.

In a juvenile ecosystem a large proportion of net primary production is preserved in increasing biomass which comprises the weight per unit area of primary producers (photosynthetic plants), a chain of consumer animals (herbivores and carnivores) and various decomposer microorganisms (fungi and bacteria). As time passes, succession leads to a climax association in which the biomass has achieved equilibrium with input (primary production) and output (respiratory loss from plants, animals and decomposers). Investigation of temperate zone ecosystems rarely reveals such biomass-stable situations (Rodin and Bazilevich, 1967) and it seems unlikely that many temperate ecosystems are fully mature. Considering the comparatively short period between the end of the last glaciation and the beginning of extensive human interference with vegetation in the northern hemisphere, this conclusion does not seem unreasonable. In tropical forest with much longer ecological history, biomass stability is probably fairly common and net primary production is equalled by total community respiration (Odum and Pigeon, cited in H. T. Odum, 1971).

The growth to peak plant biomass, for example in a forest ecosystem, is related to the relative amounts of photosynthetic and non-photosynthetic plant tissue and to the increase of leaf area index to a level where self-shading reduces integrated lower canopy photosynthesis below the compensation point. The production of secondary, non-photosynthetic tissue to support an increasing leaf area is governed by a law of diminishing returns in which every increment of photosynthetic surface demands an ever larger maintenance respiration to support it.

This type of interaction is complicated by a number of factors; for example, some plants show self-pruning of lower branches under self-shading conditions. This is characteristic of many conifers and relieves the plant of leaves which are often near or below the compensation point. Many herbaceous plants maintain a constant leaf area index by sequential death of lower leaves: note, for example, the yellowed appearance of an overgrown lawn after cutting. In other cases the canopy architecture may permit plants to carry a very large leaf area index without serious self-shading. Plants with a very steep angle of leaf insertion permit deep penetration of diffuse light into the canopy, thus many grass and grass-like ecosystems have very high productivity under eutrophic conditions.

Another complication is the multiple stratification of many forest ecosystems. Each layer takes a portion of the incident light but the sub-layers are physiologically and often morphologically adapted to low light intensities so that the efficiency of the whole system is very high. By contrast, some ecosystems carry a large, non-photosynthetic surplus of respiring tissue; for example, many xeromorphic communities of sclerophyllous or succulent species which contain collenchyma or water storage parenchyma. Such plant associations are inherently inefficient.

The efficiency of primary production may be expressed as the percentage of solar radiant input which is trapped by the photosynthetic mechanism. Maximal interception of light and minimal maintenance respiration appear to be the criteria for high efficiency. The highest transient efficiencies in nature occur in shallow marine coral reef waters and shallow freshwater springs in which there is little requirement for mechanical tissue and the diffusion path to photosynthetic cells may be no more than 2–3 cells in length. Odum (1957) recorded daily efficiencies of 5% or more for *gross* primary production by *Sagittaria lorata* and algal species in freshwater springs in Florida. Efficiencies of this order are rare in nature and are maintained only for short periods in agriculture. Alberda (1962) measured *net* primary production efficiencies of up to 4% in grass swards over the whole growing season and of 5–6% in sugar beet for short periods after the canopy became closed. Annual efficiency figures are much lower: Ovington (1961), in an energy flow study of *Pinus sylvestris* woodland, found an annual efficiency of 2% over a period of 55 years. This efficiency level is comparable with annual agricultural efficiencies and suggests that coniferous forestry may be a very efficient use of poor quality land. The very low albedo (reflectance) of needle-leaved canopies and the fact that they may photosynthesize during the winter months may be a reason for their rather high efficiency.

Annual efficiencies of most terrestrial ecosystems are well below this level. Wassink (1968) discusses the various attempts which have been made to estimate the efficiency of the earth's surface in photosynthesis. Based on Schroder's very early estimates of 0·5% for the earth's forests and 0·15% for the non-forested areas he suggests an overall terrestrial efficiency of 0·3%. The estimation of efficiencies on this broad basis is so fraught with error that the values must be treated with caution, but these figures may be trusted well within an order of magnitude.

Though the inherent characteristics of the photosynthetic organisms make a considerable contribution to productivity level, the overall control is climatic, specifically in relation to radiant input. Figure 2.9 shows the distribution of productivity on a geographic–climatic basis which may be correlated with the curves of annual energy income shown in Figure 2.10. Under the continuously high radiation conditions of the tropical zone, high temperatures are maintained and, with sufficient rainfall, productivity levels are exceptionally high. If rainfall is less, then high radiant income results in very arid conditions which support a desert or savanna vegetation with much lower production potential. The much lower radiant input during the winter months in high latitudes lowers temperatures, causing cessation of plant growth, and also limits photosynthetic energy availability. Production in the temperate zone is thus limited to half of the year and, in the sub-arctic zone, to much less than this. It should be noted that the daily radiation available for photosynthesis during the growing season in high latitudes is approximately the same as that which is constantly available in the tropics.

The environmental factors limiting plant primary productivity are the same as those which were listed as governing photosynthesis. The integrated daily input in the 0·4–0·7 μm spectral range represents the maximum available energy for potential production but this input also causes other climatic effects such as temperature change which may, themselves, be limiting.

Carbon dioxide has a fairly constant concentration of between 300 and 350 v.p.m. in the atmosphere, which is limiting to photosynthesis at high sunlight intensities, suggesting that the upper canopy of many ecosystems must be CO_2-limited when wind speeds are low and there is little turbulent mixing. CO_2-fertilization of glasshouse atmospheres produces significant increments of yield for this reason.

Air temperature is a factor which is important in controlling productivity but, because low temperature is caused by low radiant input in many cases, it is difficult to separate the two effects. When radiation levels are high there is evidence that plant leaves can photosynthesize with air temperatures near or below freezing because the leaf tissues are warmed well above air temperature (Tranquillini, 1964). In higher plants, prolonged exposure to low air temperature or to radiative cooling under clear sky leads to tissue freezing and possible damage, while freezing of the soil seriously inhibits water uptake so that both photosynthesis and plant survival may also be limited by water deficit in a high radiation environment. Some of the lower plants are more tolerant; for example, Tranquillini cites the alpine lichen *Parmelia encausta* as having a photosynthetic temperature optimum between 0 and $-10°C$ and *Stereocaulon alpinum,* a photosynthetic minimum of $-23°C$.

All of the physiological processes of the plant are, to some extent, rate-limited by temperature so that increasing air temperature is accompanied by increased rates of most plant processes. Respiration is more sensitive to

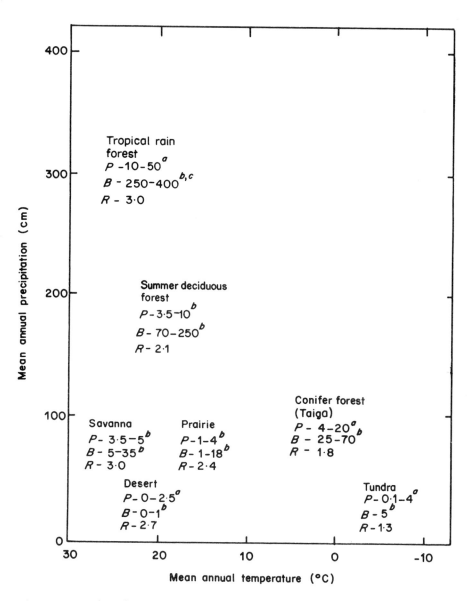

Figure 2.9 The climatic-geographical distribution of primary production, biomass and radiation input. P – primary production (t ha^{-1}); B– Biomass (t ha^{-1}); R – solar radiant input (k cal m^{-2} year^{-1} 0·3 – 3·0 μm). Production and biomass data from Whittaker (1970)[a], Rodin and Bazilevich (1967)[b] and Odum (1971)[c]. Radiation data derived from Gates (1962).

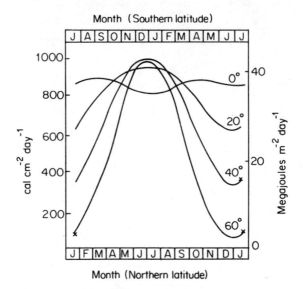

Figure 2.10 Variation in solar radiation with latitude and time of year. Reproduced with permission from R. O. Slatyer, *Plant-water Relationship*, © Academic Press, New York, 1967.

temperature than photosynthesis and, consequently, accumulation of soil organic matter tends to be less marked under tropical conditions when compared with cooler climates. Supraoptimal temperatures are rarely encountered except as a consequence of direct radiant heating but, as discussed on p. 29, this may be of considerable physiological significance, perhaps causing midday stomatal closure and inducing the midday photosynthetic slump which occurs under intensely sunlit conditions.

Soil water availability is an obvious limiting factor in arid zone primary production and is a major source of loss to world agriculture, but its effects are also felt in the temperature zone. Figure 2.11 shows the irrigation requirement during the driest year of twenty in England and Wales, countries not normally associated with agricultural water shortage. The values are calculated from potential transpiration estimates (p. 180) and predicted soil water deficits; they show, in the extreme south-east, a deficit of more than 33 cm (13 in) of rain. Wild ecosystems are less drought-susceptible than crop plants, but a deficit of this size will reduce the growth of most plants.

Plant growth, even on fertile agricultural soils, is often limited by nutrient deficiency and, most commonly, by shortage of one or more of the major nutrients: nitrogen, phosphorus and potassium. With water shortage, nutrient deficiency shares the first place as a limiting factor in world agriculture except, perhaps, for pathogenic limitation. Most wild

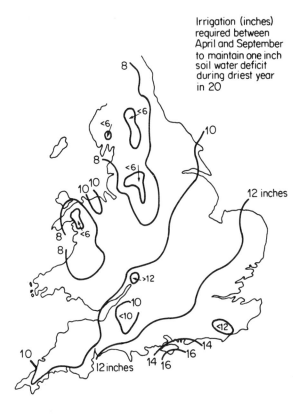

Figure 2.11 Irrigation requirement during the mean driest year in twenty for England and Wales. Modified, with permission, from W. H. Hogg, *Atlas of long-term irrigation needs for England and Wales,* Ministry of Agriculture, Fisheries and Food, London, 1967, Map C57.

ecosystems show either a surge of growth or a change of species composition after major nutrient fertilization, for example Willis and Yemm (1961) and Willis (1963) found a very large increase in productivity and an alteration in species composition after NPK fertilization of a calcareous sand dune grassland (see p. 290). Bradshaw (1969) cites the case of chalk grassland in S. England which was once thought of as summer drought-limited but now gives heavy agricultural yields after fertilization.

One of the most marked illustrations of this nutrient effect is the historic Park Grass experiment which was established at Rothamsted Experimental Station in 1856 and continues to the present day. The experimental plots were laid down to investigate the effects of different fertilizer treatments on a grassland of a neutral silt–loam overlying chalk. The original grassland, as it persists in the untreated plots, has an annual dry matter yield of 14–18 cwt acre^{-1} (1·8–2·3 t ha^{-1}) while the highest yielding plot, receiving lime and NPK, gives 69 cwt acre^{-1} (8·65 t ha^{-1}) (Thurston, 1969).

BIOMASS AND PRODUCTIVITY

The relationship between biomass and productivity is a complex function of ecosystem maturity, food chain effects and environmental factors. The various transfers of energy in food chains are inefficient, with the consequence that a large amount of photosynthetic production is required to support a comparatively small consumer biomass.

The various links in the ecosystem food chain may be represented as a sequence of trophic levels of which the first is the primary producer (autotrophs), the second is the primary consumer (herbivorous heterotrophs) and the third and fourth levels are the secondary and tertiary consumers (carnivorous heterotrophs). In parallel with this chain of organisms is another group, the decomposers or heterotrophic microorganisms.

If the various trophic levels are quantified in terms of energy flow, they may be represented as a pyramid of organization (Figure 2.12a) in which a large, annual energy fixation in primary production supports sequentially smaller energy flows at each consumer level, the remainder having been dissipated to the environment as heat generated in respiration. This loss is usually greatest at the producer–primary consumer transfer while the remaining transfers become rather more efficient.

If the pyramid is represented as relative biomass (Figure 2.12b) the relationship is usually similar to that of energy flow for terrestrial ecosystems. Occasionally, the biomass pyramid is reversed (Figure 2.12c) for example, when microorganism primary production is consumed by filter feeding aquatic or marine herbivores. Odum (1953) suggests that this reversal is a consequence of the inverse size : metabolic rate law. Small organisms have a higher metabolic rate per unit of body weight than large organisms so that a low biomass of primary producing algae is capable of supporting a larger biomass of consumers. This relationship demands continuous rejuvenation of the ecosystem (prevention of biomass accumulation) by removal of mature algal cells to maintain an optimal photosynthetic concentration in which serious self-shading cannot occur. This condition is satisfied by molluscan or crustacean filter feeding on planktonic algae or, perhaps, molluscan browsing of thin films of bottom-living algae.

Artificial culture of green algae in shallow tanks of nutrient solution, with continuous filtration or centrifugation to remove mature cells, also simulates the filter feeding situation and results in very high production and efficiency levels. Odum (1971) cites, for a continuous culture pilot plant, an algal

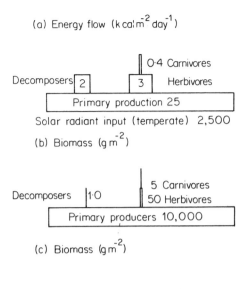

(a) Energy flow (kcal m^{-2} day^{-1})

0·4 Carnivores

Decomposers 2 | 3 Herbivores

Primary production 25

Solar radiant input (temperate) 2,500

(b) Biomass (g m^{-2})

5 Carnivores

Decomposers 1·0 | 50 Herbivores

Primary producers 10,000

(c) Biomass (g m^{-2})

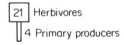

21 | Herbivores

4 Primary producers

Figure 2.12 Energy flow and biomass pyramids. (a) Energy flow in immature summer deciduous woodlands. Radiant input in the photosynthetic range (0·4 – 0·7 μm). (b) Biomass distribution in the same woodland. (c) Inverted biomass pyramid for phytoplanktonic primary producers and zooplanktonic and benthic faunal consumers in the English Channel. Data for (a) and (b) are hypothetical values based on Rodin and Bazilevich (1967), Edwards, Reichle and Crossley (1970) and Odum (1971) and (c) Harvey (1950).

production of 0·30 MJm^{-2} d^{-1} (3% efficiency) compared with 0·31 MJm^{-2} d^{-1} for sugar cane under a much higher light regime (1·8% efficiency).

The inverse size: metabolic rate law also influences the metabolism of decomposer microorganisms, thus a very small standing biomass is capable, annually, of oxidizing a great deal of organic matter (Figure 2.12a, b). The efficient functioning of the decomposers is not only of direct significance in ecosystem energy flow, but is also responsible for returning inorganic nutrients to the soil where they are available for re-use. The rapid metabolism of soil organic matter may, thus, prevent the whole ecosystem energy flow from becoming nutrient-limited at the primary production level. The mycorrhizal members of this soil community of microorganisms may be of more direct significance in recycling inorganic nutrient elements without allowing them to enter the soil pool of dissolved nutrients.

ENERGY FLOW

The concept of energy flow, the trophic–dynamic aspect of ecology, was first critically formulated in a classic paper by Lindeman (1942). This laid the foundation of more recent approaches to ecosystem analysis in which energy flow has been measured and related, as a driving force, to such other aspects of community dynamics as nutrient cycling. As Pomeroy (1970) has pointed out, this approach to the systems analysis of flows and cycles in ecology is a useful alternative to the limiting factor interpretation which has placed a constraint on the understanding of the multifactorial interactions which are so often encountered in natural systems.

Major (1969) has warned that energy flow studies may produce the same paper figures in arid zone and subarctic environments. This criticism is overcome by the proviso that the energy flow paths should be related to individual species' function in the ecosystem, in which case the approach provides a most useful, broad interpretation of ecosystem metabolism and interrelationship.

Energy flow analysis takes a sequence of thermodynamically logical steps in which the original solar radiant energy income is passed from one trophic level to the next and, at each step, a portion of the energy becomes dissipated to the environment as heat generated in respiration. As a generalization it may be assumed that the energy loss at each step represents a tenfold reduction (order of magnitude) though the early transfers are often less efficient and the later ones more efficient than this.

Figure 2.13 shows a hypothetical energy flow diagram for four trophic levels. The heavy-type boxes are symbolic of the relative biomasses of the component organisms while the flow paths indicate the relative energy flows, either as organic matter transfer (unshaded) or as radiant input or heat loss (stippled). This type of approach provides a great deal of flexibility in analysis as the individual boxes and pathways may, if necessary, be subdivided to represent species or groups of similar organisms. The model is directly analogous to the flow diagram which must be constructed in a systems analysis (p. 17) and, as such, quantifying the various flows, storage and transformations allows computer simulation of the ecosystem energetics. As shown on p. 249, the energy flow model may also be linked to the nutrient cycle, which is not only rate-dependent on energy flow but also rate-limiting if the rate of nutrient return through decomposer metabolism is inadequate or if the initial supply is deficient.

51

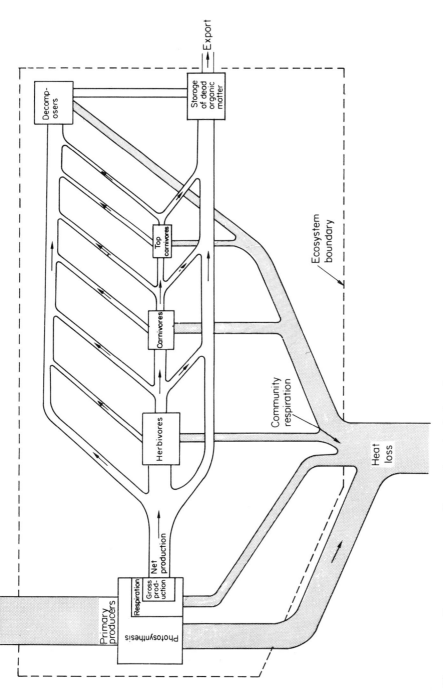

Figure 2.13 Ecosystem energy flow web. Stippled pathways: heat or radiation. Unshaded: energy transfer in organic matter. Heavy boxes represent relative biomasses. Note the one-way flow and large respiratory loss at each transfer. The diagram is not to scale as the energy flow through the top carnivores may be several orders of magnitude less than net primary production.

MEASURING PRODUCTIVITY AND ENERGY FLOW

Energy flow must be computed from the changing quantities of organic matter in the various trophic levels during the course of a year. Net primary production (P_n) may be calculated as the change in biomass $(B_2 \approx B_1)$ over a sampling period:

$$P_n = (B_2 - B_1) + L + G$$

where L is the loss by death and shedding and G is the loss to consumer organisms of all types during the sampling period. This is the most fundamental procedure for measuring primary production and requires the careful design of sampling techniques to measure the above- and below-ground components of biomass, to measure leaf and branch litterfall and, most difficult, to establish all consumer losses. These latter data must be drawn from estimates of consumer populations and metabolic rates or, in the case of large grazing mammals where capture is impossible, from observation and estimation of consumption. Newbould (1967) discusses the basic concepts and techniques of sampling for primary productivity.

Estimation of biomass change in the various consumer levels poses similar problems to the estimation of consumption. Changes in population number, population age structure and individual weight must all be considered. There are innumerable difficulties of trapping, particularly in relation to random sampling and, as with primary production sampling, there must be consideration of functional damage to the ecosystem either by excessive loss of individuals or imposed changes in population behaviour. Southwood (1966) discusses many of these problems both in relation to population sampling and to establishment of energetic data for animal communities.

Assuming that all these problems may be solved, the researcher then has available information from which the total dry weight per unit area may be calculated for all organisms at each trophic level. The biomass pyramid of Figure 2.12 is derived from data of this sort. To convert this dry weight data to energetic terms an oven-dry subsample of each component must be analysed for calorific content. The material is ground to a fine powder and either compressed into a tablet or loaded into a gelatine capsule after weighing. The sample is then loaded into a bomb calorimeter where it is ignited in oxygen, taking the precautions outlined by Lieth (1968a). The calorific value is determined from the rise in temperature in the calorimeter. For herbaceous plants, mean calorific values are fairly constant, approximating 4 kcal g^{-1}; a rather higher figure of $4 \cdot 7 \text{ kcal g}^{-1}$ seems more appropriate for woody plants while seeds are often well over 5 kcal g^{-1} $(1 \cdot 0 \text{ kcal g}^{-1} = 4 \cdot 18 \text{ kJ g}^{-1})$.

Seasonal changes of biomass may then, using these factors, be converted to energy units per unit area. With the addition of estimates of the metabolic utilization of energy amongst the consumers and decomposers, a complete energy flow sheet may be drawn-up for the whole ecosystem. This may be balanced as an energy budget by adding the solar energy input from which the primary production derives. This energy input is best measured as the difference between the annually or seasonally integrated readings of two glass-domed radiometers, one of which reads at all wavelengths up to 3·0 μm and the other which reads only infrared input, being filtered to cut off visible light below 0·7 μm.

Chapter 3

Soils

Almost any part of the earth's surface which supports vegetation also bears a covering of soil. It is thus possible to define a soil as 'the material in which plants root' though this is a limited definition as soil is formed by the long-term modification of parent geological material through a combination of biological, climatic and topographic effects, the multifactorial interactions of which lead to a great variety of potential soil types. A further strong source of variation is the feedback between organism and environment. Plant species are often habitat–specific but their presence and activity gradually produce alterations in the environment which may influence their own vitality and allow other species to invade and supersede them.

Earlier views of pedogenesis suggested that the sequence of invasions by organisms contributed to a monocyclic situation in which a soil would gradually evolve, with its associated vegetation, until it reached an equilibrium with the prevailing climate, but it is now accepted that many soils are polycyclic, their pedogenesis having been interrupted and diverted by repeated intervention of external factors. Soil formation must, then, be looked upon as a long-term process of complex interactions leading to the production of a mineral matrix in intimate association with interstitial organic material both living and dead.

The earth's crust is composed of a relatively few rock-forming minerals mainly based upon various alumino-silicates, metal silicates and variable quantities of free silica (SiO_2). In their primary form these minerals arose by the solidification of molten magma and are referred to as igneous rocks. Weathering and denudation leads to their mechanical and chemical breakdown and redistribution; the secondary products of this process are the sedimentary rocks which form from the deposition of weathering products in ocean basins (water transport) or on terrestrial surfaces (aeolian, glacial or solifluction transport). Some water-deposited sediments may be biogenic or of chemical origin; for example, most limestone rocks. A third category of rocks arises by the influence of heat and/or pressure on pre-existing igneous or sedimentary rocks: these are named metamorphic. Table 3.1 gives some examples of these varied rock types.

The alumino-silicate minerals are the most widespread constituents of igneous rocks and may be subdivided into two main groups: the feldspars (potassium, calcium–sodium and sodium alumino-silicates) and the micas (complex potassium alumino-silicates with included hydroxyl groups, often containing magnesium, iron, sodium and fluorine).

These minerals are all comparatively easily weathered and it is their breakdown products which form the fine-grained mineral matter of most soils and, because of their rather complex constitution, provide most of the mineral nutrients required by plants. The small particle size of this clay mineral gives the soil alumino-silicate fraction a large specific surface which is physicochemically very active in the nutritional and water relationships of the soil.

The coarse-grained particles of most soils are derived from free silica in the parent rocks. Silica is very resistant to both chemical and physical weathering processes, at least under temperate climatic conditions, and tends to accumulate in the soil mantle as a mechanical matrix of sand and gravel particles amongst which the alumino-silicate clays are distributed. It is the denudation of this material which leads to the formation of the sedimentary rocks and these, having already passed through one cycle of weathering process, tend to be softer and more rapidly reweathered than the hard, primeval igneous rocks. For similar reasons the metamorphic rocks have more physical and chemical coherence than their igneous or sedimentary progenitors and so resist the weathering process more successfully. Table 3.2 summarizes the weathering characteristics of the various rock-forming minerals.

THE WEATHERING PROCESS

Bare rock surfaces and developing soils are exposed to a range of physical, chemical and biological processes which lead to mechanical and chemical disruption of their components (Table 3.3). Chemical and physical processes alone give rise to abiotic crusts of weathering products which are only the raw material of soil formation. However, careful investigation shows that most bare rock surfaces do not remain free of life for very long and the physicochemical weathering processes are soon reinforced by the often potent effects of numerous microorganisms. Jacks (1965) stresses the role of the lichen symbiosis in this early phase, suggesting that lichens are able to extract nutrients which would be unavailable to higher plants. Retention of water by the thin layer of lichen, fungal and bacterial organisms on rock surfaces prolongs the period during which chemical processes can proceed, splitting the rock

Table 3.1 Soil-forming rocks and their composition

Rock-type	Example	Composition
IGNEOUS — QUARTZ RICH (acid magmas)	GRANITE	Mica, alkali feldspar, quartz in coarse, discrete crystals
IGNEOUS — QUARTZ POOR (basic)	BASALT	Fine crystalline structure of plagioclase feldspar and augite
SEDIMENTARY — Coarse texture Water-deposited	SANDSTONE SAND	Quartz grains cemented with Fe_2O_3, $CaCO_3$ or loose.
SEDIMENTARY — Fine texture Water-deposited	MUDSTONES SHALES CLAYS	Predominately finely particulate alumino-silicates with various cementation such as Fe_2O_3, $CaCO_3$
BIOGENIC CHEMOGENIC	LIMESTONE	$CaCO_3$ as organic remains or as chemically precipitated material. May be very pure as in chalk
AOLIAN	LOESS	Wind-transported deposits of medium texture. Often $CaCO_3$-rich
GLACIAL	DRIFT TILL BOULDER CLAY	Often heterogenous mixture of diverse geological material. Often clay-rich and impermeable
METAMORPHIC — Metamorphosed sedimentary	MARBLE SLATE	Metamorphosed limestone Metamorphosed shale
METAMORPHIC — Metamorphosed igneous	GNEISS	Metamorphosed quartzites and granitic rocks

Table 3.2 Rock-forming minerals, their composition and form in soil

Rock-forming mineral		Chemical constitution	Form in soil
ALUMINOSILICATES	FELDSPARS	Alumino-silicates of K, Na, Ca, Alkali feldspars are KNa forms and plagioclase feldspar NaCa	Mainly as clay-sized particles in much modified form. Very little unaltered feldspar
	MICAS	Hydroxyl-alumino-silicates of K, Na, (KMgFe), K(Mg) and K(Fe)	Similar to feldspar in producing much modified clay mineral
SILICATES	OLIVINE AUGITE GARNET PYROXENE AMPHIBOLES	Mg, Fe, Ca, Na and Mn silicates	Not usually represented in any quantity
	QUARTZ	SiO_2	Persists in soil as sand and gravel, often forming main matrix
ALUMINIUM OXIDE IRON OXIDE	(Al_2O_3) (Fe_2O_3)	Fairly low concentration in most rocks	Becomes concentrated by loss of silica in tropical wet conditions (ferrallitization). High concentrations in some tropical soils (oxisols)

Table 3.3 Weathering processes

Physical	Chemical
WETTING–DRYING E.g. Disruption of layer lattice minerals which swell on wetting	**HYDRATION** E.g. Reversible change of haematite to limonite which is accompanied by swelling and so disrupts cementation of sandstones etc. $Fe_2O_3 \leftrightharpoons Fe_2O_3 3H_2O$
HEATING–COOLING E.g. Disruption of heterogeneous crystalline rocks in which inclusions have differential coefficients of thermal expansion. Surface flaking of large boulders, particularly in arid climates, due to sun heating	**HYDROLYSIS** E.g. Silicate breakdown $K_2Al_2Si_6O_{16} \rightarrow Al_2O_3 2SiO_2 2H_2O$ Orthoclase Kaolinite K and surplus Si are washed away in solution
FREEZING E.g. Disruption of porous, lamellar or vesicular rocks by frost shatter due to expansion of water during freezing	**OXIDATION–REDUCTION** E.g. $Fe^{3+} \leftrightharpoons Fe^{2+}$ causes disruption of cementation as Fe^{2+} is much more soluble than Fe^{3+}
GLACIATION E.g. Physical erosion by grinding process	**CARBONATION** E.g. $CaCO_3 \leftrightharpoons Ca(HCO_3)_2$ leads to solution loss of limestone or disruption of $CaCO_3$ cemented rocks as the hydrogen carbonate is more soluble than the carbonate
SOLUTION E.g. Removal of more mobile components such as Ca, SO_4, Cl etc.	
SAND BLAST E.g. Erosion of upstanding rocks in arid, desert conditions	**CHELATION** Essentially a consequence of biochemical activity, various metals being dissolved as chelates with organic products of plant and microorganism activity

alumino-silicates by hydrolysis and carbonation into the simpler clay alumino-silicates. Carbon dioxide released by respiratory processes must further accelerate this type of weathering. Photosynthetic energy fixation by the algal component of the lichen will increase the available organic matter at the surface and Rogers, Lang and Nicholas (1966) have shown the appreciable contribution which the blue-green algae of the lichens may make to nitrogen fixation. Invasion by bryophytes increases the photosynthetic capacity of the living layer, more organic matter and weathered rock becoming incorporated in what must now be recognized as a thin soil layer covering the surface.

Various exudates from these organisms, other than respiratory CO_2, are likely to speed the pedogenetic process; for example, some of the lichen acids have strong chelating properties; organic acids generally are potent in dissolving mineral components, while some species of bacteria directly influence the solubility of nutrient elements and cementing compounds. Mulder, Lie and Wolderdorp (1969) cite the solubilization of phosphorus by microorganism-formed organic acids and microorganism reduction of insoluble ferric phosphates. Further examples are the bacterial reduction of manganese and iron, increasing their solubility. The cycles of sulphur, nitrogen and phosphorus are all strongly governed by microorganisms through the sizes of the organic and inorganic pools and the rates of change between soluble and insoluble, available and unavailable forms. Further discussion may be found in Campbell and Lees (1967), nitrogen; Cosgrove (1967), phosphorus; Freny (1967), sulphur and Ehrlich (1971), minor elements.

Colonization of a juvenile soil by higher plants adds yet another complication to the soil-forming process, greatly increasing the energy-fixing capacity of the surface and increasing the supply of decaying organic matter. Soluble organic compounds also diffuse into the rhizosphere zone from the roots and wash into the soil surface from leaf-drip. Deeper penetration of roots will tend to increase the depth range of the cyclic processes involving nutrient elements, soluble elements leached downward being returned to the surface by transport through the plant.

Rock weathering is therefore, for a short time, a physicochemical process but rapidly becomes biogenic with a consequent increase in the overall rate. Pedogenesis may thus be considered as a biological phenomenon by which crusts of weathered rock debris are converted to true soils comprising a complex mineral matrix in association with a great range of organic compounds; very often carrying a rich microorganism population which is a reflection of the nature of the parent material and its interaction with climate, topography, plant cover and age.

COMPOSITION OF SOIL

The components of a soil can be classified into four categories:

(i) A matrix of mineral particles derived by varying degrees of breakdown of the parent material.

(ii) An organic component derived from long- and short-term additions of material from plants, animals and microorganisms above and below ground.

(iii) Soil water held by capillary and adsorptive forces both between and at the surface of the soil particles, its amount varying with the balance between precipitation, evapotranspiration loss and drainage. Soil water is in reality a dilute solution of many different organic and inorganic compounds and forms the immediate source of plant mineral nutrients.

(iv) The soil atmosphere occupies the pore space between soil particles which, at any time, is not water-filled. Its composition differs from the above ground atmosphere as it is normally lower in oxygen and higher in carbon dioxide content. In very wet soils oxygen may be almost absent as it can only be replenished by diffusion in solution from the surface.

The relative proportions of the mineral and organic solids to the liquid-

or gas-filled pore space are determined by the particle size distribution of the mineral matrix (the soil texture) and the binding of these fundamental particles into larger units or aggregates. The pore spaces between aggregates are much larger than those between the textural particles: the aggregated condition is often described as soil *structure*. Aggregation is related to the organic content of the soil and to its degree of microbiological activity.

Soil particle size distribution is usually expressed as frequency in arbitrary size-classes: a commonly used set of classes is that of the International Society of Soil Science:

		Diameter range (mm)
I	Coarse sand	$2.0 - 0.2$
II	Fine sand	$0.2 - 0.02$
III	Silt	$0.02 - 0.002$
IV	Clay	0.002

This classification, though arbitrary, achieves a separation which is pedologically useful, the boundary between silt and clay roughly marking a transition in the properties of the material. Much physicochemical activity of soils is a function of the surface area: silt, fine sand and coarse sand have specific surfaces, respectively, of less than 1, 0·1, and 0·01 m^2 g^{-1} compared with the clay fraction which commonly ranges between 50 and 700 mg^2 g^{-1} (Fripiat, 1965). Hence the clay content determined by particle-size analysis gives a very good idea of the ion-exchanging, adsorptive and capillary activities of the soil. This physical difference is also accompanied by chemical differences. The principal components of the sand/silt fractions are quartz, unmodified silicates and alumino-silicates from the parent material (mainly feldspars, micas, pyroxenes, amphiboles, olivene, etc.), iron oxides, and a few larger particles of true clay mineral. In temperate soils quartz is usually the commonest mineral, making up 90–95% of the sand fraction in soils derived from sedimentary rocks. It may be less common in igneous soils depending on the content of the parent rock.

The clay fraction is typically different from the sand/silt, comprising an assemblage of minerals formed by weathering of alumino-silicate rock minerals. The coarsest of the clay fractions may occasionally contain a little finely particulate quartz and mica but the finer material is almost entirely true clay with some iron and aluminium oxide. The clays are layer-lattice crystalline materials based upon two subunits of silicon–oxygen and aluminium–oxygen; the details of the structure will be discussed in the next chapter. Their small size and consequent large surface area, coupled with strong ion-exchanging activity, is responsible for the importance of soil clay as the major source and reservoir of plant nutrients and in governing soil water relationships and aeration characteristics.

SOIL DEVELOPMENT—PEDOGENESIS

The comparatively recent realization that many geochemical processes involve biological activity tends to blur the distinction which was previously made between geological weathering and pedogenesis. However, the latter process is recognizable as that which develops the characteristic morphology and 'metabolism' of a soil by which it may be recognized to type in the same way that the morphology and physiology of an organism permits its specific identification. The phenotype of an organism is developed by the interaction of genetically determined raw material with the environment, while the manifested soil type analogously develops from the influence of environment on geological parent material.

Joffe (1936) subdivided soil-forming factors into two main groups, passive and active:

Passive factors.
　　1. Parent material.
　　　　(a) Physical constitution.
　　　　(b) Chemical composition.
　　2. Topography.
　　　　Influential on both macro- and micro-scale through its influence on drainage, solifluction and insolation.
　　3. Time.

Active factors.
　　1. Rainfall.
　　　　(a) Determines the direction of solute translocation (up or down) according to precipitation/evaporation (P/E) ratio.
　　　　(b) In conjunction with other factors determines depth to water table and therefore capillary water and aeration status.
　　　　(c) Indirect effect through influence on vegetation.
　　2. Temperature.
　　　　(a) Interacts with rainfall in governing P/E.
　　　　(b) Influences rates of physicochemical processes, thus controlling rate of organic turnover in the soil.
　　　　(c) Influences the growth rate of vegetation and other organisms.
　　3. Humidity/Evaporation.
　　　　Low relative humidity gives high evaporation rates and offsets the influence of high rainfall by reducing amount of downward water movement.
　　4. Wind.
　　　　Of minor importance except that high maintained wind speeds cause increased evapotranspiration and may limit vegetation growth.

5. Biosphere effects.

The activity of living organisms must be considered definitive in pedogenesis as they speed-up and modify the physical and chemical processes which would normally occur in their absence. Their narrow ecological tolerance ranges tend to amplify the environmental differences already influential in pedogenesis.

(i) Phytosphere.

    (a) Direct plant activities such as the secretion of organic acids and enzymes. Respiratory production of CO_2.

    (b) Input of organic matter after death.

    (c) Micrometeorological influence of vegetation cover.

(ii) Zoosphere.

    (a) Direct interaction between primary production and consumption by animals with consequent effect on input of organic matter to the soil.

    (b) Soil-dwelling micro-arthropods, molluscs, lumbricids, etc., have a direct influence on organic matter turnover and incorporation.

(iii) Microorganisms.

Bacteria and fungi play a most important part in the geochemical, biochemical and biophysical processes of pedogenesis.

The parent material factor strongly influences soil composition both through its chemical characteristics, some of which are handed on to the derived soil, and through its physical constitution which influences, for example, leaching rate and soil aeration. Generally speaking, silicate-rich rocks with a high content of plant nutrient elements produce soils which are eutrophic or mesotrophic and highly favourable for plant growth, while silica-rich, acid rocks tend to produce oligotrophic soils with a rather depauperate vegetation. Physically, a high silica soil will produce a sand-rich or coarse-textured soil which is prone to leaching loss of soluble nutrients. This contrasts with silicate-rich rocks which may give a high clay content and good nutrient status but provide so little coarse pore space that the soil may be badly aerated.

Topography influences soil formation through drainage and retention of water. The soils of waterlogged hollows and, on a larger scale, of valley floors and extensive ill-drained plains, often have quite different characteristics from adjacent well-drained soils on identical parent material. Such toposequences have been described as *catenas* (p. 93). Most important in this context is the aeration status of the soil in relation to decomposition of soil organic matter and to the redox characteristics of the soil. Atmospheric oxygen can diffuse freely into the air-filled pore space of a soil, but when it becomes water-filled and oxygen has to diffuse in solution, it is not able to keep pace with the respiratory demand of soil microorganisms. In consequence, the soil becomes anaerobic, its redox potential falls and its microflora is replaced by a population of anaerobes. The reduced rate of oxygen entry leads to an increase of organic matter in the soil and the fall in redox potential may convert iron to the ferrous state, increasing its solubility and mobility in the profile. Other inorganic consequences of the redox change will be described later.

Time as a factor in any ecological context is all too often neglected. The life span of many large, woody plants lies in the range 10^2 to 10^3 years and, as they may be of prime importance in the soil formative process, no soil can be considered to be in biotic equilibrium before many tree generations, representing many thousands of years, have passed by. Much of the land area in the Northern Hemisphere emerged from the late glacial period only 10,000 years ago so that it is doubtful whether its soils and vegetation can have been in any sort of equilibrium before extensive human interference became important during the Bronze Age. The situation may be different in Equatorial areas where, but for the possible interpolation of pluvial periods, the climate has probably remained stable for 1–2 million years.

The time required for soil formation appears to be widely variable as it is so strongly influenced by climatic and parent material factors. Evidence from buried soils under Bronze Age tumuli in Northern Europe suggests that extreme podsolization has taken place in many heathland areas during the past 4000 years, probably following human destruction of the original forest cover (Dimbleby, 1962) and the same author cites other evidence which suggests that podsolization can be even more rapid. Under tropical conditions, soil formation may occupy a much longer time-scale. Mohr and van Baren (1959) wrote: 'The process of laterisation in the tropics . . . must be regarded as a geological phenomenon. This is because the time factor, next to climate, is of the greatest importance in determining the final stage in the cycle of weathering'. Soil formation may thus range from being a short-term process comparable in scale to the generation time of many plants to a long-drawn-out and possibly cyclic process over periods of near geological time.

Rainfall appears to play a dominant part in pedogenesis, coupled with evaporation rate and the drainage characteristics of the soil and site. With a continuous excess of precipitation over evaporation, there is uninterrupted downward movement of water in the soil profile. Initially, soluble salts such as carbonates and more mobile ions such as sodium are leached from the surface layers, a distinct front of solute loss advancing down the profile with time. With more coarsely textured soils, or a greater surplus of percolating water, material in the clay fraction may be washed downward and redeposited below (*lessivation*); in many temperate soils this is the first evidence that *illuviation* of the *B* horizon is occurring (see p. 66). With extremes of downward water movement, coupled with soil acidity and nutrient deficiency, the iron oxide coating of the soil grains may become soluble and the iron, after translocation downward, is redeposited at a deeper level in the soil. Under these circumstances there is nearly always an accompanying transport of organic materials from the surface horizons and formation of a *B* horizon organic-rich layer. Other metal sesquioxides such as manganese and aluminium may be *eluviated* from the *A* horizon along with iron to reappear as concentrations in the *B* horizon.

Interruption of the high P/E regime by a dry season may temporarily stop the downward leaching or, in some cases, return soluble materials to the surface. If the dry period is extended it may come to dominate the pedogenic process, which will then resemble the one which would occur with a P/E ratio constantly less than one. Soluble materials are carried upward in the soil capillary matrix and, if the water table is sufficiently high, may lead to massive enrichment of the surface horizon with soluble salts (characteristically carbonates, sulphates and chlorides of calcium and sodium) and the development of 'alkali' and saline soils in arid zones.

Temperature obviously interacts strongly with the water balance effects described above. Solar radiation is the main energy source for water evaporation, hence high radiation, high temperature and low P/E tend to be correlated together. High temperatures have a further marked effect in increasing the rates of chemical and, in particular, biochemical processes. As the photochemical step of photosynthesis is rather insensitive to temperature, a rise tends to increase respiratory breakdown of organic matter more than it increases photosynthetic production. This is generally reflected in the increased organic content of soils formed at lower temperatures. Figure 3.1 shows the strong interaction between temperature, soil wetness and organic matter accumulation. In cold climates photosynthetic production exceeds the rate of decomposition in the soil but, as temperature increases, so bacteria take over from the fungi as the major decomposers and the rate increases. At about 25°C with good soil aeration there is a balance between production and breakdown, and above this temperature organic matter cannot accumulate.

If the soil is flooded, the reduced oxygen supply permits only anaerobic bacteria to function and organic matter continues to accumulate in waterlogged soils up to about 35°C. These relationships satisfactorily account for the high organic content of peat soils formed under conditions of waterlogging (temperate and tropical valley peats), high rainfall (blanket peats) or cold (tundra and alpine soils).

SOIL PROFILES

The interaction of these pedogenic factors permits the differentiation of a variety of soil types which may be defined by the nature of the mineral matrix, the vertical distribution of organic matter and the movement and

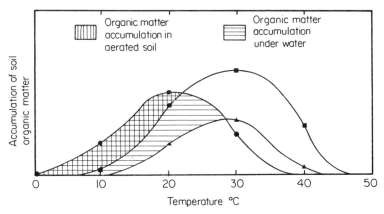

Figure 3.1 The relationship of soil organic matter accumulation to wetness and temperature. Reproduced with permission from E. C. J. Mohr and F. A. van Bahren, *Tropical Soils*, van Hoeve, The Hague, 1959, Figure 56, p. 280.

redeposition of various inorganic constituents. Soils are described and identified by reference to their *profiles*—the sequence and nature of the horizons (layers) exposed in a pit-section dug through the soil mantle. The dimensions of a definitive profile vary with soil type as some soils show little horizontal variability, whereas others may show recurrent variation over horizontal distances of several metres. The smallest three-dimensional volume of a soil needed to give full representation to such features is termed a *pedon* (Soil Survey Staff, 1960).

'A soil horizon may be defined as a layer which is approximately parallel to the soil surface and that has properties produced by soil forming processes but that are unlike those of adjoining layers' (Soil Surv. Staff, 1960). Horizons may usually be identified visually in the field but they also have chemical and physical properties which can be measured in the laboratory to confirm the diagnosis.

Horizons may be classified by reference to both position and constitution. Historically, the soil horizons of the Russian steppes were the first to be classified by the *ABC* terminology. The *A* horizons were the dark-coloured surface layers and the *C* horizons the underlying layers largely unaffected by the soil-forming process. The *B* horizons were transitional in properties and position. Adoption of this terminology in western Europe led to the concept of the *B* horizon as one into which materials were washed (illuviated) by downward water movement. This, however, raises difficulty with some tropical soils (latosols) in which the *A* horizon is recognizable but the horizon in the *B* position is enriched with sesquioxides, not by illuviation but by the leaching loss of other components (silicates and silica).

Five main groups of horizons may be defined and are designated *O, A, B,*

C and R. Their definitions are fairly broad in order to solve such problems as the terminology of sesquioxide-enriched horizons described above.

O horizons

Organic horizons forming above the surface of the mineral matrix and dominated by fresh or partially decomposed organic matter.

A horizons

Mineral horizons formed either at or adjacent to the surface and characterized by enrichment with organic matter and/or downward loss (eluviation) of soluble salts, clay, iron or aluminium and consequent enrichment with silica or other resistant minerals.

B horizons

Mineral horizons forming below the surface and having one or more of the following features. (i) Enrichment with inwashed clay (lessivation), iron, aluminium, manganese or organic matter. (ii) Residual enrichment with sesquioxides or silicate clays which has occurred other than by the removal of carbonates or easily soluble salts. (iii) Sesquioxide coatings of mineral grains sufficient to give a more intense colour than horizons above or below. (iv) Alteration of the original rock material to give silicate clays or oxides in conditions where (i), (ii) and (iii) do not apply.

C horizons

Mineral horizons below the B layer but excluding true bedrock and lacking any characteristics of the A or B horizons. Formed under conditions little affected by the pedogenic process. May or may not be the material from which the A and B horizons formed.

R horizons

Bedrock. May or may not be the material from which the A, B and C horizons formed (soil may develop in drift overlying bedrock).

It must be stressed that the O, A and B horizons are entirely pedogenic and are unrelated to geological layering; the process of horizon formation may in fact continue uninterrupted downward through a geological discontinuity such as the boundary of superficial drift and bedrock. The number and relationship of the soil horizons is governed, in particular, by the direction and quantity of water movement in the profile. With a high ratio of precipitation to evaporation (> 1) downward washing of materials predominates leading to the eluviation of surface (A) horizons and illuviation of B horizons. If the P/E ratio falls below one then capillary rise of water may occur, transporting soluble materials upward towards the surface.

Profiles may be further subdivided by numbering from the surface downward, e.g O_1, O_2, A_1, A_2, A_3, etc. Each subdivision is an integral unit and needs its own separate definition as a horizon. Transitions between horizons may be shown by the use of both respective symbols; thus A/B includes features of both an A and a B horizon. A number of other horizon characteristics may be indicated by the use of a lower case suffix: C_{ca} indicates a C horizon with calcium carbonate enrichment; A_p and A_i horizon disturbed by ploughing, and B_g a B horizon showing gleying: the evidence of reducing conditions consequent on waterlogging and poor aeration. Gleyed horizons are grey, blue-grey and green-grey due to the reduction of iron from the ferric to the ferrous condition. They may often be mottled with red-brown patches where ferric compounds have been redeposited in soil voids, root channels or around living roots (see p. 90).

SOIL CLASSIFICATION

Classifications of soil types are inevitably a little arbitrary as they constrain a natural continuum in a linear or branching system. The majority of soil classifications have been based on a taxonomic and/or genetic approach, though purely artificial classifications are possible. The earliest attempts were made in Russia at the end of the last century and were type-defined by a combination of climatic/vegetation data and soil morphology. During the past half-century the U.S. Department of Agriculture Soil Survey has been responsible for the continuous development and refining of a system much influenced by Dokuchayev's (1900) final classification of Russian soils (Table 3.4) which defined a group of zonal soils, roughly corresponding geographically with the great climatic

Table 3.4 A classification of soils (Dokuchayev, 1900)

Class A. Normal or Zonal soils			
Zones	1. Boreal	2. Taiga	3. Forest-steppe	4. Steppe
Soil types	Tundra (dark-brown soils)	Light-grey podsolized soils	Grey and dark grey soils	Chernozem
Zones	5. Desert Steppe	6. Desert	7. Subtropical and tropical forest	
Soil types	Chestnut and brown soils	Aerial soils Yellow soils White soils	Laterite or red soils	
Class B. Transitional soils			
8. Dry land moor soils or moor-meadow soils	9. Carbonate containing-soils (rendzinas)	10. Secondary alkali soils	
Class C. Abnormal soils			
11. Moor soils	12. Alluvial soils	13. Aeolian soils	

Reproduced from Soil Surv. Staff (1960).

68

zones of the earth's surface, and two other groups which depart from the nature of the zonal soils by reason of their unusual water relationships or parent material. The U.S. system in its revised form (Thorp and Smith, 1949) has been widely used in the mapping of American soils and has also been adopted, in modified forms, elsewhere. An outline of the classifica-

Table 3.5 A classification of soils: the U.S.D.A. revised system of Thorp and Smith (1949)

Order	Suborder	Great soil groups
Zonal soils	1. Soils of the cold zone.	Tundra soils.
	2. Light-coloured soils of arid regions.	Desert soils. Red Desert soils. Sierozem. Brown soils. Reddish-brown soils.
	3. Dark-coloured soils of semi-arid, subhumid, and humid grasslands.	Chestnut soils. Reddish chestnut soils. Chernozem soils. Prairie soils. Reddish Prairie soils.
	4. Soils of the forest-grassland transition.	Degraded Chernozem. Noncalcic Brown or Shantung Brown soils.
	5. Light-coloured podsolized soils of the timbered regions.	Podsol soils. Grey wooded, or Grey Podsolic soils. Brown Podsolic soils. Grey-Brown Podsolic soils. Red-Yellow Podsolic soils.
	6. Lateritic soils of forested warm-temperate and tropical regions.	Reddish-Brown Lateritic soils. Yellowish-Brown Lateritic soils. Laterite soils.
Intrazonal soils	1. Halomorphic (saline and alkali) soils of imperfectly drained arid regions and littoral deposits.	Solonchak, or Saline soils. Solonetz soils. Soloth soils.
	2. Hydromorphic soils of marshes, swamps, seep areas, and flats.	Humic Gley soils (includes Wiesenboden). Alpine Meadow soils. Bog soils. Half-Bog soils. Low-Humic Gley soils. Planosols. Ground-Water Podsol soils. Ground-Water Laterite soils.
	3. Calcimorphic soils.	Brown Forest soils (Braunerde). Rendzina soils.
Azonal soils		Lithosols. Regosols (includes Dry Sands). Alluvial soils.

Reproduced from Soil Surv. Staff (1960).

tion is shown in Table 3.5; the resemblance of this classification to Dokuchayev's 50-year-old work is immediately apparent.

In Europe, at about the same time as the revised U.S.D.A. system was published, Kubiena (1953) produced a classification of European soils, utilizing a binomial naming scheme within which many soil types resembled those of the U.S. system. The approach to identification was, however, entirely taxonomic, using factors such as soil colour, reaction, parent material and humus type in the construction of artificial keys. Biological aspects were included in the discussion of the soil types but the genetic nature of the soil was not utilized as a classifying factor.

The 1949 U.S. system suffered certain defects imposed by the use of genetic characteristics in type definition as some soils could not be placed until their pedogenesis was fully understood. By contrast, a fully taxonomic scheme using manifested characteristics such as colour, morphology, chemical constitution, etc., allows immediate classification of any newly observed soil type.

The most recent and significant advance has been the appearance of the 7th Approximation of the U.S.D.A. classification (Soil Surv. Staff, 1960), which departed radically from the 1949 system in recognizing ten, as opposed to three, soil *Orders* of which only one corresponds even roughly with the original classification. Table 3.6 shows an outline of the classification and the equivalence of the new orders to the previously named soil groups of the 1949 system; the rationale of the naming is given in the third column.

Each order is identified by a group of diagnostic characters pertaining to its various horizons: their sequence, morphology, colour and constitution.

Table 3.6 A classification of soils: the U.S.D.A. 7th Approximation (Soil Surv. Staff, 1960)

Present order	Approximate equivalents
1. Entisols	Azonal soils, and some Low Humic Gley soils.
2. Vertisols	Grumusols.
3. Inceptisols	Ando, Sol Brun Acide, some Brown Forest, Low-Humic Gley, and Humic Gley soils.
4. Aridisols	Desert, Reddish Desert, Sierosem, Solonchak, some Brown and Reddish Brown soils, and associated Solonetz.
5. Mollisols	Chestnut, Chernozem, Brunizem (Prairie), Rendzinas, some Brown, Brown Forest, and associated Solonetz and Humic Gley soils.
6. Spodosol	Podsols, Brown Podsolic soils, and Ground-Water Podsols.
7. Alfisols	Grey-Brown Podsolic, Grey Wooded soils, Noncalcic Brown soils, Degraded Chernozem, and associated Planosols and some Half-Bog soils.
8. Ultisols	Red-Yellow Podsolic soils, Reddish-Brown Lateritic soils of the U.S., and associated Planosols and Half-Bog soils.
9. Oxisols	Laterite soils, Latosols.
10. Histosols	Bog soils.

The diagnostic characteristics omit any which are seriously affected by the intensity of genetic processes, thus avoiding the problems of classifying immature and ill-understood soils. The orders, nevertheless, tend to represent the various zonal and associated azonal soils of the 1949 system.

Further subdivisions are made: into suborders; great groups; subgroups; families and series. The differentiae used are not discussed in this limited text but may be found in Soil Surv. Staff (1960).

SOILS—THE GEOGRAPHY OF THE GREAT GROUPS

The earliest classification of soil-types showed an awareness that their distribution was a generalized reflection of latitudinal variations in climate. In this awareness, pedology was closely akin to plant ecology in the Clementsian view of the climatic climax. In both fields the subsequent collection of more detailed information has shown that climatic determinism may be misleading if it is accepted on too local a basis: in both soil and plant distribution there are other strongly acting factors such as topographic water regime, physicochemical status of the parent material, grazing intensity, and anthropic effects such as extensive forest clearance for hunting and agriculture.

Current classifications make varying degrees of allowance for these interactions but there is still sufficient evidence of a zonal distribution of soils for Figure 3.2 to give a crude picture of the relationship between climatic temperature and wetness as a determining factor in the worldwide pattern of soils. In the following sections a brief description of the more widespread zonal soil-types is given.

Tundra soils

The tundra climate is exceptionally harsh with extremes of low temperature, high wind speed and, often, aridity. As a result, many soils of this zone are pedologically immature with little or no true horizon differentiation. On stony substrata such soils are named *Lithosols,* and on sands, gravels and other unconsolidated materials, *Regosols.* Soils of this description are not limited only to the tundra but may occur in other parts of the world if the climate and parent material are unfavourable to soil development. For this reason they may be considered *azonal* soils.

The true zonal tundra soils (Figure 3.3) generally have a peaty or organic-rich A horizon resting on a pale coloured gley horizon which, during the summer months, may be almost liquid, its drainage impeded by the permanently frozen soil (permafrost) beneath. The soil is strongly seasonal in its biological activity, being completely frozen for the greater part of the year. The expansion and contraction of the soil water on freezing and thawing results in the enlargement of soil voids and a general mixing of the soil (cryoturbation) which leads to portions of the A horizon

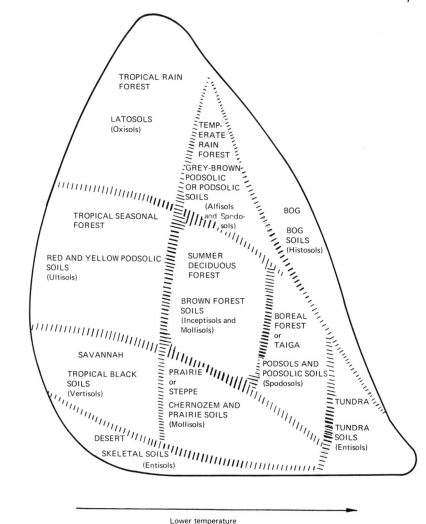

Figure 3.2 The geographical distribution of soil types in relation to climatic temperature and wetness regime. After Muckenhirn *et al.*, *Soil Sci.*, **67**, 93–105 (1949).

becoming incorporated at other levels in the profile. The same process often causes migration of clay mineral which forms coatings on stones or an indurated zone at the base of the profile where it is squeezed between the permafrost and the overlying soil on freezing.

Cryoturbation often leads to the patterning of tundra soils which is so obviously seen from the air. The surface is dissected into regular polygons of which the centres are raised and the borders depressed. One cause of this

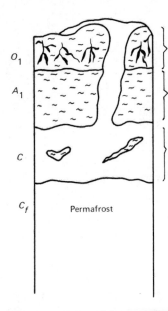

O_1 — Black-brown mat of moss and plant root remains; occasionally, as here, penetrated by C horizon material which has been ejected to the surface by mud explosion.

A_1 — Very dark grey-brown layer, often with a weak, platy structure. High organic matter content

C — Grey with brown mottles. Weak platy structure or almost fluid. Often streaked with inclusions of A_1 material which have been incorporated by cryoturbation

C_f — Permafrost — Permanently frozen during whole of year and forming a barrier to water drainage thus maintaining a very wet surface condition

Figure 3.3 A Tundra Anmoor profile. During the summer months the soil thaws to about 75 cm or less. During the winter, the whole profile is frozen. Vegetation consists of peat-forming mosses such as *Sphagnum* spp. Large areas are dominated by *Carex* spp and *Eriophorum* spp.

is mud-explosion in which semi-liquid mud is forced from the central regions of the polygons by the lateral pressure of ice forming in wedge-shaped cavities along the sunken borders. Such morphological features remain as 'fossil' features of soils which formed in the northern hemisphere in the periglacial climate following the last glaciation (Fitzpatrick, 1956).

Kubiena (1953) described these soils as *Tundra Anmoor* (swampy soils) and *Tundra Moss* (peat moor). In the 7th Approximation (Soil Surv. Staff, 1960) they belong to the Order *Entisols,* Suborder *Aquents (Cryaquents)* though, if the organic layer is very deep, as in tundra moss, they intergrade with the *Histosols.* Some less mature profiles belong to the Order *Inceptisols.*

Tundra vegetation ranges from isolated patches of lichen, mosses and occasional higher plants of Rock Tundra to the more continuous cover of moss, lichen, *Carex* or dwarf-shrub Tundra. Almost any depression may accumulate water over the permafrost, giving rise to Tundra Moor with peat-forming mosses. The whole vegetational physiognomy is markedly xeromorphic, the frequently frozen soil and shallow rooting imposed by waterlogging interacting with extremes of wind exposure and low humidity to cause potentially serious water loss. The majority of the plants are drought-tolerant (poikilohydrous mosses and lichens) or avoid water loss by their low stature, xeromorphy and short growing season. Most of the vascular plants either belong to the geophyte and hemicryptophyte life-forms or are dwarf chamaephytes with xeromorphic leaves.

The harsh climate seriously limits species diversity and some areas may be extensively dominated by a single species; thus moss tundra may be dominated by *Polytrichum* species and lichen tundra, for example, by *Cetraria* or *Cladonia* species. *Carex* tundra has a grassy aspect due more to the inclusion of *Carex* and *Eriophorum* species than to an abundance of grasses. In high summer the meadow-like appearance is accentuated by profuse flowering of various circumpolar species of prominent-flowered plants such as *Ranunculus* and *Potentilla*. Other flowering plants characteristic of high latitudes are numerous species of *Gentiana* and *Saxifraga*. *Dryas octapetala* is so typical of this habitat that its remains, in lake deposits, may be used diagnostically of past climate.

The very limited chamaephytic flora include the dwarf birches and willows: *Betula nana, Salix herbacea, S. reticulata* and *S. arctica*. In more southerly areas and slopes of warmer aspect *Betula* and *Salix* species may form a continuous thicket cover. The more heath-like tundra may be dominated by one of several Ericaceous plants, examples being *Empetrum nigrum* and *Cassiope tetragona*.

Podsols and podsolic soils

To the south of the tundra regions of N. Europe, N. America and Asia the higher annual temperatures permit an enlargement of the life-form spectrum. The dwarf-shrub cover of the southernmost tundra gives way to Boreal forest characteristically dominated by needle-leaved conifers. The soil most frequently associated with this formation is the podsol, though, as a soil type, it may also extend southward into the broad-leaved forests and heathlands of lower latitudes, a zone usually dominated by various brown and grey forest soils.

The fully developed podsol shows an extreme degree of horizon differentiation which is best seen in the *Iron-humus Podsol*. This is most often found under heathland with Ericaceous dominants; the more typical conifer forest podsol is the *Iron Podsol*. Figure 3.4 shows the interrelationship of a range of podsol types which may form according to degree of wetness, parent material, etc.

The most prominent feature of the podsol is the intense eluviation of the surface horizons with mobilization of organic matter and its redeposition at lower levels. The process is so potent that all easily soluble inorganic materials are leached downward and the cation exchange complex is desaturated (page 102) with a corresponding increase in soil acidity. This drop in pH and the removal of calcium cause a change in the microbiological pattern of organic matter decomposition leading to the accumulation of a peaty, fibrous organic layer near the surface. This material has been named raw-humus or *mor* (page 111).

The fully developed podsol profile (Figure 3.4) has a surface O_1 horizon

74

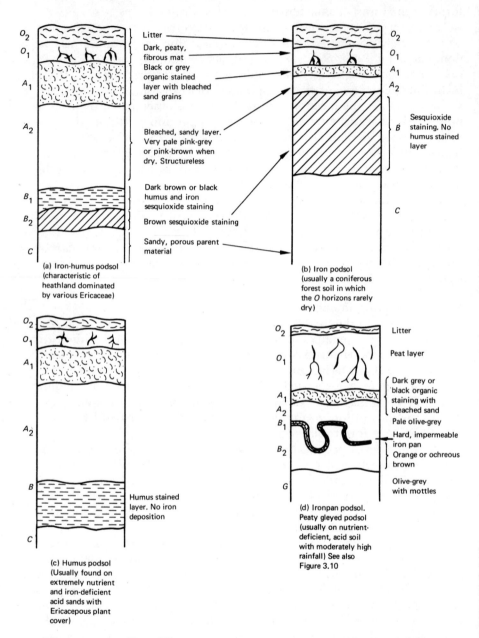

Figure 3.4 Profiles of four contrasting podsol types (see also Figure 3.10).

of undecomposed plant litter (A_{00} in other terminologies) which grades into the $O_2(A_0)$ horizon of peaty mor which is considerably decomposed but still contains microscopically recognizable plant fragments. The bottom of this layer rests on the first horizon of true mineral soil, the A_1, which is stained black or grey by organic matter. Its individual sand and larger silt particles can be seen to be bleached due to the mobilization of the ferric sesquioxide which normally coats such grains, giving them a brown coloration. Below the A_1 is the A_2 horizon which shows the same extensive bleaching of its larger particles but no organic staining. The horizon has a whitish to pale grey, ashen appearance and is a characteristic of all well-developed podsols. Bleaching due to removal of ferric sesquioxide is a diagnostic feature of *albic* horizons. With such potent eluviation it is not surprising to find that the A_1 and A_2 horizons are also often depleted of manganese and aluminium sesquioxides, of clay mineral and of more soluble compounds and ions such as carbonates, calcium, magnesium and potassium.

Below the A horizons there is a sharp transition to the illuvial zone of the B horizons in which percolating water has redeposited various materials to form the B_1 horizon, which is enriched with organic matter and ferric sesquioxide, and the B_2 horizon enriched predominantly with ferric sesquioxide. The humic and sesquioxide coatings of the mineral grains may cause cementing of the particles and 'pan' formation. Some iron pans have a rock-like consistency and are impervious to water, thus causing a superficial waterlogging of the A horizons.

The C horizon of the podsol, if it is also the parent material, is most likely to be a sand, the rock-rubble of a sandstone or acidic igneous rock such as granite. All of these parent materials weather to form coarse-textured soils which are much more likely to show full podsolization than fine-textured parent materials which impede free drainage in the profile. Podsol formation is thus strongly influenced by parent material and also by topogenic waterlogging, vegetation type and human interference. The boreal forest consequently has a mixture of soil types of which the podsol is dominant only when other factors permit. There is evidence in northern Europe that much podsolization and soil acidification is anthropogenic following the removal of the primeval forest (Dimbleby, 1962) which accounts for the occurrence of podsols in the broad-leaf forest areas to the south of the climatic podsol zone.

The vegetation of the podsol, characteristically considered to be needle-leaved forest, may also be heathland in these more southerly areas. In N. Europe, where heathland is most prominent, they are usually dominated by Ericaceous plants; for example, the central species of much European lowland heath is *Calluna vulgaris*. Under rather dry conditions *Erica cinerea* may become codominant or take over full dominance. Schimper (1898) long ago recognized this association as an edaphic one and it has been more recently realized that it is strongly anthropogenic, being maintained by burning and tree felling (Tansley, 1939). In the absence of human control the instability of the association is manifested in the ease with which it is invaded by, for example, *Pteridium aquilinum* (bracken) and subsequently colonized by *Betula* spp. or conifers. With increasing wetness there is a transition to various types of wet heath and extensive peat formation. Under high rainfall conditions ombrogenous peat may form from *Sphagnum* spp., *Eriophorum* spp. or *Molineae caerulea*. Extensive tracts of country may become covered with blanket peat which, pedogenetically, is the *O* horizon of a peaty podsol or peaty gley. Similar peat formations may occur in valley bottoms but, under these soligenous conditions, the peat is more nutrient-rich. Tansley (1939) gives an extensive account of the relationships of the various upland and lowland heath formations.

Boreal forest, for geographical reasons, like the tundra, is mainly confined to the northern hemisphere. It forms a prominent belt across N. America, N. Europe and Asia and includes a large range of locally dominant conifer species, mainly of the genera *Picea, Pinus, Abies* and *Larix*. The diversity is exemplified by contrasting the range of dominants in these continents. In the east of N. America *Picea glauca, P. mariana, Larix laricina, Abies balsamina* and *Pinus banksiana* may variously be dominant, but further west the climatic change results in their replacement by *Pinus contorta* and *Abies lasiocarpa*. In Western Europe different species occur, *Pinus sylvestris* and *Picea abies* commonly dominating. Further east, and in Asia, *Larix sibirica, Abies sibirica* and *Pinus obovata* are prominent.

The range of terminology of the podsols is very wide: in addition to the *humus, humus-iron* and *iron podsols* already discussed, Kubiena (1953) deals with two types of gleyed podsols (*Gley* and *Molken Podsols*) and *Peat podsols* which occur under very wet conditions. In the 7th Approximation (Soil Surv. Staff, 1960) the podsols are mainly included in the order *Spodosols*. *Podsolic Brown Earths* are commonly the product of leaching and subsequent podsolization of acid brown soils (page 77) on coarse-textured parent materials. Their formation is often associated with mismanagement under agriculture or forestry and their existence may be related to the genetic continuum which exists between soils of the brown earth type and the true podsols.

The processes involved in podsolization commence with leaching and decalcification and proceed to the formation and migration of clay mineral; formation of free oxides and destruction of clay mineral; mobilization of sesquioxides and their deposition in the *B* horizon (Bunting, 1965). The fully developed podsol is usually very acid and its cation exchange complex is strongly desaturated (page 102). It is difficult to specify precise pH ranges but Pearsall (1952) identifies pH 3·8–4·0 as the upper limit for mor humus formation and most podsol horizons are near or below this pH.

In the northern hemisphere the processes leading to podsolization probably started when the land-surface emerged from periglacial conditions at the end of the last glaciation and accelerated when man began extensive deforestation in the late Neolithic or early Bronze Age (Dimbleby, 1962). Soils such as the brown podsolic are probably intermediates in a continuum of pedogenesis. There is now evidence to suggest that certain plant species hasten the podsolization process; for example, the Ericaceous heath which so often succeeds cleared forest on acid brown soils (Grubb et al., 1969). Other plants recognized as soil acidifiers are some conifers and, amongst the broad-leaved tree species, *Fagus sylvatica* and *Quercus* spp. in Britain (Mackney, 1961) and *Populus trichocarpa* in Alaska (Crocker and Dickinson, 1957).

Brown earths or brown forest soils

The brown forest soils are characteristically those of the broad-leaved summer forest formation, the most productive of temperate zone ecosystems. The high productivity is reflected in the great biological activity of these soils in which microorganisms and the soil fauna play a very considerable part in pedogenesis.

Like the podsol, the brown forest soil is as much a function of parent material as of climate. Most typically it forms on silicate-rich parent materials: of the sedimentary rocks, the shales, mudstones, clays, finer sandstones and also impure limestones. On igneous substrata it is most often found on base-rich materials such as basalt and other deep-seated lavas.

The brown forest soil is maintained in a constant state of horizon mixing by the activity of its prolific earthworm fauna. As a result the *B* horizon rarely shows sharp delimitation from the *A,* other than in its lower organic content and difference in colour and structure. It is often designated (*B*) to indicate its indeterminate nature. Any leaching which occurs is usually counterbalanced by biological mixing but in some cases (*Sols bruns lessivés*—leached brown soils) some downward washing of clay (lessivation) has occurred and, on more siliceous parent materials, leaching of carbonates may have lowered the pH into the acid range (*Sols bruns acides*—acid brown soils).

The central group of brown forest soils is of high cation saturation (above 50%) and has a pH near 6. These are, however, variable characteristics since calcareous brown soils, on limestones, may be of pH 7 or above, and have high calcium saturation of the exchange complex, while some leached brown soils may be well below pH 6 and have a cation saturation of less than 35%. The acid brown soils, forming on siliceous parent materials, podsolize easily and have a base saturation and pH range overlapping with that of the podsols. Pearsall (1952) suggested that pH 4·8–5·0 was the lower limit for the mull humus formation (page 111) characteristic of brown forest soils but their pH range extends upwards above pH 6·5, which he considered the lower limit for the distribution of calcicole plants.

The *A* horizon of the nutrient-rich brown forest soil is markedly structured as a result of earthworm activity. The worm casts, particularly those voided underground, contain organic matter which has been comminuted in the animal's alimentary tract and forms a rich substrate for fungal and bacterial growth. The binding effect of the hyphae coupled with the influence of bacterial gums stabilized the soil in the casts into persistent structural aggregates which, in some particularly rich soils, may form a large proportion of the *A* and (*B*) horizons, giving them a soft, spongy and well-aerated nature.

The action of the worm population keeps the surface of the soil more or less clear of organic debris so that O_1 and O_2 horizons are minimal or non-existent. In some of the more acid soils there is formation of *O* horizons (**Figure 3.5**) as the worm population falls with increasing acidity. Worms appear to be absent below a pH of c. 4.8 (their activity is instrumental in mull humus formation so that this limitation is further evidence for Pearsall's suggestion noted above). Soil Surv. Staff (1960) attribute the nature of

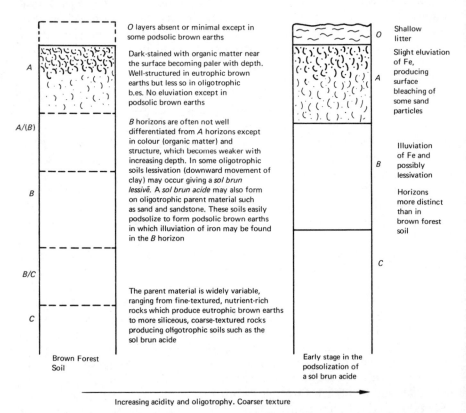

Increasing acidity and oligotrophy. Coarser texture

Figure 3.5 Brown Forest Soils or Brown Earths: their profiles and relationships.

the upper horizon of the brown forest soil to the activity of earthworms in pulling leaf litter into their burrows, thus incorporating it uniformly into a considerable depth of mineral soil. This is comparable with the effect of a grass sward in which the roots are responsible for returning a large proportion of the photosynthesate into the soil. Both brown forest soils and grassland soils are characterized by a superficial *mollic* horizon; a thick surface layer of organic enrichment, dominantly saturated with bivalent cations, strongly structured and having a narrow C/N ratio (see p. 120). These soil types are consequently recognized as belonging to the order *Mollisols*. Many brown forest soils have been included in the order *Inceptisols* as their diagnostic horizons are thought to form rather quickly and do not show significant illuviation, eluviation or extreme weathering.

The vegetation of the brown forest soils is typically a broad-leaved deciduous forest, though many areas have been taken by man for agriculture. Under arable management large amounts of organic matter must be added if the soil is to retain its originally well-structured condition while heavy inorganic fertilization is needed to compensate the loss to the crop. There is a lesser problem in using these soils for grass production as much organic matter is returned to the soil by the root system, but inorganic fertilization is required and some care has to be taken to avoid damaging the soil structure by overstocking and 'poaching' by cattle trampling, or by the use of heavy machinery.

The natural forests of these soils are physiognomically similar in all parts of the world. There is usually a single stratum of dominant trees and, below this, sparse strata of undershrubs and herbs. Few climbers or epiphytes are present so that the general aspect is much less luxuriant than that of tropical forest.

Five general types are recognized (Polunin, 1960).

1. Oakwoods of west and central Europe. These are dominated by *Quercus robur* and/or *Q. petraea* with a number of associated species such as *Fraxinus excelsior, Populus* spp., *Betula* spp., *Ulmus* spp. and *Alnus glutinosa*. These vary in abundance according to soil wetness, pH and nutrient status. A number of shrub species such as *Corylus avellana* and *Crataegus monogyna* are common and the ground flora may vary from a luxuriant cover of herbaceous plants on rich soils to tussock grass cover of species such as *Deschampsia caespitosa* on moist or heavy clay soils or to a *Pteridium aquilinum–Rubus fruticosus* association on light sandy soils.

2. Forests of N. America, E. Asia and S.E. Europe. These are more varied in their species, the wide range including species of *Quercus, Fagus, Betula, Carya* (Hickories), *Juglans, Acer, Tilia* (Basswood), *Fraxinus, Ulmus, Liriodendron* (Tulip tree), *Castanea* and *Carpinus*. The underflora of these forests resembles its western European counterpart in its wide variety of herbaceous, grass or Pteridophyte cover. In some areas bordering the boreal forest region conifers may occur as codominants with the deciduous species.

3. Beech forest. Widespread forest areas in Europe may be dominated by *Fagus sylvatica* which forms such a dense, shade-casting canopy that the shrub and herb layer is usually either very sparse or absent while the soil is characteristically covered with a thick brown mat of the slow-decaying beech leaves.

4. Southern Beech. An association of the southern hemisphere dominated by *Nothofagus* spp. closely crowded in the tree layer with relatively few undershrubs having a luxuriant carpet of Bryophytes and ferns.

5. Variants due to topographic waterlogging, containing *Salix* or *Alnus* spp. as dominants and having a rich underflora of hygrophyllous herbs. According to soil nutrient status and acidity the underflora may range from a wet heath *Sphagnum* spp. or *Molinia caerulea* association to a lush, rich fen flora under eutrophic conditions. Under these circumstances the soil will be of a humic gley type (page 91) rather than the brown forest soil common to the rest of the formation.

Chernozemic and subhumid grassland soils

In central continental areas the gradient of decreasing precipitation in the deciduous forest region leads to a greater tree spacing and the development of a park-like vegetation. Further reduction in rainfall prevents tree survival and the forest vegetation gives way to the subhumid plainland grass associations: the prairies of N. America, the steppes of E. Europe and Asia, the pampas of S. America and the grasslands of S. Africa and Australasia.

These grasslands are essentially midcontinental in their distribution and their associated soils are much influenced by climatic conditions. The annual precipitation range lies between 1000 mm and less than 350 mm. At the lower end of this range there is a transition to semi-desert conditions and, also, rainfall effectiveness may be reduced by the high intensity of summer storms. The upper limit overlaps the precipitation region of summer deciduous forest but the continental climate with continuous sub-freezing conditions during the winter months and the high evapotranspiration in the midsummer months is unfavourable to tree growth.

Leaching is not extreme as the winter-frozen soil prevents continuous downward water movement for a large proportion of the year while high evaporation rates induce an upward capillary movement during the summer.

The climatic range for grassland as a climatic/biotic climax is thus fairly wide and includes soils which range from near relatives of the podsolic type to poor arid soils in which biological activity is limited by persistent drought. The soils thus span the range from highly leached soils with spodic horizons of R_2O_3 and organic carbon enrichment to soils in which a past or present low P/E regime has caused secondary accumulation of calcium carbonate to form a calcic horizon at some level in the profile.

The range of soil type, while reflecting latitudinal climate, is also governed by parent material and soil hydrology; consequently, sandy materials may become leached to thin, red podsolic soils while adjacent lime-rich loess, limestone, basalts and dolerite form the chernozemic soils so typical of the region. In river valleys, strips of forest cover may penetrate deep into the open grasslands and have wetter varieties of mollisols ranging from brown forest soils to various gleys. This characteristic geographical pattern has been almost entirely eliminated in N. America and Eurasia as the majority of rich grassland soils have been cultivated for cereal growth, and the former grassland persists only in relict sites.

The modal chernozemic profile (Figure 3.6) is dominated by a dark-coloured A horizon of extreme depth (to 1·5 m) and high organic content.

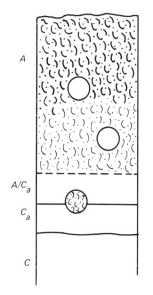

O horizons absent

A horizon black near surface grading through dark to light grey with depth. Very deep (often several metres). Very well structured with spongy fabric. Contains infilled mammal burrows (crotovinas)

A/C_a

C_a

Below this level contains free $CaCO_3$ and effervesces with HCl

Contains precipitates of $CaCO_3$

C

Most characteristically deep loess but most easily weathered calcareous rocks can produce suboptimal chernozems

Figure 3.6 Chernozem profile. The formation of this soil, in addition to requiring a calcareous parent material, is a function of the P/E balance which governs the depth to the calcareous layer (A/C_a and C_a). Under rather higher P/E regimes various brown prairie and steppe soils are formed which have similar profile morphology to the chernozems but show some downward translocation of materials to form a B horizon and are usually devoid of free $CaCO_3$.

The horizon may be calcium carbonate saturated and has a pH range from 6·5 to 8·0 or more and rests directly on a highly calcareous C horizon. At some level in the profile there is a zone of calcium carbonate enrichment (C_{ca} or B_{ca}) the depth depending on annual P/E balance. In N. America the depth of this horizon increases with precipitation on a west–east gradient.

The chernozem is extremely biologically active with a high earthworm population and an accompanying collection of soil-dwelling mammals whose burrows are so extensive as to form a diagnostic part of the profile with their infilled remnants. In the dark A horizon they often appear as 'white eyes' containing the contrasting material of the calcic C horizon. Having deep, mollic A horizons, these soils, like the brown forest soils, are classified among the mollisols (Soil Surv. Staff, 1960).

Under rather more humid conditions the black chernozem soils are replaced by those which have been named *Brunizems* (*Prairie soils*) and *Reddish Prairie soils* in N. America and *Degraded* and *Leached Chernozems* in Eurasia. In all of these soils some downward translocation of materials may be detected in the formation of a B horizon and a brown coloration formed by weathering production of iron oxides. It is possible that these soils may formerly have carried a forest cover before its removal by human activity. On coarser parent materials under these conditions *Thin Red Podsolic* soils may also be formed.

With an annual precipitation below 350–400 mm the Chernozems give way to the *Chestnut soils* of N. America and Eurasia and with further reduced precipitation these intergrade with various *Arid Brown Desert* and *Semidesert soils*. Under these drier conditions there is a transition from full grass cover to sparse grass or desert scrub with the inclusion of more xeromorphic species.

In Eurasia the Chernozems are dominated by grasses including the genera *Festuca, Koeleria* and *Stipa* and, in N. America, by *Boutoloua, Agropyron, Andropogon, Sorghastrum, Koeleria, Stipa, Poa, Panicum* and others. Associated with these is a large range of flowering herbs which, in certain aspects, may appear to dominate the grassland.

As a major vegetational region of the earth, these grasslands were not only the reflection of climatic factors but also showed the strong biotic pressure of the grazing mammals formerly so abundant on the plainlands of the midcontinents. The role of fire in these habitats was probably equally important in maintaining the open nature of the habitat and it may be that these grasslands could be regarded as deflected climax (plagioclimax) associations which were prevented from achieving the full climatic status for the region. Man has, however, now destroyed much of the vegetation but the soils remain as identifiable entities, albeit modified by cultivation.

Mediterranean and subtropical soils

Soils of these regions are much influenced by the length of the dry season and the general aridity of the environment. In this they resemble the grassland soils but, of course, do not experience a cold winter climate in which biological and chemical weathering come to a standstill. For this reason most of the varied soils occurring in the region show considerable weathering effects compared with the relatively unweathered state of the modal chernozem.

Arid subtropics: Terra rossa *and* Terra fusca

In the drier regions of the Mediterranean type climate *Terra rossa* (red soils) are characteristically developed on limestone or dolomite karst and on some other basic rocks. The terra rossa have a constitutional, but not genetic, affinity with the tropical latosols, their red colours being caused by the enrichment of the surface layers with iron oxides consequent on decalcification. Dehydration of the iron oxides during the intensely dry summer months produces a vivid red coloration of the soil which often contrasts with the whiteness of the fairly pure limestone from which it forms.

The profile (Figure 3.7) constitutes a poorly developed A horizon of low organic content, grading through an A/B transition to a red B horizon which then shows a sharply defined junction with the parent limestone below. The texture may vary from clayey to sandy according to the nature of the impurities in the limestone or to the ingress of wind-blown material. Morphologically the profile resembles the limestone rendzinas of more temperate climates and Stace (1956) suggests that there is a continuum of characteristics between these two soil types.

In Europe the vegetation of the terra rossa is often sparse, the scattered low trees, shrubs and herbs forming what is termed in France a *garigue,* which represents the remains of a formerly more luxuriant forest cover dominated by *Quercus ilex* and *Pinus halpensis.* The change has mainly been wrought by man and his voracious companion, the goat. Erosion has subsequently stripped much of the soil mantle, producing the characteristic karstic features of the landscape, bare limestone outcrops contrasting with the vegetation-covered hollows.

Siliceous rocks or wetter areas have a more biologically active brown soil covering of *Terra fusca* with a greater organic content than the terra rossa and in which earthworm activity produces a greater mixing. On limestone parent materials the two soil types appear to be interconvertible, the removal of forest cover promoting the change to terra rossa while reestablishment of a closed vegetation cover protects the soil surface from intense summer insolation and may permit a terra rossa to revert to the terra fusca condition (Bunting, 1965).

The sclerophyll *maquis* vegetation of the terra fusca is much richer than the garigue and contains a wide variety of shrubs which were probably the subdominants of former forest cover. They include such plants as *Olea europea* (Olive), *Cystus* spp., *Myrtus communis, Rosemarinus officinalis* and many others. The original dominants in the Mediterranean area were *Quercus*

84

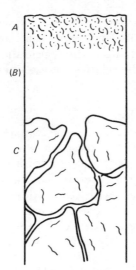

Very shallow *A* horizon

(*B*) horizon is vivid red in terra rossa due to dehydration of its high iron oxide content. In terra fusca the colour is ochreous or red-brown

Terra rossa on purer limestones. Terra fusca on less pure limestone and/or under higher rainfall or higher P/E conditions

Figure 3.7 *Terra rossa* and *Terra fusca* profiles. In both cases the *A* horizon is very shallow (< 10 cm) and is deficient in organic matter. The vivid red of the terra rossa and the ochreous colour of the terra fusca are more strongly shown in the (*B*) horizon. The high iron oxide content is a consequence of the leaching loss of calcium carbonate. The *A* and upper (*B*) horizons are decalcified with a slightly acid or neutral pH contrasting with the temperature zone rendzinas which form on similar parent materials and have similar profile structure.

suber and *Pinus maritima* (*pinaster*). Similar sclerophyllous vegetation and soils may be found in the Cape region of S. Africa, in W. California (chaparall), in central Chile and parts of S. Australia. The Australian areas are, in part, dominated by a much more luxuriant forest vegetation of *Eucalyptus* spp. than is found elsewhere, but the S. African vegetation shows similar grazing damage to the garigue, while the chaparall vegetation is characteristically exposed to repeated burning.

Humid subtropics: red and yellow podsolic soils

These soils reflect an intermediate situation between the soil formative processes of the temperate zone and the tropical latosol region. They contain more iron and aluminium than the temperate podsols but less than the ferrallitized tropical soils. They tend to be associated with geologically old land surfaces or with younger deposits of already highly weathered materials (Soil Surv. Staff, 1960).

Profiles are usually well drained and acid with thin O_1 and O_2 layers, a dark A_1 horizon containing organic matter and grading into a bleached A_2 which then shows a sharp boundary with the red or yellow B horizon (Figure 3.8). The rather sparse literature on these soils suggests that they are widespread in subtropical climates but are very variable and associated

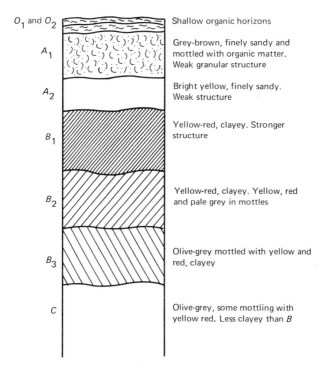

Figure 3.8 A typical Red Podsolic profile formed under a mixed deciduous and pine forest. Most of the profile is well below pH 5, is strongly unsaturated and the B horizons are rich in Fe and Al oxides. The B horizons are also, typically, more clay-rich than the A or C layers.

with a wide range of vegetation types. Their former vegetation was usually a summergreen forest, but in many cases this has been removed by man. Richards (1952) describes the transition between tropical rain forest and drier habitats; evergreen forest giving way first to summergreen forest followed by savanna forest, thorn forest and finally arid scrub. The red/yellow podsolic soils belong usually to the wetter part of this transition but Richards describes a great diversity of vegetation and soils for the region on a worldwide basis. The 7th Approximation (Soil Surv. Staff, 1960) recognizes most of them as suborders of the *Ultisols,* so named because of their high degree of weathering.

Latosols and other tropical soils

Soil formation in the west tropical regions has long been a topic of controversy, particularly in relation to the *Latosols.* The interaction of high temperature and rainfall causes the weathering process to proceed with unrivalled intensity, mantles of weathering products in some cases reaching depths of 300 m. The term latosol is used to denote a whole range of soils and weathering crusts which have most of their free silica, silicates and bases removed by weathering, leaving a residuum consisting of iron and aluminium oxide, some free quartz particles and 1 : 1 lattice clays, enriched with oxides of titanium chromium and nickel.

The weathering process is described as *ferrallitization* and it causes the development of very deep and uniform profiles with a red or pink coloration and a low organic content, the high temperature causing organic matter to be destroyed faster than it is accumulated. These soils all belong to the order *Oxisols* (Soil Surv. Staff, 1960), their diagnostic characteristic being the possession of an *oxic* horizon. Such an horizon is defined by its extreme state of weathering and oxidation associated with the sesquioxide enrichment and silicate impoverishment described above (Figure 3.9).

The earlier term, laterite, has been used to describe such soils but in its original sense it referred to those layers of ferrallitized materials which harden irreversibly on exposure to air and sun, and may be used as building materials (Latin *later* = brick). The 7th Approximation refers to the material as *plinthite*; it does not occur in all latosols but is common, in the unhardened state, in wet ground-water latosols and, as a hard desert crust, in those soils which have a prolonged dry season. In the latter case it formed, almost certainly, under a wetter climate or with a wetter soil water regime and is present today as a 'fossil' structure. In many areas such a layer forms an extensive, eroded *cuirass* over the land surface (Mohr and van Baren, 1959) but this rarely occurs under forest cover. Shallow hard pan layers, at least, seem to form after felling and make these oxide-rich soils agriculturally difficult unless precautions are taken to maintain a protective plant cover.

The characteristic vegetation of the latosol is evergreen *Tropical Rain Forest* though, in drier areas or with human interference, this may be replaced by summergreen forest or even savanna. Latosols characteristically occupy geologically old land surfaces, and Richards (1952) suggests that in some areas the vegetation has existed uninterruptedly since 'a very remote geological period'. The forest is floristically very rich and this, in part, may be due to its ancient nature,

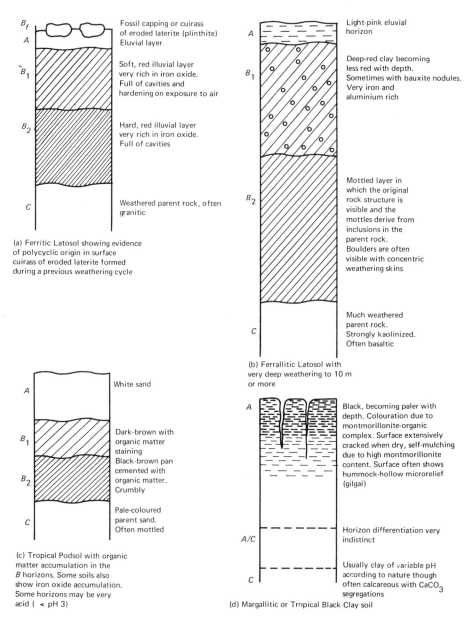

(a) Ferritic Latosol showing evidence of polycyclic origin in surface cuirass of eroded laterite formed during a previous weathering cycle

B_f / A — Fossil capping or cuirass of eroded laterite (plinthite) Eluvial layer

B_1 — Soft, red illuvial layer very rich in iron oxide. Full of cavities and hardening on exposure to air

B_2 — Hard, red illuvial layer very rich in iron oxide. Full of cavities

C — Weathered parent rock, often granitic

(b) Ferrallitic Latosol with very deep weathering to 10 m or more

A — Light-pink eluvial horizon

B_1 — Deep-red clay becoming less red with depth. Sometimes with bauxite nodules. Very iron and aluminium rich

B_2 — Mottled layer in which the original rock structure is visible and the mottles derive from inclusions in the parent rock. Boulders are often visible with concentric weathering skins

C — Much weathered parent rock. Strongly kaolinized. Often basaltic

(c) Tropical Podsol with organic matter accumulation in the B horizons. Some soils also show iron oxide accumulation. Some horizons may be very acid (< pH 3)

A — White sand

B_1 — Dark-brown with organic matter staining

B_2 — Black-brown pan cemented with organic matter. Crumbly

C — Pale-coloured parent sand. Often mottled

(d) Margallitic or Tropical Black Clay soil

A — Black, becoming paler with depth. Colouration due to montmorillonite-organic complex. Surface extensively cracked when dry, self-mulching due to high montmorillonite content. Surface often shows hummock-hollow microrelief (gilgai)

A/C — Horizon differentiation very indistinct

C — Usually clay of variable pH according to nature though often calcareous with $CaCO_3$ segregations

Figure 3.9 Four typical soil profiles. Tropical soils are widely variable according to parent material, relief and climate but generally fall into these four categories.

evolution and selection having continued for very long periods without serious environmental change. In its ecological relationships it is also unique, again perhaps because of its extreme age and selection for optimal niche occupancy (p. 307). Despite having a soil which is intensely leached of nutrient elements, the vegetation is rich and of great biomass; most of the ecosystem nutrients are locked up in the living cover and must be recycled with considerable efficiency to maintain the climax status of the forest. Human interference and felling have a catastrophic effect in removing this pool of nutrients and the regeneration of a secondary forest is consequently a very long process, which often leads to a secondary climax of poorer ecological status than the original with smaller dominant trees and a very dense, tangled shrub and creeper undergrowth, contrasting with the more open floor and multiple stratification of the primary forest.

The strongest characteristic of mixed tropical rain forest is its multitude of codominating tree species and its multiple strata with perhaps three distinct tree layers. The uppermost layer has a discontinuous canopy but, towards the lower layers, overlap increases. Below these layers are shrub and herb strata which show an equivalent richness of species composition. The association is also complex on a broader scale as the distribution of tree species varies according to the catenary mosaic of topogenic soils. The diversity of dominant species prevents description other than this, but an extensive account may be found in Richards (1952).

The occurrence of other soil types in the tropics is related to the nature of the parent material and to drainage. Under conditions of impeded drainage near base-rich or little-weathered rocks, dark, clay-rich *Margallitic* soils develop, but under similar drainage conditions on acid rocks, *Ground water Laterites* are formed. More extreme waterlogging gives rise to some form of swamp soil. With free drainage and on siliceous, coarse-textured parent materials *Tropical Podsols* are developed.

Margallitic soils

These are the *Tropical Black earths, Cotton soils* or *Regur* (Figure 3.9). Dominated by montmorillonite clay, their volume changes on wetting and drying make them 'self-ploughing' and for this reason they are classified in the order: *Vertisols* (Soil Surv. Staff, 1960) and show microrelief phenomena such as *gilgai* (micro-basins and micro-knolls of a few cm to 1 m, or so, in relief). Though their formation is associated with drainage impediment, they normally occur in areas of lower rainfall than many latosols and appear to occupy the zone climatically intermediate between these and the red/yellow podsols. Their natural vegetation is either grassland or savanna.

Tropical podsols

These are not dissimilar to the temperate zone podsols and occur on siliceous coarse-textured parent materials. The *A* horizon often consists entirely of excessivly bleached quartz sand of extreme acidity (Figure 3.9) which, despite the high temperature, permits the accumulation of a thick O horizon of mor-like humus. The 7th Approximation includes them in the great group *Thermaquods* of the *Spodosols* and also suggests that some should belong to a thermic suborder of the *Humods*. The podsol profiles described by Mohr and van Baren (1959) are generally developed on excessively drained substrata and they discuss the possibility that they were formed under conditions of greater climatic wetness and that some of the laterite cuirass areas are in fact 'fossil' illuvial (B) horizons of eroded palaeogenic podsols.

Richards (1952) indicates that bleached sands are widespread in the tropics and describes their vegetation generally as tropical rain forest, summergreen forest or savanna.

Swamp soils

Richards describes two general types of tropical swamp soil developed in areas where the ground water rarely falls more than a few cm below the surface. They are, firstly, organic-rich soils with an ombrogenous peat cover and, secondly, organic-deficient silt/clay textured soils rich in 1 : 1 lattice clays and silicic acid. The formation of peat in such high temperatures must involve factors other than waterlogging, the major one being extreme nutrient deficiency, a suggestion which is supported by the extremely acidic nature of the drainage waters. The peat cover is often lens-like in section, resembling the raised bog association of temperate climates. The vegetation is a 'moor forest' variant of the normal tropical rainforest association.

A third type of swamp soils occurs under Mangroves on the sea coast and fills the same ecological niche as temperate zone salt marsh. The soils are dark-coloured with a high organic content and the deeper layers are black with precipitated ferrous sulphide derived from sulphate reduction (p. 195).

VEGETATION OF THE ZONAL SOIL TYPES

Descriptions of the vegetation associated with the various soil types have, of necessity, been kept to a minimum in this account, particularly in the case of the widely diverse tropical and subtropical forms. Very early source descriptions may be found in Schimper's classic work *Plant geography upon a physiological basis* (1898) and more recent plant geographies by Dansereau (1957) and Polunin (1960) give useful accounts of the vegetational zones. Oosting (1956) gives a brief general description of the N. American vegetation types which are also dealt with more extensively and with beautiful illustrations in Gleason and Cronquist (1964). The classic account of the tropical vegetation is that of Richards (1952). All of these works may be recommended as an addition to this very limited chapter.

AZONAL SOILS

The azonal soils or *Entisols* of the 7th Approximation are incipient soils in which pedogenesis has hardly begun. They show no obvious profile differentiation and thus have no apparent affinity with the neighbouring zonal soils. They are classified, according to parent material, into *regosols, lithosols* and *alluvial* soils. Regosols are formed on soft, unconsolidated materials such as dune sand, loess and glacial till, while the lithosols are associated with stony substrata and include many desert and tundra soils and soils of unstable slopes. Some thin rendzinas (p. 93) also fall into the lithosol category as do *rankers* with an OA_1 horizon resting directly on nutrient-deficient, siliceous parent rocks.

Alluvial soils are very variable and are classified by texture, ranging from the finest clay/silt soils through sand and gravel to boulder deposits. Many *marsh soils* are of clay/silt texture of either freshwater or marine (saltmarsh) origin. The waterside meadow or *warp* soils of many temperate zone river valleys are often rich, having a strong biological activity and being maintained in a pedogenetically young condition by repeated input of silt. Another characteristic alluvial form is the *gyttja* soil, initially formed under water in eutrophic conditions, and consisting of clay with much organic matter, coloured black by the precipitation of ferrous sulphide.

INTRAZONAL SOILS

These are soils in which a local factor dominates over the zonal climate and vegetation in pedogenesis. There are three normal variants of hydromorphic, calcimorphic and halomorphic types. Many classifications have considered these to be distinct soil types but the 7th Approximation presents most of them as deviants from the adjacent zonal soil forms.

Hydromorphic soils

The most widespread variant in humid climates is the *Gley,* which has a part of the profile waterlogged and consequently in a reducing rather than an oxidizing condition. The detailed consequences of waterlogging are discussed in Chapter 7 but the immediate morphological effect is seen in soil colour. Iron in the ferric condition gives most soils their yellow, brown or red colorations which contrast strongly with the blue-grey and grey colours of the ferrous compounds in waterlogged soils. This subdued coloration is often intermingled with a mottling of yellow or red-brown where local oxidation occurs adjacent to air-filled voids or to living roots.

Another consequence of the lowered oxygen tension and reduced oxygen diffusion rates in wet soils is that organic matter decays rather slowly, either causing organic enrichment of the A horizon or build-up of superficial peat.

The origin of the gley characteristic may be a high ground water level or the retention of surface water by a superficial impervious layer in the soil, the latter situation forming *surface water gleys (pseudogleys)*.

The gley soils show enormous variation in response to degree of wetness and nutrient/acidity status. Under eutrophic, high pH conditions *mull gleys* occur and have an A horizon not dissimilar to those of adjacent brown forest soils. Increasing wetness brings waterlogging nearer to the surface, causing fen-like conditions to develop and the accumulation of a surface layer of nutrient-rich fen peat (Figure 3.10). Increasing acidity and nutrient deficiency reduce rates of organic decomposition and produce a range of soils having affinities with the podsols: these are the *humic gleys, peaty gley podsols (half-bog soils)* and *bog gleys* with a thick peat cover (Figure 3.10).

The acid gley soils best show their relationship to the podsols in the sequence of soils commencing with the peaty gley podsols which have a thin iron pan in the upper part of the B horizon, but otherwise are similar in profile to the normal podsol (Figure 3.10). Their A_2 horizon is, however, grey or greenish-grey in colour with none of the pale pink tinge associated with well-drained podsols. The iron pan, which is hard and black or brown in colour, underlies this layer but, below it, the remainder of the B horizon is bright reddish-brown and well-aerated. Crompton (1956) suggests that superficial waterlogging mobilizes iron which reprecipitates as a pan, in the B horizon, when it comes into contact with this better aerated layer. The peaty gleys are consequently typical of areas with high rainfall and low evaporation in which the surface soil remains very wet during most of the year. As the pan thickens it may increase this effect by impeding drainage. Thus there is a hydromorphic continuum of soils through the podsol, peaty gley podsol with iron pan (or iron-pan podsol) and bog gley underlying *ombrogenous (blanket) peat*. The bog gley often has an iron pan and may not differ much from the peaty gley podsol except for its far greater thickness of peat covering.

If these soils are engendered by high rainfall they must be considered to be zonal rather than intrazonal in nature. There is, however, some suspicion that blanket peat and peaty gley podsols with iron pans may be anthropic in the sense that their development has followed the prehistoric removal of tree cover and subsequent degradation of habitat caused by over-intensive grazing (Crompton, 1968).

Some other peat soils are topogenous rather than ombrogenous; thus *basin peat* formation is characteristic of the *valley bog* which is fed with oligotrophic, acid drainage water. These peats are in capillary contact with ground water and are more nutrient-rich than the rain-fed blanket peats. In high-rainfall areas both valley bog peat or equitrophic fen peat may grow vertically until they are out of contact with the ground water, become entirely ombrogenous, and create the typical *raised bog* habitat.

The various gley soils are classified in the 7th Approximation (Soil Surv. Staff, 1960) as aquic suborders of the associated zonal orders. The deep

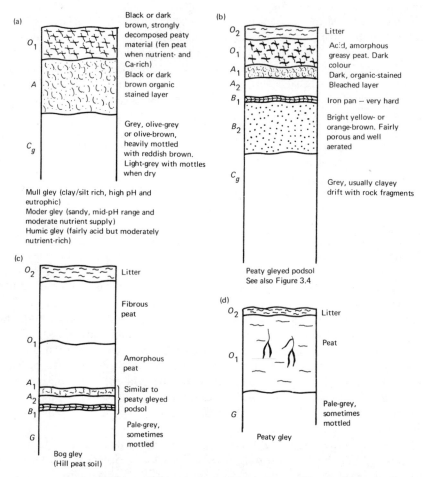

(a)

O_1 Black or dark brown, strongly decomposed peaty material (fen peat when nutrient- and Ca-rich)

A Black or dark brown organic stained layer

C_g Grey, olive-grey or olive-brown, heavily mottled with reddish brown. Light-grey with mottles when dry

Mull gley (clay/silt rich, high pH and eutrophic)
Moder gley (sandy, mid-pH range and moderate nutrient supply)
Humic gley (fairly acid but moderately nutrient-rich)

(b)

O_2 Litter
O_1 Acid, amorphous greasy peat. Dark colour
A_1 Dark, organic-stained
A_2 Bleached layer
B_1 Iron pan — very hard
B_2 Bright yellow- or orange-brown. Fairly porous and well aerated

C_g Grey, usually clayey drift with rock fragments

Peaty gleyed podsol
See also Figure 3.4

(c)

O_2 Litter
Fibrous peat
O_1 Amorphous peat
A_1
A_2 Similar to peaty gleyed podsol
B_1
G Pale-grey, sometimes mottled

Bog gley
(Hill peat soil)

(d)

O_2 Litter
Peat
O_1
G Pale-grey, sometimes mottled

Peaty gley

Figure 3.10 The influence of wetness, acidity and nutrient status on the formation of gley soils. (a) Under nutrient-rich conditions a mull gley is formed. With declining nutrient status and increasing acidity either moder gley or humic gley soils develop. These three types have similar profiles and differ mainly in chemical characteristics, humus form and vegetation. (b) In circumstances conducive to podsolization, moderately high rainfall leads to the formation of peaty gleyed podsols while very high rainfall induces the formation of hill peat (ombrogenous peat) giving (c) bog gley soils overlaying a thin iron pan soil similar to that of a peaty gleyed podsol. (d) Shallower peat soils often form directly overlying a gley horizon to form peaty gleys.

peat soils are considered to be different in nature and are classified in a separate order, the *Histosols*.

The vegetation of the acid gleys varies according to wetness and oligotrophy. The humic gleys often have a swamp woodland of *Salix* and *Alnus* spp. or a poor marsh vegetation of herbaceous hydrophytes. The peaty gleys usually have a surface mat of peat-forming mosses, including *Sphagnum* spp., with a cover of grasses such as *Molinia caerulea*. Under slightly drier conditions the vegetation of the iron pan podsols may be of Ericeae such as *Calluna vulgaris* and *Vaccinium myrtillus*. The bog gleys, much wetter soils, generally support peat-formers such as *Sphagnum* spp., *Molinia caerulea* and *Eriophorum* spp. In basin peat areas some indicators of nutrient enrichment such as *Juncus* spp. may appear, but generally the vegetation of valley bog, raised bog and blanket bog is superficially similar, conditioned by nutrient deficiency, acidity and waterlogging. Because these characteristics are continuous and partially independent variables, the range of intermediate soil types and vegetations is very large.

The variation of soil type with topographic wetness is expressed in the concept of the *catena*. First developed in relation to tropical soils, it is useful wherever topography generates a pattern of repeating soil types. Such toposequences may arise on the same or on differing parent materials and reflect not only soil wetness but also the influence of evaporation, insolation and erosion. Figure 3.11 shows a number of catenary soil sequences of various types.

Calcimorphic soils

Soils formed on relatively pure limestone in the temperate zone are generally *Rendzinas*. They are usually rather immature, thin soils with an A horizon resting directly on the C horizon limestone, or perhaps on a C_a horizon under drier conditions (Figure 3.12).

The A horizon is typically black, dark grey or dark brown in colour and ranges from a few cm to c. 35 cm in depth. The dark coloration is caused by organic matter in the presence of high concentrations of calcium carbonate; there are often free fragments of limestone parent material in this horizon which have been incorporated by earthworm activity. The modal rendzina has a mull or mull-like humus but there is a rather wide range of types and humus forms (Kubiena, 1953). These in which the A horizon is mollic (p. 79) belong, like the chernozems and brown forest soils, to the *Mollisols,* of which a separate suborder, the *Rendoll,* is reserved for the rendzinas (Soil Surv. Staff, 1960).

The rendzinas are most often associated with pure limestone such as Cretaceous chalk, Jurassic Oolitic limestones, Carboniferous limestone and other similar deposits. On clay-rich materials, for example the Liassic limestone/shale deposits, deeper calcimorphic brown soils may form and ultimately decalcify and show lessivation of clay. With increasing

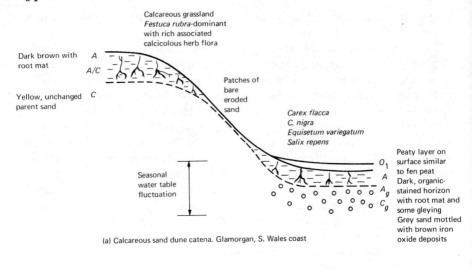

Calcareous grassland
Festuca rubra-dominant
with rich associated
calcicolous herb flora

Dark brown with A
root mat
A/C

Yellow, unchanged C
parent sand

Patches of
bare
eroded
sand

Carex flacca
C. nigra
Equisetum variegatum
Salix repens

O_1 Peaty layer on
surface similar
to fen peat
A Dark, organic-
stained horizon
A_g with root mat and
C_g some gleying
Grey sand mottled
with brown iron
oxide deposits

Seasonal
water table
fluctuation

(a) Calcareous sand dune catena. Glamorgan, S. Wales coast

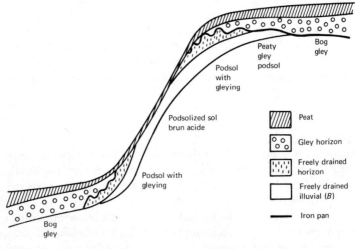

Peaty
gley
podsol

Bog
gley

Podsol
with
gleying

Podsolized sol
brun acide

Peat

Gley horizon

Podsol with
gleying

Freely drained
horizon

Freely drained
illuvial (*B*)

Iron pan

Bog
gley

(b) British upland catena characteristic of high rainfall areas with
oligotrophic parent materials

Figure 3.11 Two examples of soil catenas or toposequences.

temperature there is a transition between the rendzinas and the terra
rossa/terra fusca soil complex.

Rendzinas may be highly alkaline, their pH values ranging between c.
7·5 and 8·3, reflecting the high calcium carbonate content. Within this pH
range the solubility of many nutrients is much reduced, for example, iron
deficiency and lime-induced chlorosis are typical of such soils. As a consequence, many plant species of these soils are *calcicoles,* either obligate or

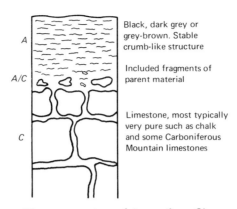

A

A/C

C

Black, dark grey or
grey-brown. Stable
crumb-like structure

Included fragments of
parent material

Limestone, most typically
very pure such as chalk
and some Carboniferous
Mountain limestones

Figure 3.12 A rendzina soil profile.

facultative, while *calcifuge* plants are usually excluded from the association. Further discussion of this problem is presented in Chapter 8.

The native vegetation of most rendzinas is woodland; in N. Europe probably of *Fagus sylvatica* or, with increasing rainfall and harder parent limestones, *Fraxinus excelsior*. In central Europe alpine rendzinas are formed under a vegetation of cushion grass or grass heath, but in the more oceanic climate of Britain and N.W. Europe the extensive chalk and limestone grasslands dominated by *Festuca rubra* and *F. ovina,* or by taller grasses such as *Zerna erecta* or *Brachypodium pinnatum,* are biotic plagioclimaxes maintained by rabbit- or sheep-grazing. With lessened grazing pressure there is a subsere through a limestone scrub of *Swida* (*Thelycranea*) *sanguinea* and *Crataegus monogyna* to a woodland dominated by *Fraxinus excelsior* and/or *Fagus sylvatica*. The limestone grasslands are particularly rich in herbaceous rosette and creeping species which give them a characteristic colourful appearance during spring and summer and comprise a range of calcicoles of varying degrees of exclusiveness and a few more ubiquitous plants.

Halomorphic soils

Most halomorphic soils are formed in arid zones by surface accumulation of soluble salts related to high ground water levels, impermeable substrata and high evaporation rates, thus having an originally hydromorphic character. Kubiena (1953) describes primary salt soils, which are gleys in which a range of salts have crystallized, at the surface, by evaporation and capillary rise under a low P/E regime. Secondary salt soils occur under the same climatic conditions by superimposed waterlogging, for example as a result of overirrigation or leaking irrigation canals and have caused serious agricultural loss in areas of extensive irrigation such as the Indus Basin in the Punjab.

The most salt-rich saline soil is the *Solonchak* which has a salt-saturated pale grey *A* horizon overlying a mottled gley layer. In the spring the soil may be flooded but, as the season advances and the soil dries, it turns white and patchy efflorescences of salt crystals may occur at the surface. This effect and the pale colour of the soil are responsible for the alternative name of *White Alkali soil*. The salts are usually chlorides and sulphates of sodium, calcium and magnesium with a soil pH of above 8·0 due to the presence of some carbonates. If very large amounts of sodium carbonate are present, the pH may rise to 10·0 or more.

The saline soils contrast with the carbonate soils or *Solonetz* in which the predominate salt is sodium carbonate. The soil organic matter is very dark in colour, causing these soils to be called *Black Alkali soils*. They are not salt-enriched at the surface, as is the solonchak, the sodium in the upper part of the profile being in exchangeable form and only appearing as a carbonate accumulation lower in the profile.

The native vegetation of these soils is strongly halophytic and includes a high proportion of species drawn from the family Chenapodiaceae; for example, the Great Basin Desert of the Western U.S.A. has extensive areas of salt desert which may be dominated by Shadscale (*Atriplex* spp.) and Winter Fat (*Eurotia lantana*). With higher salt content *Salicornia* spp. become dominant. Species of *Atriplex*, *Salicornia*, *Halimione*, *Sueda* and *Salsola* may be found on saline soils whether of salt deserts or of temperate zone salt marshes in many parts of the world.

Chapter 4

Chemical and physical properties of soils. The root environment

The soil is a three-phase system of solid, liquid and gaseous components, each of which has its own physical and chemical properties, in an equilibrium, or transient-state, relationship with the others. The liquid and gaseous phases are, in small soil volumes, fairly homogeneous. By contrast the solid phase is heterogeneous, comprising a range of different sized inorganic particles of silica, silicate clay, metal oxides and other minor components, all in varying degrees of association with different types of organic matter. The nature of this association and the characteristics of the organic matter vary greatly with soil type.

In addition to these components, each soil has a distinctive flora and fauna of Bacteria, Fungi, Blue-green algae, Algae, Protozoa, Rotiferae, Nematoda, Oligochaetae, Mollusca and Arthropoda. This assemblage of organisms not only reflects the present status of a soil, but also affects the pedogenetic process by modifying the course of organic decomposition, by influencing chemical processes and by altering the physical structure of the soil. The interaction of an earthworm population with soil fungi and bacteria, for example, causes the strong aggregation of many mollic soil horizons.

THE MINERAL MATRIX

The larger soil particles, frequently of silica (SiO_2), are direct weathering residues of quartz, or other forms of silica, in the parent material. They have a small specific surface and little physical or chemical activity, but often form a structural matrix which houses the other components of the soil.

The major chemical and physical characteristics of the soil derive from the finer textured particles: the silicate clays. Their particle size is defined as being below 0·002 mm equivalent diameter (p. 60), thus they have a high specific surface, colloidal properties and, because of the nature of their crystalline structure, have surface charges which impart strong ion-exchanging properties. They are also the source of most of the inorganic plant-nutrient elements. In conjunction with organic matter the clay fraction is the seat of most of the soil physical and chemical properties.

STRUCTURE OF CLAY

Grim (1953) gives a detailed account of clay mineral structure and attributes the basic generalizations to Pauling (1930). Electron micrographs show that clays are made up of flat plate-like or needle-like crystals in which X-ray diffraction studies have shown atomic layer lattices involving two types of structural subunits (Figure 4.1). One consists of two sheets of close-packed oxygen atoms or hydroxyl units in which aluminium (or magnesium or iron) atoms are embedded in octahedral coordination, each aluminium atoms thus being equidistant from six O— or OH— positions. The other subunit is a sheet of silica tetrahedra in which each silicon atom is equidistant from four oxygens.

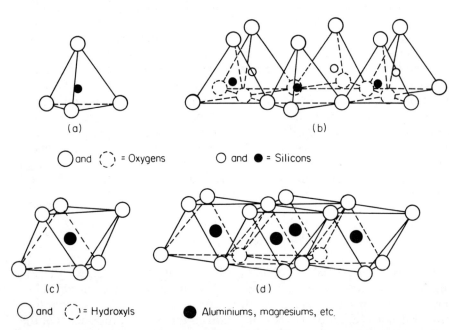

Figure 4.1 The structure of the clay lattice subunits. (a) Single silica tetrahedron. (b) sheet structure of silica tetrahedra. (c) single octahedral unit. (d) sheet structure of octahedral units. Reproduced with permission from R. E. Grim, *Clay Mineralogy*, McGraw-Hill, London, 1953, Figure 1, p. 43.

Soils contain a wide variety of silicate clays but a few common examples may be cited. Highly weathered soils such as the oxisols and ultisols usually contain *kaolinite* while soils of less extreme weathering, for example, the mollisols, often contain *vermiculite* and *montmorillonite*. Many immature soils have *mica* clays which are usually direct derivates of mica in the parent material. Many other minerals may occur in the clay fraction, for example the oxisols often have high proportions of ferric and aluminium sesquioxide. Despite their high specific surface these non-silicate clays show much less chemical activity than the layer lattice clays.

The silicate clays mentioned above have layer lattices which are based on the aluminium octahedral sheets and silicon tetrahedral sheets bonded together to form a 1 : 1 lattice with one aluminium and one silicon sheet or a 2 : 1 lattice with a central aluminium sheet sandwiched between two silicon sheets. Kaolinite is a 1 : 1 clay in which the unit cells of the lattice are hydrogen bonded together (Figure 4.2) and do not swell on hydration as water molecules cannot enter between the lattice layers. By contrast, the 2 : 1 lattices of vermiculite and montmorillonite show strong swelling on hydration as they have, respectively, magnesium and calcium ions in the interlayers, associated with varying amounts of water of hydration. The thickness of the unit cell may vary from 1·0 to 1·5 nanometers according to the degree of hydration. The 2 : 1 mica clays have potassium ions in the interlayers, without water of hydration, and do not swell on wetting. Soils such as the vertisols which show very marked volume changes and crack formation during the wetting–drying cycle have a high proportion of swelling clay, usually montmorillonite.

CATION EXCHANGE

If a solution of a neutral salt such as potassium chloride is flushed through a soil the outflowing liquid is usually found to be depleted of potassium, enriched with other metal cations and often reduced in pH due to enrichment with hydrogen ions. These changes have been wrought by the *cation exchange complex* of the soil, consisting mainly of the clay fraction and the organic matter. The cation exchanging property is caused by the presence of unsatisfied negative charges at the surface of the clay and organic particles; metal cations and hydrogen ions are bound at these sites, sufficiently loosely to show exchange with ions in the bathing solution. Table 4.1 shows the *cation exchange capacity* (CEC) of various clays and Table 4.2 that of different soils. It may be seen that the CEC of sandy soils is associated mainly with their organic content but that of finer textured soils is a function of clay content. The ion-exchanging behaviour of organic matter is discussed later (p. 119).

Silicate clays have ion-exchanging properties for three reasons: (i) broken bonds at the edges of aluminium/silicon units produce unsatisfied

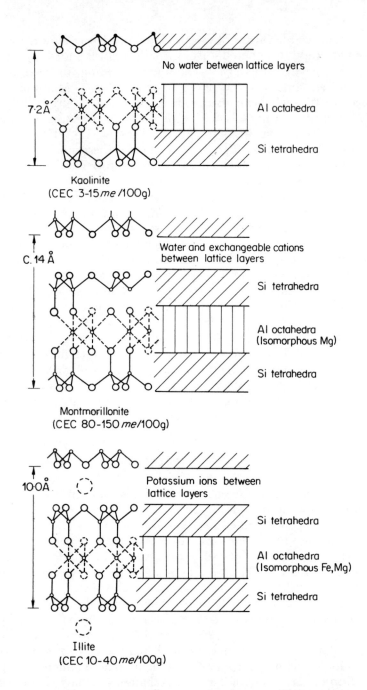

Figure 4.2 The crystal structure of 1:1 and 2:1 lattice clays. Kaolinite is a non-swelling 1:1 clay which does not take in water molecules between the lattice layers. It has no

charges; (ii) some exposed hydroxyl units may lose hydrogen ions in exchange for metal cations under certain pH conditions and (iii), usually most important, individual atoms in the lattice may be isomorphously substituted; trivalent aluminium for quadrivalent silicon and divalent iron, calcium and magnesium for aluminium. The substitution does not disrupt the lattice sufficiently to alter the crystal structure but it does result in a surplus negative charge which causes cation binding at the crystal surface. Montmorillonite has a high cation exchange capacity which is mainly due to considerable isomorphous substitution of magnesium for

Table 4.1 The cation exchange capacity of some clay minerals

Clay	C.E.C. (meq/100 g)
Kaolinite	3–15
Montmorillonite	80–150
Illite (mica-type clays)	10–40

Table 4.2 The cation exchange capacity of different soils

Soil	Horizon	% Organic C	% Sand/Silt/Clay	C.E.C. (meq/100 g)
Sandy podsol	A_1	1·12	96/3/1	5·8
	A_2	0·03	96/3/1	1·0
	B	2·54	89/4/7	27·5
	C	0·02	93/2/5	1·4
Prairie soil. High-	A	2·01	44/20/36	27·0
clay content	B_1	1·13	42/20/38	29·0
	B_2	0·63	44/22/34	26·2
	C	0·14	47/24/29	24·5

Figure 4.2—(continued)

isomorphous replacement and, consequently, a low CEC. Montmorillonite has a 2:1 lattice which, in the expanded form, holds a layer of water between the lattice layers. In the dehydrated condition the lattice closes to a unit spacing of about 9.3Å. The cation exchange capacity of montmorillonite is high because aluminium is isomorphously replaced by magnesium and a large exchange surface is available between the lattice layers. Illite is a micaceous, non-swelling clay which structurally resembles montmorillonite but has its lattice layers bound together by non-exchangeable potassium ions and cannot expand by absorption of water. Though it shows extensive isomorphous replacement of aluminium by iron and magnesium it has a low exchange capacity as the interlattice layers are not available for exchange. After R. E. Grim, *Clay Mineralogy,* McGraw-Hill, London, 1953, Figure 2, p. 44.

aluminium. By contrast, kaolinite has a much lower CEC, deriving only from broken bonds and hydroxyls, there being no isomorphous substitution within the lattice.

The presence of ion exchangers in the soil is of great importance both in pedogenesis and in the soil–plant nutritional relationship. Most metallic elements which are taken up by growing plants are absorbed as cations but they exist in three forms in the soil: (i) as sparingly soluble components of mineral or organic material; (ii) adsorbed onto the cation exchange complex and (iii) in small quantities in soil solution. Free cations in solution are easily leached from the soil if the P/E ratio is high but the exchange complex forms a reservoir of nutrients which are not so easily lost in this way, though maintaining a continuous supply to the soil solution by slow equilibration. Hence soils may exist, under high rainfall conditions, which maintain a steady supply of nutrients to the plant cover without becoming rapidly depleted of nutrients by the leaching process. The plants also act as bio-cyclers in this relationship, the root systems extracting nutrients from deeper horizons and thence returning them to the soil surface in litter. As decomposition proceeds the liberated cations return to the exchange complex of the surface layers.

Despite their nutrient-retaining behaviour many soils ultimately do become acidified by desaturation of the exchange complex; metal cations being replaced by *exchangeable hydrogen*. The acidification of a neutral salt solution on passage through a soil is caused by the flushing of exchangeable hydrogen ions from the exchange complex. There is, however, some difficulty in interpreting this observation as the leachate from an acid soil almost invariably contains aluminium amongst the displaced cations. Aluminium ions rapidly react with water to form insoluble aluminium hydroxide and hydrogen ions; it is thus difficult to know whether the increased acidity of the leachate derives from exchangeable hydrogen or exchangeable aluminium in the soil. Discussion of this difficulty may be found in Black (1968). It is conventional to express the *cation saturation* of a soil as the percentage of the cation exchange capacity occupied by metal cations. Cation exchange measurements are usually made at a standardized pH as the behaviour of some ion binding sites is pH dependent.

Most techniques for measuring cation exchange capacity, individual exchangeable cations, cation saturation and exchangeable hydrogen have depended on various exchange displacement methods in which a high concentration of a single salt is leached through a soil sample so that the exchange complex becomes totally saturated with a single cation species and the originally adsorbed cations appear in the leachate. Neutral ammonium acetate has been most widely used; metal cations may then be measured in the extract and cation exchange capacity determined by distilling the adsorbed ammonia from the soil sample and measuring it either volumetrically or colorimetrically. Exchangeable hydrogen may be estimated from the difference between the total metal cation value and the cation exchange capacity.

Exchangeable hydrogen or exchange acidity may also be measured directly, for example by leaching with a buffer solution of barium chloride and triethanolamine, followed by titration to estimate the exchange acidity resulting from displacement of both hydrogen and aluminium ions. Details of these techniques may be found in Black (1965), Jackson (1958) and Piper (1944). The origin of cation exchange activity in a number of different mechanisms, depending on dissociation of different components of the system, leads to a pH sensitivity; thus cation exchange capacity, exchangeable hydrogen and exchangeable metal cations should all be measured at a specified pH (often pH 7). It might be more desirable in plant nutritional studies to measure these values at the prevailing soil pH.

Soils also show a small anion exchange activity arising from unsaturated positive charges on clay and organic particles, but anion exchange plays little part in plant nutrition or pedogenesis because the majority of anionic nutrients (e.g. N, P and S) are present as insoluble organic compounds which slowly liberate the nutrient anions to the solution without massive adsorption on an exchange complex. This problem is discussed more fully in Chapter 8, p. 236.

SOIL ACIDITY

Soil acidity is associated with the presence of hydrogen and aluminium ions on the exchange complex and the existence of an equilibrium solution of hydrogen ions in the interstitial water of the soil. As an intensity factor it may be defined by the conventional physical chemical concept of hydrogen ion activity expressed as pH.

pH is defined as the negative logarithm of hydrogen ion activity where activity is understood to mean effective concentration. The product of hydrogen ion and hydroxyl ion activity is constant for dilute aqueous solutions and equals 10^{-14}. In pure water the hydrogen and hydroxyl ion concentrations are equal and have the value 10^{-7} g ions/litre, hence the pH of pure water is 7. Increasing acidity raises the H^+ ion concentration, lowers the OH^- ion concentration and lowers the pH value. Increasing alkalinity raises the OH^- ion concentration with a corresponding reduction in H^+ concentration and thus increases pH value. Soil pH is now usually measured electrometrically using a glass electrode referred to a calomel half-cell.

The chemical definition of pH is only valid for simple, aqueous solvent–solute systems. Measurements of pH in suspensions of surface-charged solids, though giving repeatable results, must be interpreted with caution. Readings taken with the electrodes in the supernatant liquid sometimes differ by more than 0·5 pH units from readings with electrodes in the sediment. Further differences appear if the electrolyte concentration of the soil solution is altered, for example the addition of a neutral salt such as potassium chloride usually reduces pH, presumably by displacement of hydrogen ions from the exchange complex. The electrode positional effect is probably due to a junction potential arising at the boundary between the calomel electrode and the suspension: potassium ions may enter the cationic atmosphere of individual soil particles while the chloride ions are largely excluded and move only through the external solution. The potassium ions thus move forward, as a front, faster than the chloride ions. Olsen and Robbins (1971), using suspensions of ion exchange resin, obtained results which support this theory. With the further complication that potassium may exchange with ions of different valency from adjacent soil particles, the junction effect leads to aberration in the pH reading compared with that in simple solution (Black, 1968).

The indicated soil pH may thus be strongly affected by the position of the calomel electrode but little by the glass electrode. The consensus of opinion is that the pH reading obtained is a reflection of the mean pH of the liquid surrounding the soil particles. This pH is susceptible to dilution effects and therefore, for consistency of reading, various standardized soil : water ratios have been used, for example 1 : 2·5 or 1 : 1, while some workers have attempted to simulate the field situation more closely by using a stiff paste of soil and water. After adjusting the water content of a soil sample it is necessary to allow some time for pH equilibration as fairly slow processes such as cation exchange and/or carbonate dissolution are involved.

Addition of water to solutions of fully ionized electrolytes causes a pH shift toward neutrality; alkaline solutions decrease and acid solutions increase in pH. In soils, the consequences of dilution are much more unpredictable as the addition of water reduces the electrolyte concentration in the soil solution, thus causing further dissociation of cations from the exchange complex. It also, temporarily, alters the carbon dioxide concentration in solution which, in calcareous soils, may strongly influence pH. Addition of water also increases the thickness of water films on soil particles, so modifying the local influence of adsorbed ions on pH and also altering the junction potential effect described above.

Despite the difficulties of interpreting soil pH values they show strong correlations with soil type, vegetation type, profile horizon and, agriculturally, with crop growth, lime requirement and mineral nutrition. Natural soils usually have pH values between about pH 3·0 and 8·4, the upper value being the calcium carbonate equilibrium with atmospheric carbon dioxide concentration and the lower value, the soil solution equilibrium with a highly hydrogen-saturated soil. More extreme values do occur in unusual soil types: some alkali soils with high sodium carbonate content reach values of pH 10·0–10·5 while drained gleys may produce sulphuric acid by oxidation of ferrous sulphide, their pH falling to 2·0 or below. Such low pHs may also be found in spoil heap soils derived from sulphide ores or from coal spoil with a high ferrous sulphide content.

Many years ago Salisbury (1925) showed that almost all of the calcium carbonate in dune sand was lost by leaching within 250 years in the prevailing rainfall climate of 80 cm/ann and, during this period, the pH fell from 7·8 to 5·0. This rate of change may be lessened on finer textured parent materials but, even so, with an excess of precipitation over evaporation, decalcification is a pedological fact.

Jenny and Leonard (1934) investigated soils of the loess zone of the central USA on a transect spanning a precipitation gradient from 30 to 100 cm/ann and showed that the depth to free carbonates was strongly correlated with rainfall (Figure 4.3). After decalcification, further leaching leads to the desaturation of the exchange complex and an increase in the acidity of the soil solution, due to the increased ratio of hydrogen to metal

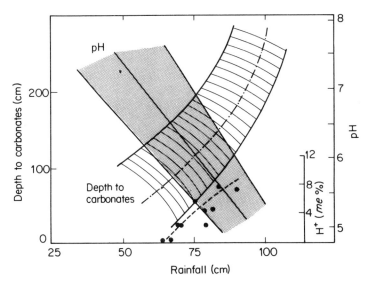

Figure 4.3 The depth of the decalcification front, soil pH and titratable hydrogen in relation to annual rainfall in the central USA. ● = Exchangeable hydrogen. Data of Jenny and Leonard (1934).

cations ionizing from the exchange sites. Figure 4.3 shows that titratable hydrogen was first detected in this study at an annual precipitation level of about 60 cm and with a soil pH of about 7. Increasing rainfall to 90–100 cm/ann is accompanied by a fall in soil pH to 5·0–5·5 and a rise in titratable hydrogen to 6–8 *me*/100 g. The main source of titratable hydrogen is likely to be dissociation from an unsaturated exchange complex.

Leaching, in pedogenesis, may therefore be considered as a process in which an advancing front of decalcification or cation desaturation passes down the profile, accompanied by a fall in soil pH. Natural soil profiles are not always, however, more acid near the surface than in the deeper horizons as other factors are confounded with the leaching process. Biocycling of elements returns cations to the soil surface in litterfall and delays or even reverses the leaching loss; faunal soil mixing limits downward movement while annual P/E and seasonal distribution of precipitation have a very strong influence on leaching. High evaporation rates may actually enrich the surface soil with solutes carried upward in capillary water and produce pedocal soils of very high pH. Figure 4.4 in-

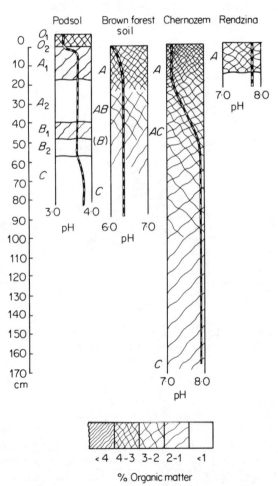

Figure 4.4 Organic matter and pH profiles of several contrasting soil types.

dicates the range of pH values which may be encountered in natural soil profiles and also shows the very steep profile pH gradients which may be found when there is a marked build-up of organic surface layers (O horizons). The steepness of these gradients is perhaps superficially concealed by the pH notation in which one unit represents a tenfold change in H^+ ion concentration.

Pearsall (1952) delimited various boundary pH values which are ecologically significant. Plants regarded as calcicoles usually occur above pH 6·5 and contrast with the extreme *calcifuges* of heath and moorland soils below pH 3·8–4·0. Soils above pH 6·5 are generally cation-saturated and have a subclass of *calcareous* soils containing free calcium carbonate, while soils below pH 3·8–4·0 are strongly desaturated and have a considerable content of exchangeable hydrogen.

These pH limits are also reflected in the nature of the soil organic matter: raw humus or *mor* is associated with soils of below pH 3·8 while *mull* is characteristic of the more cation-saturated soils of pH 4·8–5·0 and above. Between 3·8 and 4·8 intermediate organic matter forms occur. Soil organic matter is discussed in more detail later in this chapter.

Many differences between soils of differing pH are due to the processes of 'soil metabolism' which vary strongly with soil pH. Between pH 5·0 and 8·0 + both bacterial and fungal decomposition proceed rapidly, but below about pH 5·0 bacterial activity is reduced and fewer fungal species are found. There is also a change in fauna, the numerous worms and snails of higher pH soils being replaced with a less diverse population of arthropods: mainly mites and springtails.

Limitation of bacterial activity by acidity and accompanying calcium deficiency slows the rate of organic decomposition so that acid soils tend to accumulate a thick, superficial mat of undecomposed organic matter: the O_1 and O_2 horizons so characteristic of spodosols. Increasing wetness also inhibits oxygen diffusion in the soil, encouraging anaerobiosis and slowing decomposition. Hence the histosols or bog soils may accumulate layers of acid peat many metres in depth.

Long-term waterlogging also interacts with soil pH, tending to reduce the pH of alkaline soils and increase that of acid soils so that most anaerobic soils are usually in the range pH 5·0–7·0. In alkaline soils the reduction of pH may be caused by accumulation of carbon dioxide and, possibly, by release of organic acids from microorganism metabolism. Acid soils tend to be increased in pH probably by the conversion of inert ferric sesquioxide to the more basic ferrous hydroxide (Greene, 1963). Wetness and soil pH are thus inextricably bound together and may be subject to seasonal oscillations, peat bogs, for example, becoming more acid as they dry out and oxidation replaces the prevailing reducing condition. Calcareous soils, by contrast, increase in pH as they dry and reach a pH equilibrium with a lower carbon dioxide concentration. These effects may also occur on a microscale in the structural aggregates of the soil which may be oxidized on their surfaces but reduced within the bodies of the aggregates (Greenwood, 1961; Crampton, 1963).

Figure 4.5 presents some of the generalizations which may be made, relating pH wetness, soil processes and vegetation. The accumulation of organic matter associated with both low pH and increased wetness is also reflected in a greater C/N ratio, while the inhibition of bacterial activity limits the nitrifying process so that acid or waterlogged soils contain little

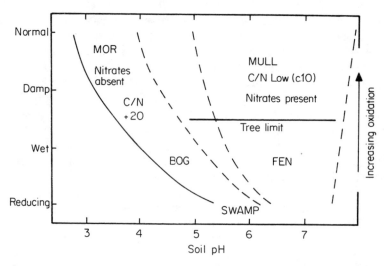

Figure 4.5 Soil characteristics in relation to wetness and pH. Reproduced, by permission, from W. H. Pearsall, The pH of Natural Soils, *J. Soil Sci.*, 3, 48, Figure 4 (1952).

or no free nitrate. A strong pattern of vegetational distribution is also imposed by acidity and wetness. As soil parameters these two factors and their interaction are possibly the most important in defining the nature of the ecosystem. An ecologist, armed with the knowledge of soil pH and long-term wetness, could essay a fairly accurate prediction of soil type, vegetation and fauna for a given geographic area.

Some of the reasons for the overriding importance of these two factors are discussed by Greene (1963). The value of soil pH can, in most normal soils, be considered as an index of its exchangeable cation saturation as most soils between about pH 5·0 and 6·0 pass from approximately 25% to 75% saturation. Under waterlogging conditions there is a further relationship between pH and redox potential which is governed by the equilibrium between ferric sesquioxide and ferrous hydroxide. Fe_3O_4 acts as a mild oxidant and, while it remains present in the soil, redox potential does not fall to a level at which the soil can become seriously toxic by production of sulphide or excessively high ferrous iron concentrations. During the reduction process, ferrous hydroxide is produced, which is weakly basic and causes a rise in soil pH.

Interference with mineral nutrition is also characteristic of both wetness and pH changes. Soluble iron and aluminium, produced at low pH or low redox potential, may remove phosphates from solution by precipitation (vivianite (blue ferrous phosphate) may occur in the lower horizons of bog soils). Increased solubility of iron, manganese and aluminium may also cause toxicity effects. Nitrogen fixation and mineralization may be limited by low pH or anaerobic conditions while dentrification seems to be encouraged by waterlogging (Greene, 1963). Low pH is particularly characteristic of cation-unsaturated soils and nutrient deficiencies, particularly of divalent cations such as calcium and magnesium, may arise, due both to their limited amount and to impairment of uptake. Hewitt (1952) has discussed the influence of soil pH on plant growth, mainly mediated through mineral nutrient effects. These are discussed more fully in Chapter 9 and in relation to the calcicole–calcifuge problem in Chapter 10.

Soil pH, as a criterion of hydrogen ion concentration, is an intensity factor reflecting degree of acidity, but soils also vary quantitatively in acidity according to the degree of unsaturation of the exchange complex. This is important in considering the agricultural implications of soil acidity, particularly lime requirement.

For agricultural purposes it is normal to adjust soil pH to about pH 6·5. Addition of lime ($Ca(OH)_2$) or ground limestone ($CaCO_3$) initially raises the pH of a soil by neutralizing the free hydrogen ions in the soil solution. This, however, promotes further ionization of hydrogen from the un-saturated exchange complex and the pH cannot rise to 6·5 until all of the exchangeable hydrogen has been neutralized and replaced by metal cations. Thus the lime requirement is not only a function of original soil pH but also of total exchangeable hydrogen. Acid soils with a high clay or organic matter content have a large reservoir of exchangeable hydrogen compared with sandy soils with a low cation exchange capacity. The sandy soil has a low *buffering capacity* and is easily changed in pH by addition of lime compared with the well-buffered clay soil (Figure 4.6).

SOIL ORGANIC MATTER

The organic matter derived from surface-living plants and animals and from soil-dwelling micro- and macroorganisms, is a key factor differentiating the mineral 'crust of weathering' from true soils. The nature of the organic matter is governed both by input and by 'soil metabolism', consequently vegetation, climate, parent material and topography all have a strong influence.

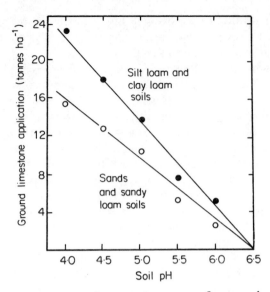

Figure 4.6 The requirement of ground limestone (CaCO₃) to adjust a soil to pH 6·5. Data of Jackson (1958).

The initial input is in the form of surface litter and subterranean dead root material but, after encountering various soil processes, this organic matter may be grossly modified and has, in the past, been referred to as *humus*. Current practice is, generally, to replace this term by 'soil organic matter'.

The first step in the complex process of its formation is the comminution and transport of the original plant fragments, usually by the soil fauna. The earthworm, *Lumbricus terrestris,* is an example of an animal which is spectacularly successful in transporting and incorporating litter which it pulls into its deep burrows (1–2 m) and devours on the spot (van der Drift, 1965). The remains of this ground and digested litter are incorporated in casts which are voided both at the surface and at all depths in the burrow system. The very rapid disappearance of litter in deciduous forests and its uniform incorporation throughout the soil profile pays tribute to the great efficiency of the earthworm population in this process. Some other worms, millipedes, Isopods, Dipterous larvae, slugs and snails, though less mobile, play an early role in the comminution of the litter but leave their faecal pellets in the surface litter. These animals also carry fungal and bacterial spores, rapidly inoculating the whole of the litter mass, and prepare the way for a succession to various microarthropods and other organisms which feed within the decaying tissues on fungal hyphae and other microorganisms. The microarthropods include springtails (Collembola) and mits (Acari). Other agents in clearing of microorganism colonies are Protozoa, eelworms (Nematoda) and Enchytraeid worms. Extensive description of the ecology of these organisms may be found in van der Drift (1965) and Burges and Raw (1967). The importance of these animals in soil development cannot be overstressed; in many soils a large proportion of the organic material is derived from the joint activity of the soil fauna and microorganisms. The mineral horizons of many soils also have abundant populations of animals, the larger members active in both modification of organic matter, and in increasing soil pore size.

After the initial breakdown, commenced by the soil fauna, microorganism attack increases rapidly and the original material is soon cleared of its labile components such as soluble sugars, polysaccharides, proteins and fats, leaving a residue of more resistant lignin and lignin derivatives. This second stage in decomposition takes place throughout the body of the soil, at least in those soil types in which a large earthworm population is available to transport litter quickly into the soil.

The composition of that part of the soil organic matter which is in a long-term steady-state relationship with input-breakdown is not only a reflection of litter chemistry, but is also influenced by the nature of the microbiological resynthesis. For example, the bulk of the soil polysaccharide is made up of bacterial cell wall material, while the more stable 'humic acids' are probably polyphenolic condensation products of microorganism metabolism. The resistance of the more stable organic matter to breakdown is evidenced by carbon-14 dating studies; for example Paul, Campbell, Rennie and McCallum (1964) found that humic acid from a grassland soil has a mean age of about 1000 years.

Humus Forms

In temperate climates the range of organic matter types reflects the gross differences between soil groups; moderate rain coupled with a nutrient-deficient, well-drained parent material tends to produce podsolization and *mor* humus while finer textured, nutrient-rich rocks with broad-leaf forest cover develop brown forest soils and *mull* humus forms. These two terms were originally used in the late 1800s by Muller (Howard, 1969) to describe the organic matter of beechwoods on, respectively, oligotrophic and eutrophic parent materials. Muller's classification is still widely used though problems do arise in applying it to other than cold temperate habitats.

Mor, or raw humus, is characteristic of the true podsol and forms at a low pH (below 3·8–4·0) in a nutrient-deficient milieu. Acidity and lack of calcium limit bacterial activity, thus slowing decomposition and causing the accumulation of a deep O_1 litter layer. Earthworms are absent, thus preventing the litter from being mixed with the mineral soil. It decays slowly, *in situ*, to form the O_2 layer of true mor. This is a compact, rather tough, black or brown peaty material in which the individual fragments are microscopically recognizable as plant material. Most of the plant roots are confined to this horizon or to the surface of the A_1 and it also contains large quantities of dark-coloured fungal mycelium. The discontinuous distribution of organic matter in the profile is quite characteristic, the majority occurring in the O horizon and a small amount of finely particulate material washing into the A_1 horizon. The remainder of the profile has little organic content except for a layer in the B_1 horizon which, probably, is carried down and redeposited in a colloidal form, perhaps associated with iron.

The formation of mor, though partially a function of climate and parent material, is also related to plant cover and human or other biotic interference. There is, in fact, a whole syndrome of soil acidification which has been interpreted as soil deterioration caused by specific types of plant cover and, at least in western Europe, attributable to human destruction of broad-leaf forest and its replacement by heathland (Dimbleby, 1962).

The podsolization process appears to be encouraged by a number of plant characteristics, for example, high fibre (lignin) content, low nutrient content (particularly calcium) of litter and, possibly, high content of various phenolic compounds which may inhibit microorganism activity. These plants form an acid litter layer with few bacteria or earthworms and have been called 'mor formers'. The commonest examples appear to be a number of genera of the Ericaceae and various conifers (Kubiena, 1953). The evidence for soil changes, induced by different species, is not particularly well documented and relies heavily on data from forestry stands. Grubb, Green and Merrifield (1969) have shown, fairly conclusively, that *Calluna vulgaris* and *Erica cinerea* may cause rapid soil acidification. Unexpectedly they also found that *Ulex europaeus* caused acidification at much the same rate as *C. vulgaris,* causing the top centimetre of mineral soil to fall from pH 5–6 to (3·5) 4–5 in 10 to 12 years. They also review the rather sparse literature of the topic.

A number of archaeological investigations of 'fossil' soil profiles buried under tumuli or other artifacts have revealed considerable change of both vegetation and soil type (Dimbleby, 1962; Crampton, 1968). Generally, the buried Bronze Age soils show less podsolization than their present day counterparts and exposed tumulus surfaces have developed podsols or podsolic soils similar to those of the surroundings. Many of these sites show pollen analytical evidence of Bronze Age tree cover while the present day soils have developed under Ericaceous heath. It is tempting to postulate that the growing technology of the Bronze Age permitted, for the first time, widespread tree clearance and initiated the spread of heathland, thereby accelerating the process of acidification by encouraging mor-forming species.

Mor is microbiologically poorer than mull, supporting a smaller bacterial population and a fungal population which, though prominent in biomass, is often limited in species numbers. Figure 4.7 shows the distribution of fungi and bacteria in soil samples taken from a sand dune sere and ranging, with soil age, from pH 6·8 to pH 4·3. In the early stages of

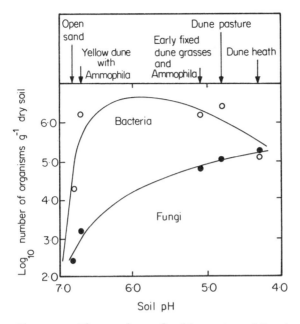

Figure 4.7 The numbers of soil bacteria and fungi in relation to soil pH during a sand dune succession. Note that the numbers of organisms are plotted on a log scale. Soil samples were taken in the depth range 5–15 cm. Data of Webley, Eastwood and Gimingham (1952).

colonization there is a marked increase in the populations of both groups, but, once the succession has reached acid dune heath, bacterial numbers decline sharply (Webley, Eastwood and Gimingham, 1952). Warcup (1951) also found that numbers of soil fungi were similarly increased by falling soil pH and that different species tended to be represented in various parts of the pH range.

Parallel with this situation, the soil fauna show the same decline in species number with falling soil pH, the most important loss being the earthworms which are responsible for soil mixing and incorporation of organic matter in mull humus soils. Barley (1961) suggests that pH 4·5 is the boundary for the larger, deep-burrowing worms; below this pH the typical surface accumulation of mor humus is encouraged. A few worms do occur below this pH but they are small and remain near the soil surface (Satchell, 1967). Those animals which do occur in mor are mainly Arthropoda such as larvae of Diptera and Coleoptera (flies and beetles), Diplopoda (millipedes), Collembola (springtails) and Acari (mites) (Raw, 1967). Some mor soils have coprogenic elements, their structural components being bound together by faecal materials from the soil animals.

An intermediate humus form, *moder,* is cited by Kubiena (1953) as being synonymous with 'insect mull' in which various arthropod droppings are micromorphologically prominent. Moder appears to characterize the intermediate pH range between true mull and true mor formation.

The process of podsolization and its apparent relationship to certain plant species has attracted the attention of numerous workers who have tried to show that various organic compounds derived either from the living vegetation, the litter layer or the O_2 mor layer, are responsible for the intense eluviation of the A horizon. Deb (1949) and Bloomfield (1953a) reviewed the previous literature and noted that organo-iron may move, as a humus-protected sol or as a complex organic ion. Bloomfield, in this and in a number of other papers (1953b; 1954a, b, c) showed that soil sesquioxides could be mobilized by aqueous extracts of the leaves of many conifers and some broad-leaved species. He also noted (1965) that the activity of leaf extracts in mobilizing sesquioxides correlated with their content of polyphenols but, since both broad-leaved species and conifers showed this activity, postulated that the residence time of polyphenols in the litter layer may be crucial and might be affected both by enzymatic effects in the decay process and by the palatability of the litter to the soil fauna. The problem is also further complicated by the possible implication of soil microorganisms in the mobilizing process (Aristovskaya and Zavarzin, 1971).

Mull humus characterizes the richer brown forest soils and some other mollisols of the temperate zone. It is associated with soil pH values above 5·0, an abundance of divalent cations and a well-developed earthworm fauna. A grey, brown-grey or blackish material, it is diffusely incorporated amongst the soil mineral particles by biological mixing. Microbial decomposition is so rapid that no plant or microorganism remains are recognizable; it has an amorphous, colloidal nature. The association with soil particles is so intimate that it cannot be separated by normal, mechanical means and is believed to be physicochemically bound as a 'clay–humus complex'. Kubiena (1953) suggests that the complex is microbiologically stable and is responsible for the persistent good structure of agricultural soils derived from broad-leaved forest or grassland areas. The natural vegetation of mull soils is a luxuriant reflection of the eutrophic nature of the material, is rich in species and, by contrast with the vegetation of mor soils, has many deep-rooting plants which are active in nutrient cycling and produce easily decomposable litter.

The rapid disappearance of litter and its incorporation into both A and (B) horizons of molisols is associated with intense earthworm activity. Many years ago Darwin (1881) noted that earthworms in pasture soils required only thirty years to produce a surface layer of worm casts six inches or more in depth; an annual turnover of some 20 tons/acre. This represents less than half of the material ingested and voided by worms as many of the casts are produced below ground. These casts contain the ground-up remains of plant litter, already attacked by cellulase enzymes, homogeneously mixed to a paste-like consistency with soil to form a perfect substrate for microorganism growth. As a result, fungal hyphae and bacterial gums bind the casts into stable aggregates which persist for long periods and form the bulk of the material in the A and (B) horizons. Such soils are very porous and have a spongy consistency underfoot. The earthworm mixing process is almost certainly responsible for the formation of the clay–humus complex.

The rate of removal of litter and its relationship to mull/mor formation is affected by the differential palatability of various plant species to earthworms. Satchell (1967) noted that the well-recognized podsolizers, the conifers and the beech, are the least palatable, the oak which can occur on podsols is intermediate, while the well-recognized mull formers are the most palatable. This observation has considerable significance in the light of Bloomfield's (1965) suggestions that leaf litter with a long residence time will be most active in mobilizing iron with polyphenolic leachates. Satchell (1967) also notes that destruction of worm populations with toxins leads to surface accumulation of organic matter in normally mull soils.

The nature of soil organic matter

Soil organic matter represents the equilibrium between input, originating from primary photosynthetic production, and the degradative and resynthetic processes associated with soil-dwelling organisms of all kinds. Most naturally occurring organic compounds may thus be found, at some time, in soil.

The input is largely of carbohydrate, lignin compounds, fats and proteins with smaller quantities of, for example, free amino acids, alkanes, terpenoids, carotenoids, flavonoids, alkaloids, polyphenols, resins and others. Most of these compounds are easily decomposed and have a short residence time in the soil; for example, carbohydrates are rapidly metabolized by both animals and microorganisms and even the more stable cellulose component is prone to fungal attack and to digestion by lumbricids and other animals which secrete cellulase or have cellulose-attacking gut flora.

Minor constituents

Carbohydrates thus occur in low concentrations in the soil, much of it being polysaccharide, secondarily derived from microbial products such as bacterial cell walls. Soils rarely contain more than 0·25–0·3% of polysaccharide, though exceptions may be found in peat soils, where it may form 10% or more of the organic matter. Soluble sugars do occur but only in minute quantities, probably as equilibria with enzymatic conversion processes. Lipids are generally broken down quickly but the cuticular waxes may persist even to the extent of forming fossil deposits. Heavily cutinized pollen grains may, for example, be used in the dating of soils (Dimbleby, 1961).

Proteins and amino acids are also rapidly attacked by soil organisms though chromatographic studies show that most amino acids are present at very low concentrations. The bulk of the nitrogeneous material becomes bound, in the form of amino–N, either to the lignic materials as 'ligno-protein' or, by adsorption, to the clay surfaces. This bound nitrogen appears to be very stable, only a few per cent being mobilized during each growing season though it has been suggested that this may be an apparent effect, due to an insufficiency of organic energy-supplying materials for the nitrifying bacteria. The nature of the soil organic nitrogen fraction is still a subject of dispute.

Binding of inorganic nutrient elements in an unavailable form also occurs with phosphorus and sulphur. As with nitrogen, the bulk of the soil content of these elements is in an organic form and only slowly mobilizable by microorganisms. Cosgrove (1967) suggests that one-half to two-thirds of the soil phosphorus is organic and that the major constituent is inositol hexaphosphate derived from plant, animal and microorganism myoinositol polyphosphate (phytin). Phospholipids, nucleic acids and sugar phosphates entering the soil are all rapidly dephosphorylated, hence one of the limiting factors in phosphorus availability is the rate of dephosphorylation by microbial phosphatases of inositol hexaphosphate. Likewise, soil sulphur is predominately organic and not freely available to plants (Freny, 1967). Under aerobic conditions it is slowly converted by microorganisms to sulphate, the form in which it is absorbed by plants (see Chapter 8, p. 229).

Organic acids may accumulate in soils, produced both by plant roots and by microorganisms. The commonest are those involved in the tricar-boxylic acid cycle but other aliphatic acids, sugar acids and aromatic acids do occur. The plant growth substances, such as indole acetic acid, also contain carboxylic acid groups while specialized aromatic acids are synthesized by the lichens. Some workers have attributed considerable pedogenetic significance to such acids as they may mobilize inorganic materials by acidification–solubility effects or by chelation. Certain other minor organic constituents of soil may also be implicated in soil-forming processes, for example by mobilization of sesquioxides. Bloomfield's work, cited above, suggests that polyphenols may be important in this role (Stevenson, 1967).

Most of the remaining minor constituents are found in trace quantities and the soil enzymes are of considerable interest in this group. In the past, various workers have treated the soil as an 'organism' and described processes of 'soil metabolism'. Generally this is a reflection of overall microbiological activity in the soil but the presence of free enzymes does suggest that some organic transformations may be considered as whole-soil processes. The most difficult problems in this field are the investigation of the origin of the enzymes and the localization of their functional sites in the soil. They may arise as microbial extracellular enzymes, by autolysis of microbial cells and from soil animals, plant roots or residues. A large number of different enzymes have been isolated from soils (Skujins, 1967) and it seems likely that most of them are adsorbed on the surfaces of colloidal soil particles; as noted above much of the soil amino–N appears to be bound in this way.

Major constituents—humic complexes

By far the greatest proportion of the organic matter, in most soils, is in the form of humic complexes comprising a mixture of diverse phenolic polymers (Hurst and Burges, 1967). Many of the humic subunits are similar to those in lignin and the chemistry of both compounds has been difficult to elucidate, not least because soil organic matter is tightly bound to other soil colloids while lignin is usually bound to structural polysaccharides. In both cases severe extraction procedures have to be used with invariable chemical alteration of the lignic or humic molecule.

The traditional extraction technique is to use dilute sodium hydroxide after removing fats, waxes etc. with benzene–methanol. The alkali extracts a large proportion of the organic matter and, on acidification, produces a dark-coloured precipitate of *humic acid*. The supernatant liquid, on drying, gives a yellow-brown *fulvic acid* while the small residue from the alkali extraction is termed *humin*. Felbeck (1971) notes that these are by no means homogeneous fractions and suggests modifications to provide sharper distinctions. Hurst and Burges (1967) conclude that there are no great chemical differences between the humic and fulvic fractions except that the latter are of lower molecular weight. On analysis there is no distinction between their degradation products.

Oxidative degradation of humic acid produces a high proportion of phenolic products while reduction produces a range of units similar to those encountered in microbial breakdown of lignin. They include various phenylpropane-derived units such as hydroxy–cinnamic and ferulic acids, substances also found as precursors in the biosynthesis of lignin (Isherwood, 1965). Reductive cleavage also produces 1.3.5-substituted rings based on the parent molecule, phloroglucinol. These are most likely to be derived from flavonoids but whether these are of direct plant origin or produced by microbial biosynthesis is not known.

Hurst and Burges (1967) summarize the ideas concerning humic acid synthesis, suggesting that the molecules grow by stepwise addition of phenolic units liberated by microorganism metabolism and that cross-linkages gradually develop. At any time the molecule contains monomers in all states of binding, ranging from loosely attached new additions to tightly condensed core units. The wide variety of degradation products suggests that there is no great rigidity of configuration; 'They may best be regarded as polycondensates of those phenolic units immediately available in a particular microarea of the soil. These units are principally derived from lignin and plant flavonoids of the overlying vegetation'. Figure 4.8 shows some of the structures which may be involved in the humic complexes.

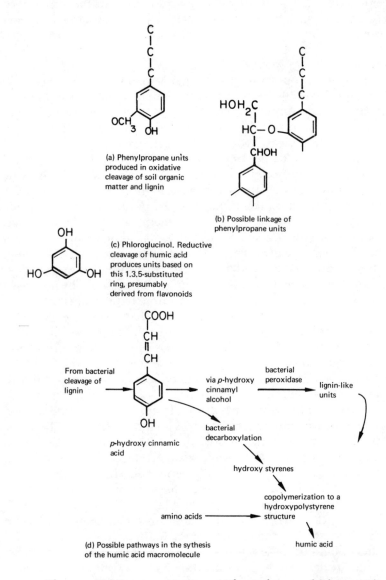

(a) Phenylpropane units produced in oxidative cleavage of soil organic matter and lignin

(b) Possible linkage of phenylpropane units

(c) Phloroglucinol. Reductive cleavage of humic acid produces units based on this 1.3.5-substituted ring, presumably derived from flavonoids

(d) Possible pathways in the sythesis of the humic acid macromolecule

Figure 4.8 Some structures and syntheses which may be associated with the soil organic matter macromolecule. Suggestions for the pathways shown in (d) are derived from Isherwood (1965) and Finkle (1965).

Properties of the humic complexes

Soil organic matter is the critical component responsible for converting a dust-like or mud-like mixture of compacted mineral matter to a structurally aggregated material with considerable pore space which provides aeration under most conditions. Unstructured soils may become anaerobic even with fairly low water contents as the capillary spaces of the matrix are very small. The organic macromolecules presumably bind clay particles with bondings such as R—COOO—Ca— clay which can bridge adjacent clay particles, forming a network of clay particles and macromolecules: the clay–humus complex.

Different soils contain varying amounts of organic matter ranging from < 1% in raw soils such as lithosols and regosols to 80–90% in organic histosols. The content is not uniform with depth for, as stressed above, nutrient-rich soils of high biological activity show strong mixing of organic matter into both A and B horizons while oligotrophic, leached soils have a marked stratification of organic matter, most of it lying on the surface and a little penetrating to the B horizon. Table 4.1 shows organic profiles of two contrasting soil types and the associated variation of cation exchange activity.

The data of Table 4.2 show that cation exchange capacity is not strongly influenced by the organic component of high-clay soils but in sandy soils it may be the main source of exchange. Carboxylic acid, hydroxyl and imide groups may all lose a hydrogen ion in exchange for a metal cation. The participation of the various groups varies with pH and with the cations involved. Most soil organic matter has a cation exchange capacity of 100–400 $me/100$ g and thus contributes about 1–20 $me/100$ g of exchange capacity to normal soils with an organic content of 1–5%. In some ombrogenous peat soils the organic exchange activity is of extreme importance as it permits the removal of low concentrations of nutrient cations from percolating rainwater; the only source of plant nutrients in such soils.

The foregoing discussion has been related to temperate zone soils which have featured most prominently in the pedological and ecological literature. Processes of organic matter accumulation and decomposition appear to be similar under tropical conditions, being related to the continuum of earth's surface temperature and wetness conditions. Generally speaking, there is a tendency for the degradation of organic matter to outstrip its production as temperature increases, hence, from poles to equator, there is a gradient of decreasing accumulation of soil organic matter. Superimposed upon this is a wetness effect: high water content slows the diffusive access of oxygen to soils and reduces the rate of organic decomposition. Figure 3.1 outlines these temperature and water relationships. Soil nutrient content and pH also influence this relationship, increased acidity and oligotrophy reducing the bacterial population and encouraging organic accumulation. The general consequence is that well-drained tropical soils are often of low organic content and show little litter accumulation but, under wetter or nutrient-deficient conditions, tropical peats or podsols may form.

Many workers have noted that the ratio relationships between soil organic carbon and some other elements appear to have pedogenetic significance. Carbon/nitrogen ratios in particular have been widely examined and endowed with varying degrees of significance. The C/N ratio is variable from soil to soil but a number of generalizations may be made. When plant litter first enters the soil it is of high C/N ratio, often exceeding 50 : 1, though the nitrogen-rich litter of some Leguminosae may be as low as 20 : 1 (Burges, 1967). The early stages of decomposition rapidly remove carbohydrate and fats so that the ratio closes, approaching a value of about 10 in calcium-rich mull humus. By contrast, the mor humus of podsols has a much wider ratio which is usually 20 and may reach 50. The highest values are often found in the surface horizons of podsols as they may contain elemental carbon, the charcoal remnants of intentional or accidental burning. Inflated values of this type are not pedologically significant as charcoal carbon is virtually inert in soil.

The relative constancy of C/N ratios in many different soil types is also reflected in similarities of elemental constitution, functional group content and cation exchange behaviour of the humic material (Kononava, 1961). It is not suggested that the humic complex has the same composition in all soils but it is likely that there is a basic similarity of constitution and properties which is the consequence of microorganism metabolism and plant composition rather than of soil type and chemistry.

Methods of measuring and characterizing soil organic matter

Soil organic matter content is usually expressed as a percentage of total dry weight and often as organic carbon content, a more meaningful expression when the composition of the organic matter is not known. Organic carbon may be measured by wet oxidation with a dichromate–sulphuric acid mixture ($Cr_2O_7^{2-}$) or by combustion of dry samples. Most wet oxidation techniques rely on the reduction of the chromic ion by the organic matter and titration of the excess oxidant. Allison (1965) gives an account of two variants of this technique and also describes the combustion method in which carbon is determined gravimetrically. Both wet oxidation and combustion techniques may also be used in conjunction with a carbon dioxide absorbing and measuring arrangement. The normal $Cr_2O_7^{2-}$ oxidation measures only organic forms of soil C while dry combustion measures, in addition, elemental C. Soils which contain free carbonates require special precautions when gravimetric or CO_2 absorption analysis is used.

Techniques for the extraction and fractionation of organic matter may be found in Black (1965) and more general reviews in Felbeck (1971), Schnitzer (1971) and Scharpenseel (1971). The detailed techniques presented by Black *et al.* (1965) also include those for characterising the soil fauna and microflora.

SOIL TEXTURE

The International Society of Soil Science classification of soil particle size distribution (texture) was outlined in Chapter 3. Stones and gravel above 2·0 mm diameter are excluded from the textural classes, of which there are four: coarse sand (2·0–0·2 mm), fine sand (0·2–0·02 mm), silt (0·02–0·002 mm) and clay (0·002 mm). Separation of the size fractions is preceded by oxidation of organic matter and removal of inorganic cementation to break down structural aggregates. The fine fractions, silt and clay, are removed from the coarser material by a sedimentation technique and the coarse fraction dried and graded by sieving. Stones and gravel are retained by a 2·0 mm perforated plate sieve, the coarse sand is retained on a 70 mesh wire sieve (0·23 mm per opening) while the fine sand passes through the 70 mesh. The silt and clay fraction is graded according to sedimentation velocity in a liquid column; it is assumed that clay and silt both have the same particle density and hence their sedimentation rates will be influenced only by relative diameter. Sedimentation may be followed gravimetrically either by taking pipette samples from the column or by using a suitable hydrometer to follow the liquid density change as the particles settle. Details of the procedures may be found in Day (1965).

Soil texture directly influences soil–water relationships, aeration and penetrability through its relationship with interparticle pore space. These factors are, of course, also influenced by the degree of structural aggregation discussed below. Indirectly, there is a further relationship with soil nutrient status as the clay fraction is the main source of many plant nutrients and of cation exchange activity. Sandy soils thus tend to be inherently nutrient-deficient and, because of their high porosity, to lose nutrients by leaching in humid climates.

Figure 4.9 shows an arbitrary textural classification of natural soils which may be used with laboratory particle size analyses or, more crudely, with 'finger tests' in the field. It also serves as a framework to relate particle size to water availability: the highest values are associated with a uniformly moderate particle size (silt) or with a balanced size range (silty and sandy loams). Low values occur in predominantly sandy soils in which the interparticle spaces are so large that they are mainly air-filled at −0·3 bar (approximately field capacity; see Chapter 5). Soils with a high clay content generally have small pore sizes from which much of the water is not extracted at a water potential of −15 bar (permanent wilting percentage). Salter and Williams (1967) have used this approach to estimate available water in the field. In addition to influencing water availability, particle size also influences water infiltration rate during rainfall. Coarse, open-

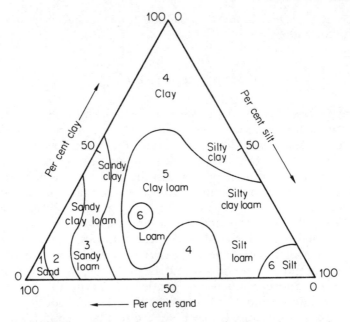

Figure 4.9 Soil texture and water availability. The triangle shows the range of soil textural types which can be related to variation in sand, silt and clay content. The triangle is contoured in cm water available per cm of soil in the water potential range −0·3 to −15 bar. 1= 0·00–0·05; 2 = 0·05–0·10; 3 = 0·10–0·15; 4 = 0·15–0·20; 5 = 0·20–0·25; 6 = 0·25–0·30 cm water. Data from Black (1968)

textured soils manifest high and maintained infiltration rates but soils with a high proportion of silt- and clay-sized particles may produce a surface 'pan' when exposed to raindrop beating or surface run-off. Panning is also related to the degree of aggregation as it is caused by the blockage of soil pores with suspended clay/silt particles which are liberated from the surface only if the soil is poorly structured. Protection by a vegetation canopy and good soil structure plays a major part in stabilizing soil surfaces against panning, run-off and soil erosion. The relationship between water matrix potential, water retention and pore size is more fully discussed in Chapter 5.

The relationship between cation exchange capacity and particle size distribution is shown in Figure 4.10, which indicates a positive correlation between CEC and clay content. The wide scatter of points derives from the variable organic matter content of the samples and the unspecified nature of the clay mineral. The samples are all from the lower parts of profiles and have comparatively small organic contents. Clay content also correlates well with the supply of many plant nutrients; for examples, Figure 4.10 also shows exchangeable potassium plotted against clay content for the same samples.

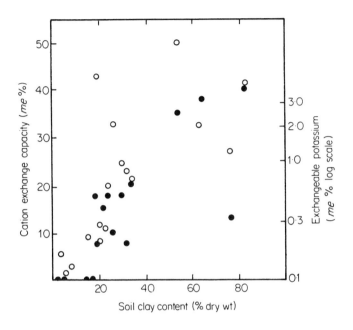

Figure 4.10 The relationship between soil clay content, cation exchange capacity and exchangeable potassium. Data from Soil Surv. Staff (1960).

STRUCTURE AND POROSITY

In the preceding sections it has been stressed repeatedly that the action of organisms modifies the raw mineral matrix of the soil by binding particles together to form aggregates. These are more or less well defined structural units which may be much larger than the textural particles, ranging from a few mm to several cm in diameter. The activity of burrowing animals, effect of root penetration, and cracking, due to shrinkage in high-clay soils, all interact to create air spaces between the aggregates. These spaces are much larger than those which exist between the textural particles and greatly modify the soil as a medium for plant growth and habitat for soil-dwelling organisms.

Structure may be investigated by separating and size grading stable aggregates (crumbs) or by measuring the pore space of undisturbed samples. Aggregate size analysis is usually undertaken by passing a field sample through a nest of suitable sized sieves with a standardized washing procedure (Kember and Chepil, 1965). The sizes of sieve mesh have usually been chosen to suit specific investigations; there has been little standardization. Soil porosity may be measured by (i) relating the bulk density of a field sample to the mean density of the matrix particles and calculating the pore space by difference; (ii) measuring the pressure : volume relationship of the sample in an air pressure chamber and using the Boyle's law relationship to calculate air-filled porosity; (iii) using a suction plate (Chapter 5) to withdraw water from the sample over specified ranges of water potential and estimating the equivalent air-filled pore volumes from the amounts of drainage water (Vomocil, 1965).

Russel (1971) notes that, in addition to shrinkage cracks, root decomposition channels and faunal burrows, plant roots may also propagate cracks by localized drying of the soil and also by forcing their way into pores of smaller diameter than the growing point. Pores formed by all of these processes may be stable and long-lived or may collapse by crumbling of the pore wall or slumping of whole aggregates under wet conditions. Russel discusses the stability of pores in relation to organic matter and microorganism activity, concluding that roots have a direct influence in nourishing rhizosphere bacteria whose capsular slimes and gums stabilize the crumbs. Allison (1968) made this point very strongly, suggesting that the rhizosphere zone provides, simultaneously, nearly ideal conditions for both aggregate formation and aggregate stabilization by incorporation of bacterially synthesized macromolecules. The efficacy of grassland cover in promoting aggregation is almost certainly due to its rapid and prolific root production. Allison also suggested that the coating of the aggregates with clay skins (cutans) and inorganic cementation by ferric and aluminium oxides, the latter particularly in tropical soils, also aids aggregate stability.

Satchell (1967) discusses the role of earthworms in creating pore space and aggregate stability. After casts are voided, fungal hyphae develop in them, giving an initial stability which is short-lived but probably replaced by cementation with bacterially produced polysaccharides. Kubiena's (1953) statement that nearly all aggregates in forest mull soils 'are earthworm casts or residues of them' is frequently true. As the annual turnover of A horizon soil through the earthworm gut is fairly small, this would suggest that the stabilization must endure for fairly long periods. Jacks (1965) commented: 'it may be justified to make the rather sweeping generalisation that in the two great plant ecologically distinguished worlds of the soil, the movement of plant roots is the major structure-forming factor in grassland soils, and the movement of animals in forest soils'.

The effect of worms in aerating soils is probably not immediately great as their annual production of burrows, as measured by surface voiding of worm casts, is not more than 5–10% of the soil in the main rooting zone, this soil often already having a pore space of 40–60%. Though the burrows are not a large annual contribution to total pore space they must also be regarded in the light of the potentially long residence time of aggregates—there is not much evidence relating long-term build up of pore space to earthworm activity but the spongy earthworm mull soils referred to by Kubiena (1953) suggest that they have considerable influence. The worm burrows are also important because they open to the surface and provide large-diameter channels for water infiltration.

The structuring of soils is of considerable importance in relation to root penetration. Wiersum (1957) showed that roots would not pass through rigid pores of less than about 0·15–0·20 mm, roughly equivalent to the mean root diameter. The upper value corresponds to the pore size of an unaggregated sandy soil with a uniform particle diameter of 0·8 mm. Growth can only occur freely in such soils if the particles can be forced apart by the extending roots: this requires plenty of available pore spaces to absorb the deformation. Taylor and Ratliff (1969) have shown that both root growth and plant yield are reduced as soil strength (resistance to deformation) increases. There is some difficulty in relating the influence of pore size and soil strength to plant root behaviour as they are confounded with soil aeration characteristics when the soil is wet. Wiersum's data were, however, derived from experiments with sintered glass filter plates and sand beds which contained some air-filled pore space; it is unlikely that oxygen deficiency affected these results. Other workers, e.g. Barley (1963) have overcome the aeration problem by applying lateral pressure to beds of simulated soil. This varies the mechanical strength without altering pore space or water content. Useful reviews of these topics are presented by Graecen, Barley and Farrel (1969) and Eavis and Payne (1969).

Much investigation of structuring and pore space effects has been confined to agriculture and forestry but, as might be expected from the role of soil organisms in developing structure, there is a strong correlation between natural soil types and the degree and nature of structuring. Bryan (1971) presents data for a number of soils which show that the proportion of large structural aggregates is greater in eutrophic, high-calcium soils than in the more acid, nutrient poor sol brun acide and podsols (Table 4.3).

Table 4.3 The relationship of aggregate size distribution to soil type (after Bryan, 1971)

Aggregate size range (mm)	Per cent of aggregates falling within defined size range			
	Calcareous brown soil	Sol brun acide	Iron podsol	Humus–Iron podsol
0.5–1	6·75	7·5	9·75	12·0
1–2	18·0	15·0	21·0	24·0
2–3	12·0	9·75	9·0	12·5
3–6	39·0	31·5	21·0	15·0

SOIL AERATION

As noted above, aeration is closely related to water content and, in the context of anaerobic soils, will be discussed in Chapter 7. Various workers have noted, for aerobic soils, that the nitrogen + argon content remains close to that of the external atmosphere (79% v/v) while oxygen and carbon dioxide vary in complementary proportions to make up the remaining 21% (Van Bavel, 1965).

Experimental investigation of root growth in relation to aeration suggests that quite low levels of oxygen may be tolerated, but the accompanying increase in carbon dioxide may inhibit growth, water uptake and nutrient absorption in some species when it exceeds 4–5% in the soil atmosphere. Some studies also show that the CO_2 level must exceed the O_2 level for toxic effects to occur (Sheikh, 1970). The duration of exposure is also important, the toxicity of CO_2 often remaining hidden in short experiments. Most experimental work has been undertaken either by flushing nutrient solutions with the requisite gas mixtures or by forcing the gas through the soil or some other rooting medium; neither of these techniques fully simulates the natural root environment.

Van Bavel (1965) describes both suction and diffusion techniques for sampling the soil atmosphere in the fields. In the former case a sample is taken by inserting a hypodermic needle to the required depth and withdrawing a sample while the latter method permits diffusive equilibration of a buried sample chamber before the gas is extracted. The suction technique has been criticized as it sometimes appears to cause changes in composition, presumably due to differential gas density/transport effects. Using methods of this sort, most workers have found appreciable O_2 contents in the majority of soils unless they were almost entirely water-saturated. As a water table is approached, from above, the O_2 content gradually falls and the CO_2 content rises proportionately. Normal, dry soil rarely has a CO_2 content of more than $0 \cdot 5$–$1 \cdot 0\%$ or an oxygen content of less than c. 20%. Between about 15 to 30 cm above the water table, according to soil texture and structure, the CO_2 content reaches c. 5%. This corresponds with a very wet soil condition: a water potential of $-0 \cdot 03$ bar, much wetter than field capacity. With even wetter conditions the CO_2 and O_2 concentrations may become equal at about 10%.

These effects are closely dependent on soil texture and structure. Coarse-textured or well-structured soils have higher gaseous diffusive transfer rates than fine-textured or poorly structured soils under wet conditions. This is because they contain many large pores which remain gas-filled. The water-filled pores of the fine-textured soils form a potent barrier to gaseous diffusion; oxygen, for example, diffuses ten thousand times more slowly in water than in gas. The relationship is reversed, quite often, in dry soils as the finer textured soils often have a greater total pore space and provide a larger gas-filled cross-sectional area for diffusion. The major part of the gaseous transfer in the soil is diffusive: Kimball and Lemon (1971), for example, showed that the effects of wind turbulence caused only a slight change in the rates of surface gas exchange even in coarse-textured soils. Changes in atmospheric pressure and temperature must result in some gas exchange but probably not fast enough to influence the equilibrium between diffusive movement and microorganism metabolism.

INVESTIGATION OF ROOT GROWTH AND DISTRIBUTION

Despite the inherent difficulties of technique, studies of root distribution have been made since the early years of plant ecology. Amongst them is a remarkable example: Weaver's (1926) classic monograph on the root systems of N. American crop plants and native indicator species. Weaver made his observations by dissecting out and mapping roots in the walls of soil pits but he cites earlier workers who used washing-out techniques after stabilizing the roots by thrusting closely spaced pins through an isolated soil monolith. This 'pinboard' technique has been widely used to gain a macroscopic view of root system distribution which may be related to various soil physical and chemical conditions (e.g. de Roo, 1969).

Extraction of the whole root system gives some impression of its condition at the time of sampling but it is not a suitable technique for displaying the dynamic aspects of growth: the time course of root initiation, the extension and ageing of roots and the exploitation and reexploitation of rooting volume. The most satisfactory approach to this problem has either been the use of glass-sided containers or the installation of glass panels in trench walls in the field. Though the rooting conditions are not entirely natural, a great deal of information has been gained by measurement of root extension and time lapse photographic studies of the distribution of roots and soil organisms. One of the most extensive investigations reviewed by Rogers and Head (1969), with an underground laboratory installation (Rogers, 1969), has given information on the seasonality of root growth in perennial woody species, the length of life of corticated and decorticated roots and the influence of various environmental factors. The quantity of organic matter returned to the soil by sloughing of the cortex approximates to half the dry matter of the young root and as it occurs at a root age of only 2–3 weeks, represents a very large input. Disruption of the decaying cortical tissues is largely due to the activity of soil animals: small worms and arthropods. The experiments have also given information on root competition and specific replant effects possibly associated with allelopathy (see Chapter 10). Newbould (1968) notes that root extension data may, by expressing them as a percentage of existing root, be converted to dry weight gain if they can be related to root weight per unit area measured by other techniques.

The measurement of root weight has been discussed by Lieth (1968) and Newbould (1968). Samples may be removed by cutting out soil blocks of known volume, or with coring tools, and the roots removed by gentle washing and sieving procedures. These techniques are fraught with difficulties: the sampling is laborious and, because the variability is usually high, demands extensive replication while the washing procedure caused the loss of some fine rootlets. To reduce the first problem, various investigators have designed powered core-sampling devices, most of them using hammer-driven tubes (e.g. Wellbank and Williams, 1968).

The high replicate variability is usually a reflection of the heterogeneity of the soil environment which is discussed in Chapter 10 in relation to the niche concept in competition. The biogenic nature of soil formation must lead to spatial variability even when parent materials are very uniform. Consider, for example, a young dune soil: when the first plant colonists appear they will cause a localized accumulation of organic matter, a localized specialization of the microflora and fauna and, on death, the localized release of mineral nutrients which were gathered from a large soil volume. Some of these nutrients, such as nitrogen, phosphorus and sulphur, are not rapidly mobilized from organic material and, thus, a three-dimensionally variable mosaic of chemical conditions is imposed on the soil and influences root growth on both a macroscopic and a microscopic scale. The existence of 'pattern' in vegetation and the 'clumping' of some species due to the convergence of rhizomes of different individuals, discussed by Kershaw (1963), the mutual exclusion of rooting volumes noted by Rogers and Head (1969) and the localized, long-term accumulation of nutrients by plants (Goodman and Perkins, 1959) all have relationship to this small-scale heterogeneity which makes root study, in the field, so difficult.

Table 4.4 Vertical distribution of root biomass and root tips in an acid brown soil under *Fagus sylvatica* (after Meyer and Göttsche, 1971)

Horizon	Depth (cm)	Root biomass per 100 ml soil (mg)		Root tips per 100 ml soil
		2–5 mm diameter	0·5–2 mm diameter	
O_2	2–3·5	99·9	342·4	6259
O_1	3.5–5	39·3	226·7	2523
A	5–7	150·4	70·0	961
B	13–17		46·9	439
B	19–23	50·1	44·8	309
B	29–33		28·7	328
B	35–39	60·3	27·7	346
B	43–47		13·5	194
B	52–56		8·8	123
B	61–65	7·3	2·9	67
B	72–76		3·1	75
B	83–87		0·7	13

In addition to the horizontal and vertical variation in root exploitation of soil volume, there is also considerable variation in the size and age of roots found in different parts of the soil. The distribution of the finest, youngest absorptive roots may differ from that of older roots and also changes rapidly with time. Meyer and Göttsche (1971) investigated root biomass distribution in a *Fagus sylvatica* stand and found that fine roots (2–0·5 mm diameter) and active root tips were most abundant in the superficial organic layers of an acid brown soil but roots of 5–2 mm diameter reached maximum development in the *A* and upper *B* horizons of the mineral soil (Table 4.4). They also noted that the comparative content of active root tips was a sensitive index of soil condition; for example, the maximum number of root tips per 100 ml of soil, in the *A* horizons of *F. sylvatica* stands, varied between 556 in a eutrophic brown earth and 46,600 in a podsol, presumably reflecting the need for additional absorptive area in the nutrient-deficient soil.

Chapter 5

Plants and water deficit: physiological aspects

Man's early attempts at agriculture were probably more often doomed to failure by drought than by any other factor; plant husbandry in the Old World first developed in the arid zone of south-western Asia (Whyte, 1965) and there is little evidence of a radical climatic change since the dawn of prehistory (Butzer, 1965). Thus the first physiological and ecological problem to be attacked by the human race was that of water shortage, the remains of irrigation works being amongst the earliest artifacts of the arid zones. The struggle to understand and alleviate the effects of water deficit continues today at both theoretical and applied levels and attracts worldwide interest in response to the serious depletion of water resources in many regions.

The need of plants for water and the consequence of its shortage pose one of the most complex physiological problems in which experimental isolation of the plant from its normal environment is of doubtful value. Plant–water relationships must generally be considered ecologically as a field problem of microenvironmental physics or submitted to laboratory attempts at environmental simulation.

SOIL–PLANT–ATMOSPHERE CONTINUUM (SPAC)

The rooting zone of the soil, the plant body and the lower layer of the atmosphere behave as a continuum in relation to water transfer and must be considered as a whole in any complete analysis (Phillip, 1966). Soil water movement takes place in the liquid phase by capillary flow and in the gaseous phase by molecular diffusion or mass flow. In moist soil capillary transfer predominates but it is possible that vapour transfer may be important in rather dry soils (Cowan and Milthorpe, 1968). Water transport in the plant occurs by liquid-phase viscous flow in channels or by diffusive transfer through membranes. Final escape to the atmosphere is by vapour-phase molecular diffusion across the intercellular spaces of the mesophyll to the stomatal pores. From this point turbulent diffusion in the leaf and canopy boundary layers completes the transfer to the atmospheric sink, but molecular diffusion may become important under protected or very calm conditions.

The flow-path through the SPAC is necessarily complex, a sequence of 'series-parallel' resistances and capacitances complicating the theoretical analysis of the system. Many of the resistance and capacitance values are difficult to measure empirically and show large changes of value according to water-content of the pathway. For this reason it is difficult to establish a simple 'Ohms law' analogue of the SPAC and the interpretation of certain limiting conditions remains in doubt. The main components of the SPAC are shown in Figure 5.1.

Solar radiation is the primary energy source for the water transport process in the SPAC, the sink being the latent heat change in the evaporation of water at the mesophyll cell surface and at the soil surface. Some of the energy for evaporation is drawn from environmental longwave radiation, sensible heat transfer in the SPAC and water-vapour deficit in the free atmosphere, but these all depend secondarily on the effects of absorption of solar radiant income (Chapter 2).

SOIL AND PLANT WATER:
TERMINOLOGICAL DEVELOPMENT

During the growth of any science the concepts and terminology which are developed may have to be changed to accommodate new advances. The terminology of water deficit in soils and plants has suffered repeated modification and alteration of usage, and now appears confusing to students and occasional readers.

Various suggestions have been made for a unified thermodynamic terminology, amongst the earliest being those of Schofield (1935) and Edlefsen (1941). From 1960 onwards, following the publications of Taylor and Slatyer (1960) and Slatyer and Taylor (1960), the term 'water potential' (ψ) has gradually been adopted. Water potential is a function of the difference between the chemical potential of water in the specified system and that of pure, free water at the same temperature and elevation. The difference between the chemical potential of a substance in two parts of a system determines the direction in which the substance will spontaneously diffuse and the magnitude of the difference will determine the rate of diffusion. Water potential is expressed as energy per unit volume (or mass) and has units dimensionally equivalent to pressure ($J\ kg^{-1} \equiv N\ m^{-2}$ or, in c.g.s. units, $ergs\ g^{-1} \equiv dyne\ cm^{-2}$). Consequently, it may be presented in the more

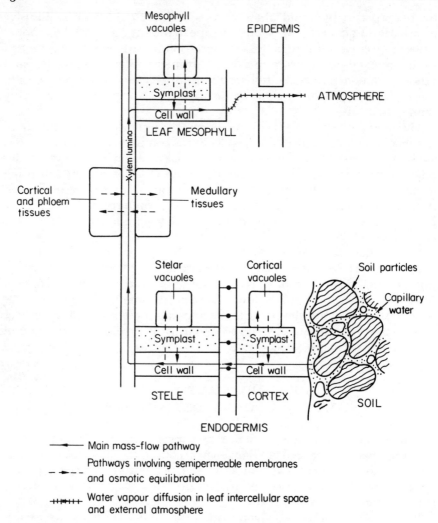

Figure 5.1 Flow paths in the soil–plant–air continuum (SPAC). After Weatherly (1969).

familiar units of pressure, the bar or the atmosphere. For convenience, the bar will be used in the following discussion. The equivalences are: 1 bar = 0.99 atm = 10^6 erg g^{-1} = 10^5 N m^{-2} = 1017 cm of water. The energy status of water in any part of the SPAC and the direction of movement may be specified by water potential which is defined as

$$\psi = \frac{(U_w - U_w{}^0)}{\bar{V}_w}$$

where: ψ = water potential (J kg^{-1});
U_w = chemical potential of water in the system (J mole^{-1});
U_w^0 = chemical potential of pure, free water at the same temperature and elevation as the water in the system (J mole^{-1});
$\bar{V}_w$ = partial molal volume (m^3 mole^{-1}).

It is an intensive property of the system (analogous to temperature), transfer taking place from regions of high to regions of low water potential.

A range of terms and units has been used in the description of solution effects, such as freezing point depression, boiling point elevation and development of osmotic pressure across differentially permeable membranes. Similarly, other terms have been used to describe capillary tension effects. The use of water potential permits a complete unification of the treatment as

$$\psi = \frac{(U_w - U_w^0)}{\bar{V}_w} = \frac{RT \ln (e/e_0)}{\bar{V}_w}$$

where: e = vapour pressure of water in the system (mm Hg);
e_0 = vapour pressure of pure, free water at the same temperature and elevation as the water in the system (mm Hg);
R = Universal Gas Constant (erg degree^{-1} mole^{-1});
T = temperature (°K).

The relative vapour pressure (e/e_0) is a function of both solute and capillary (matric) effects. Pure, free water is arbitrarily defined as having zero water potential under standard conditions: the addition of a solute or the presence of a porous capillary matrix will reduce the potential. For this reason unsaturated soil water potentials (ψ soil), solute (osmotic) potentials (ψ solute) and plant water potentials (ψ plant) are negative, while turgor pressures of cell contents (ψ turgor) are positive.

During the development of studies in plant–water relationships various terms have been used to describe the tendency of cells to take up water. This has long been recognized as deriving from the balance between the osmotic effect of the cell contents and the multidirectional hydrostatic pressure (turgor pressure) within the cell. It has been termed suction force, suction pressure, suction potential or diffusion pressure deficit, and is defined as

$$\begin{matrix} SP \\ (or\ DPD) \end{matrix} = OP - Tp$$

where OP= osmotic pressure of cell contents and TP = turgor pressure. This definition assumes the behaviour of the cell as an ideal osmometer. In the thermodynamic terminology the equation may be rewritten in the form of relative water potentials

$$\psi \text{ cell} = \psi \text{ vacuolar fluid} + \psi \text{ turgor}.$$

The change of sign for turgor pressure in this equation results from the negative values of osmotic potential and cell water potential.

If the values of ψ cell are considered in integrated form for the whole plant (ψ plant), this is some measure of the driving force available for water uptake from the soil. However, the dynamic nature of the process must be considered in this context as the balance between ψ plant and ψ soil always represents a transient or steady-state response to the overall transpiration stream which is limited by various resistances in the SPAC as well as by the available energy gradient. Because of the variable nature of some of the SPAC resistances, ψ plant is rather too general an expression; the water potentials of specific tissues and organs give a more comprehensive picture of the situation in the plant body. Detailed discussions of the state and terminology of plant and soil water are presented in Slatyer (1967) and Taylor (1968).

TERMS USED TO SPECIFY SOIL AND PLANT WATER STATUS

Soil

Soil water potential (ψ soil)
This is an expression of the total reduction of water potential in the soil which is attributable to matric and solute effects. External pressure and gravitational effects must also contribute to ψ soil which may thus be defined as:

$$\psi \text{ soil} = \psi \text{ matric} + \psi \text{ solute} + \psi \text{ pressure} + \psi \text{ gravity}.$$

Field capacity (F.C.)
This is the water content of an undisturbed soil (% oven-dry weight) after saturation by rainfall and followed by the cessation of gravitational drainage. In practice it involves sampling about 48 hours after rain. The technique is valid only for soils which are out of capillary contact with a water table. For various reasons, to be discussed later, the field capacity percentage may be considered as a soil 'constant' and is equivalent to

values of ψ matric ranging from -0.1 to -0.3 bar according to soil type. As its determination is rather awkward, attempts have been made to simulate F.C. values by equilibrating soil samples with suction-plate water potentials of -0.1 or -0.3 bar or by centrifuge-draining samples at 1000 g for 30 minutes. The water content in the latter case has been called the *moisture equivalent* (M.E.). There is, however, no universal relationship between suction-plate or M.E. values and field capacity. It should be noted that soil water content would be more usefully stated as per cent by volume but this involves the complication of a bulk-density sampling.

Permanent wilting percentage

Where field capacity is used to specify the upper limit of soil water storage, permanent wilting point may similarly express the lowest level of easily available soil water. The determination is made by growing test plants in watertight containers of soil until several leaves are formed. At this stage the soil surface is sealed and the plants left until wilting occurs. If recovery cannot be induced by placing the plants in a water-saturated atmosphere, the soil water percentage is then measured. For many years PWP was accepted as a soil 'constant' which differed little with the species of test plant, but the development of techniques for measuring *in situ* soil water potentials has revealed that the water potential corresponding to the permanent wilting percentage may range from (approx.) -10 to -20 bar. The apparent discrepancy arose because a comparatively small change of volumetric water content may induce a very large change in water potential in this region of the characteristic curve. The mean water potential value of -15 bar has, however, been widely identified with the PWP.

Slatyer (1957) criticized the concept of PWP as a single-value constant on the grounds that loss of turgor will be a reflection of the water potential of the leaf tissue which may, with species and environment, range between -5 and -200 according to the osmotic potential of the leaf tissue sap at the time when wilting occurs. The limited changes of soil water potential around the PWP appear almost insignificant compared with the possible range of leaf values.

The water which may be extracted from soil between the field capacity percentage and the permanent wilting percentage has been described as the *available water*; this is, volumetrically, a function of the pore size distribution and hence of soil texture and structure. The term is used with the implication that water beyond the PWP is not easily available for plant growth and also suggests the concept of 'equal availability' which was first put forward by Veihmeyer and Hendrickson (1927) and generated a controversy which has only recently approached a settlement (page 151). The whole concept of available water is far less valuable than that which relates volumetric water content to soil water potential; instrumentation under field or controlled environment conditions now permits much greater flexibility in this type of study.

Plant water potential (ψ plant)

Plant water potential is a measure of the mean water potential of the cells in all tissues of the plant; the cell water potentials deriving from the balance between solute potential and turgor pressure in living, vacuolated cells, and from the negative hydrostatic pressure in the non-living cells of the xylem, and in the microporous structure of cell walls. Whole-plant water potential is a less meaningful concept than the water potential of individual tissues of organs; for example, leaf water potential, which is a controlling factor in water loss to the atmosphere via the stomatal pores. Leaf water potential is one of the characteristics governing stomatal aperture and is also likely to establish an equivalent hydrostatic potential throughout the adjacent xylem, the resistance to flow in tracheids and vessels being comparatively low.

The water potential induced in the xylem system in equilibrium with the leaf water potential correspondingly has its influence on the extra-stelar tissues of the root. Under flow conditions the equilibrium value at the root surface will depend on the resistance of the pericycle–endodermis–cortex pathway. Under steady-state flux conditions the gradient of water potential through the SPAC is not uniform but shows a series of abrupt changes according to the distribution of resistances in the pathway. At any one time the highest resistance may act as a limiting factor in the transport system. Heath (1967) noted that resistances on the atmospheric side of the mesophyll evaporating surfaces are 'protective' while those below this are 'harmful', in that they may induce water deficits in the tissues of the plant.

Van den Honert (1948) expressed the relationship as a catenary transport function which may be rewritten in the water-potential terminology as:

$$\text{Rate} = \frac{\psi_1 - \psi_2}{R_{1-2}} + \frac{\psi_2 - \psi_3}{R_{2-3}} + \frac{\psi_{n-1} - \psi_n}{R_{(n-1)-n}}$$

in which ψ_1, ψ_2 etc. represent the component potentials and R_1, R_2 etc. the component resistances of the pathway. Figure 5.2 shows a hypothetical range of potential existing in the complete SPAC.

Relative water content

When an isolated plant organ or tissue is placed in water it will absorb water until it reaches 'full turgidity'. The relationship between the initial and the turgid water content has been expressed in various forms; water deficit (Stocker, 1929), saturation deficit (Oppenheimer and Mendel, 1939)

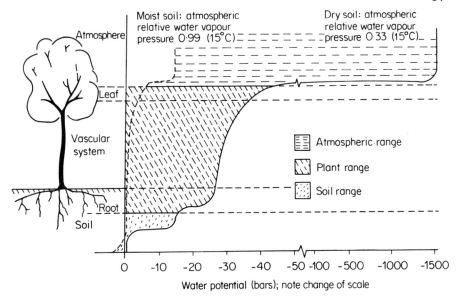

Figure 5.2 A hypothetical range of water potential in the SPAC under different conditions of soil wetness and atmospheric water vapour pressure. Note that the soil water potential becomes positive only when it is below a water table. The values may be considered representative of temperate zone conditions.

and relative turgidity (Weatherly, 1950) are examples. Weatherly's relative turgidity has recently and more accurately been described as relative water content (Ehlig and Gardner, 1964; Slatyer and Barrs, 1965) and is defined as

$$R = \frac{\text{Fresh wt.} - \text{Dry wt.}}{\text{Turgid wt.} - \text{Dry wt.}} \times 100.$$

Relative water content is a plant water characteristic which is much simpler to measure under field conditions than water potential and it is also possible to make specific calibrations so that relative water content may be converted to water potential (Slatyer, 1961) though there are some difficulties arising from the change in the relationship with plant age (Millar, Duyser and Wilkinson, 1968).

BEHAVIOUR OF WATER IN SOILS

The physical description of the status and movement of water in the non-homogeneous, three-phase matrix of a water-unsaturated soil is a problem which has exercised soil physicists for many yeas. Briggs (1897) visualized the retention of water in terms of the force provided by the curved capillary menisci between soil particles and proposed the classification of soil water as hygroscopic, capillary and gravitational. Hygroscopic water is held by the surface forces of the soil particles, capillary water by the interparticle meniscus effects and gravitational water is surplus to this holding capacity and drains from the soil. The range of water potential corresponding to the draining potentials of the various capillary pores in the soil includes the optimum region for plant growth.

The capillary retention concept was further developed by Buckingham's (1907) treatment of water flow in terms of *capillary potential* in which the flow of water through soil is considered analogous to the flow of heat or electricity through a conductor. Buckingham was one of the first workers to treat soil water in terms of its relative energy status and the term capillary potential is synonymous with matric potential (ψ matric).

The characteristic relationship between soil water content and water potential may be ascertained by the use of suction plates, pressure membrane apparatus or vapour equilibration methods (Richards, 1965). It is of a hyperbolic form (Figure 5.3) and is strongly influenced by soil texture; at

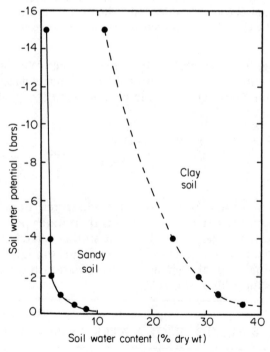

Figure 5.3 Water characteristic curves of a sandy and a clay soil.

Table 5.1 Some methods of measuring soil water status. See Gardner (1965), Richards, C. A. (1965) and Richards, S. J. (1965)

Technique	Principle	Application
		F – field and controlled environment L – laboratory S – spot reading C – continuous reading
(i) Tensiometer	Equilibration of soil matric potential with hydrostatic pressure in porous pot linked to a pressure gauge (limited 0 to −1·0 bar range)	F S C. Useful in monitoring water potential of undisturbed moist soils or experimental containers. Direct water potential reading
(ii) Suction plate	Equilibration of sample soil matric potential with imposed suction through a water-saturated capillary plate (limited 0 to −1·0 bar range)	L S. Construction of soil water potential v. water content characteristic curves
(iii) Pressure plate on pressure membrane	Equilibration of sample soil matric potential with imposed gas pressure through a water saturated capillary plate or membrane	As (ii) but covers whole soil water potential range
(iv) Neutron scatter	Detection of backscatter of slow neutrons by hydrogen nuclei. Fast neutron source and slow neutron detector lowered into borehole. Changes in water content are mainly responsible for changes of hydrogen content. Difficult in swelling soil	F S or C. Integrated value for a volume of soil. Needs calibration against water potential or content
(v) Gamma ray absorption	Density of soil changes with water content. Source and detector lowered into adjacent boreholes. Difficult in swelling soil	F C C or S. Needs calibration as (iv)
(vi) Resistance blocks	Buried porous block containing electrodes. Electrical resistance of unit is calibrated against water content. Not very accurate in moist soils	F C S or C. Needs calibration against soil water potential or content
(vii) Vapour equilibration	Soil samples are vapour equilibrated with solutions of known vapour pressure (water potential). Gravimetric determination of water content. Time-consuming	L S. Construction of soil water potential v. water content characteristic curves
(viii) Gravimetric	Soil samples are weighed, dried at 105°C, reweighed. Damage to site by repeated sampling	F L S. water content

lower potentials a fine-textured soil generally contains a greater amount of water than one of coarse texture as there are more contact points between particles. The radii of curvature of the menisci are thus smaller and more energy is required to drain each pore. Table 5.1 gives information on methods of measurement of soil water status.

Haines (1930) developed a concept of water retention and movement in a geometrically ideal soil based on the nature of the pore space in open- and close-packed arrangements of spherical particles. His experiments with glass spheres revealed the effect of the past history of wetting and drying cycles on the characteristic curve of water content against potential. There is a marked hysteresis in the cycle, the soil containing more water at any potential during the first drying than during wetting (Figure 5.4). Even at zero potential it is possible for air to remain entrapped during the wet- ting process. The hysteresis exists because the draining of pores is limited

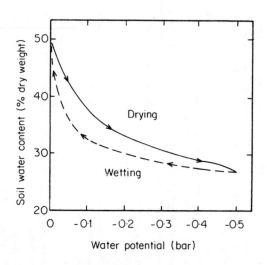

Figure 5.4 The hysteresis of the water characteristic curve of a soil during wetting and drying.

by the neck size, the neck radius determining the potential which is necessary to empty the whole pore. The neck radius likewise governs the positive potential which is needed to refill the pore, hence the tendency for more water to be held during the drying process.

Buckingham's analogy with electrical flow cannot be pursued too far as the conductivity of unsaturated soil is not independent of applied poten- tial. In *saturated* soil flow is governed by Darcy's law which states that the quantity of water passing unit cross-sectional area in unit time is proportional to the gradient of hydraulic head (potential):

$$v = -K\psi_{grad}$$

where: v = water flow velocity (m s^{-1});
K = hydraulic conductivity (m s^{-1});
ψ_{grad} = gradient of water potential expressed as head of water (m m^{-1}).

Under standard conditions K is a property of a particular soil but, once the soil is unsaturated, the value of K is no longer constant but becomes a function of water potential (Figure 5.5). This is a consequence of the emptying of pores and the resultant decrease in cross-sectional area of the flow-path. The sharp inflection of the conductivity : water potential curve explains the relative constancy of the field capacity percentage and the abrupt cessation of gravitational drainage after rain. Early work of Veihmeyer and Hendrickson (1927) illustrates the lack of mobility of water in unsaturated soil;

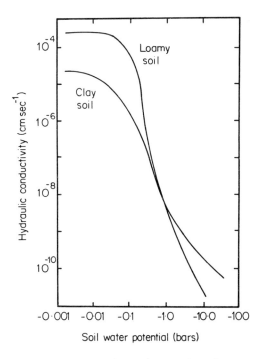

Figure 5.5 The relationship between hydraulic conductivity and soil water potential in a sandy and a clay soil.

Figure 5.6 shows the results of an experiment in which a slab of wet soil was sandwiched between two slabs of dry soil for 144 days. The two water distribution curves show the extreme slowness of capillary movement under these conditions. For the same reason water often penetrates soils non-uniformly, the downward movement of 'wetting-fronts' interacting with the localized removal of water by roots to establish complex moisture profiles which may be detected by direct gravimetric sampling as in

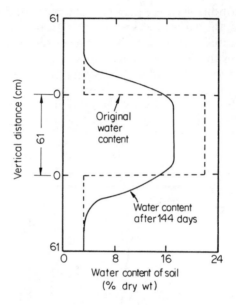

Figure 5.6 Unsaturated water flow between a slab of wet soil (field capacity) and two adjacent slabs of dry soil over a period of 144 days. After F. J. Viehmeyer, *Hilgardia,* **2,** 125–291 (1927).

Specht's (1957) study of an Australian sclerophyll heath (Figure 5.7a), or by arrays of measuring instruments as in Etherington's (1962) work in a natural stand of *Agrostis tenuis* (Figure 5.7b). Youngs (1965) gives an excellent theoretical account of these aspects of water movement in soils.

The low conductivity of soils at water potential less than -1 to -2 bar has prompted the suggestion that very steep gradients of water potential may develop around rapidly absorbing roots, with the consequence that water uptake may become self-limiting even in fairly moist soils (Richards and Loomis, 1942; Weatherley, 1951; Macklon and Weatherley, 1965; Etherington, 1967). There is no unambiguous evidence for this view, which

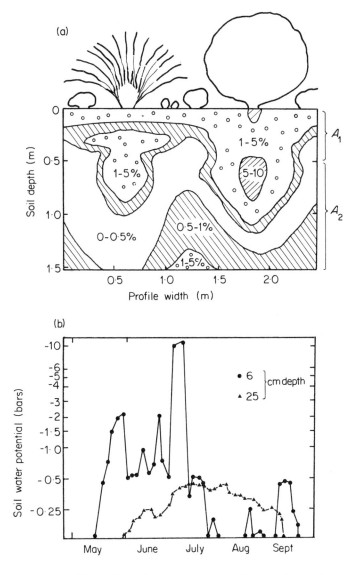

(a)

(b)

Figure 5.7 (a) Soil moisture pattern under an Australian sclerophyll heath vegetation after 0·96 inches of rain had fallen when the profile was near its wilting point throughout. Stem flow and canopy drip has caused a marked irregularity in the rewetting of the profile. Reproduced with permission from R. L. Specht, Dark Island Heath, IV. Soil moisture, *Aust. J. Bot.*, **5**, 137–150, Figure 2 (1957).

(b) Development of soil water potential profiles during the summer months in S. E. England. Records for four depths in a sandy soil with grass cover. Etherington (1962).

has been criticized on both theoretical and experimental grounds by Newman (1969a and b). For example, in Macklon and Weatherley's experiments *Ricinus communis* plants, growing in soil or in water, were exposed to different relative humidities; the plants in soil showed a marked decline in leaf water potential with high transpiration rate, whereas the leaf water potential of the water-rooted plants was unaffected. The authors explained the difference by postulating the formation of drying zones around individual roots, but Newman suggests that the resistance may be to movement of water into the whole rooting zone rather than into individual rhizospheres. The settlement of the controversy awaits experimental measurement of water potentials in the soil adjacent to roots; technically a difficult problem. Plants in the field, under conditions of water stress, may also obtain further water supplies during the extension of their roots into untapped parts of the soil. This was elegantly demonstrated by Davies (1940) who grew *Zea mays* in a long, rectangular soil container fitted with tensiometers at different distances from the plant, and by Majmudar and Hudson (1957) who showed that lettuce plants could grow to maturity on a soil originally at field capacity without any further irrigation, their root systems gradually extending further into moist soil.

The interdependence of soil water content and hydraulic conductivity poses an experimental problem as it is not possible to adjust soils to a predetermined water potential by adding water to the surface. If this is done a small portion wets to field capacity and the rest remains unchanged. Some workers have injected water in localized, small amounts (Whitehead, 1965) but this only succeeds in establishing a three-dimensional mosaic of high and low potentials. For this reason most experimenters have chosen to use 'water regimes' in which the whole soil mass is returned to field capacity after drying to a predetermined water potential. Using small volumes of soil, a few recent attempts have been made to control water potential using suction plates of various types, for example, Etherington (1962) used plates of porous polyvinyl chloride, and Babalola, Boersma and Youngberg (1968) encased thin slabs of soil in semi-permeable membranes so that external osmotica could be used to establish equilibrium potentials. None of these methods is suitable for the growth of large plants.

PLANT TRANSPIRATION AND THE INFLUENCE OF ENVIRONMENTAL FACTORS

The primary energy source for transpiration and evaporation is solar radiation, and for this reason water loss closely parallels the diurnal and seasonal fluctuations of solar radiant income. Other factors which control evaporation are temperature, relative humidity, wind speed, and longwave reradiation from all parts of the environment, but these are all either directly or indirectly a function of solar radiation.

Under conditions of moderate wind speed the transpiration rate of an individual leaf will be strongly correlated with the mean stomatal aperture (Figure 5.8). Very low wind speed or still air causes an increase in the depth of the leaf or canopy boundary layers, thus raising the resistance to water transfer into the free atmosphere. Under these circumstances stomatal

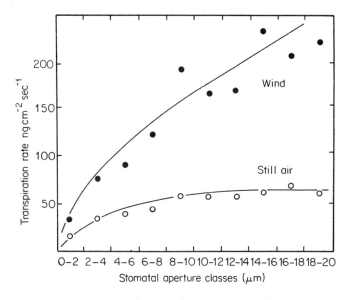

Figure 5.8 The regulation of transpiration by stomatal aperture in *Zebrina pendula* leaves under moving and still air conditions. The dashed lines show predicted behaviour based on calculated diffusive resistances. Data of Bange (1953).

aperture no longer limits transpiration rate but it should be noted that such still conditions are rarely encountered in the field. For this reason, those factors which influence guard cell opening also have a marked effect on total ecosystem transpiration.

The subject of stomatal behaviour has been extensively reviewed (Meidner, 1965; Meidner and Mansfield, 1968; Zelitch, 1969) with the general conclusion that reduced CO_2 concentration in the leaf intercellular space and increased light intensity favour opening, while high CO_2, low light, water stress and (usually) high temperature cause closure. Endogenous rhythm effects generally reinforce the normal pattern of diurnal opening established by the above factors. Most plants show stomatal opening during the day and closure at night, but an exception is found amongst some succulents which open at night and close during the day (Meidner and Mansfield, 1965): many of these plants show dark-fixation of CO_2 and organic acid accumulation during the night. Another departure from the simple diurnal pattern is the occurrence of midday closure in some plants, probably as a consequence of high leaf temperature inducing respiratory production of CO_2 and its release into the intercellular space of the leaf (see p. 29). The diurnal pattern of transpiration is usually strongly influenced by these various stomatal activities.

PLANT WATER BALANCE

At the commencement of the diurnal transpiration cycle a disturbance of water potential is propagated through the cell walls of the leaf mesophyl to the lumina of the xylem elements. The cohesive nature of the water in these elements permits the transmission of this lowered water potential to the surface of the roots via the root endodermis and the root cortical cell walls; the vacuoles of the living cells in this pathway equilibrate with this potential gradient rather more slowly. Current thinking suggests that the transpiration stream is a physically driven mass-flow process within the microporous structure of the cellulose cell walls and the lumina of xylem elements. Transfer across semipermeable membranes need not be involved other than in the endodermis where the Casparian strips prevent radial flow of water along the cell walls (Figure 5.1), and even here the passage cells may provide some wall pathway.

The physical demands of such a system are (i) a lowering of water potential at the leaf surface, of sufficient magnitude to account for the rise of water in the tallest of trees, and (ii) the existence of continuous root to leaf water columns showing sufficient tensile strength not to rupture under these large hydraulic potential gradients. There is no problem in explaining the lowering of leaf water potential caused by transpiration; from a thermodynamic point of view, evaporation of water into air at the temperatures and humidities which prevail in the leaf canopy provides more than sufficient energy. The second requirement, the maintenance of uninterrupted water columns, was first considered by Dixon and Joly (1894) and Dixon (1914), and led to the development of the cohesion/tension theory of water rise which is widely accepted. Recent work by Milburn and Johnson (1966), who used a sensitive microphone and amplification equipment to detect the sound pulse from cavitation in xylem elements, has shown that the water columns break down quite freely even in herbaceous plants. This poses a problem in the acceptance of the cohesion/tension theory concerning the mechanism by which the water columns might be regenerated, but it does remain the best overall explanation of the water transport process.

The flow resistance of the SPAC is not negligible and tissue water deficits will build-up during periods of heavy transpiration; these decline slowly during the following night. Such deficits have been attributed to root resistance, and Kramer (1938, 1940) showed that de-rooted plants or plants with roots killed in boiling water had much lower internal resistances and accrued much smaller transpirational deficits. It is possible, however, in soil, that the drying zones previously suggested (page 142) may cause much larger resistances than those of the roots.

Such changes in plant water deficit (and consequently water potential) are a reflection of the long-term lack of equilibrium between plant and environment. Figure 5.9 shows the diurnal course of leaf water potential in *Capsicum frutescens* during a period when soil water was becoming depleted

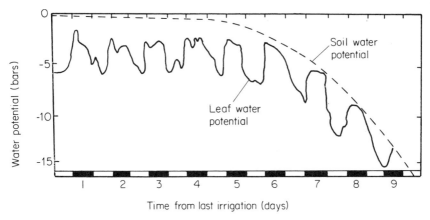

Figure 5.9 Diurnal changes of leaf and soil water potential of a pepper plant rooted in a clay loam soil. Reproduced with permission from W. R. Gardner and R. H. Nieman, *Science*, **143**, 1460–1462, Figure 1 (1964). Copyright 1964 by the American Association for the Advancement of Science.

(Gardner and Nieman, 1964). The irregularity of the ψ leaf curve shows the daily reduction followed by slow nocturnal recovery upon which is superimposed a short-term fluctuation due to variable cloud cover. The ability of a plant to adjust to very sudden changes of light intensity and evaporating conditions may be as important as its ability to cope with long periods of intense, unchanging radiation. Under transient state conditions it is difficult to predict flow rates from the known water potential gradients as many of the resistances in the SPAC are potential-dependent.

Large changes of water potential cause variations in the turgidity of un-thickened cells and, in lignified tissues, the reduction of water potential may cause a shrinkage as the water columns, which are under tensile stress, adhere to the walls of the lumina and cause contraction of individual elements. Diurnal fluctuations in the diameter of various plant organs have been measured, for example recording dendrometer traces for tree trunks show daytime contractions which provide indirect evidence for the existence of very large negative water potentials in the xylem (Figure 5.10). Kramer and Kozlowski (1960) and Kozlowski (1964) present reviews of this type of work.

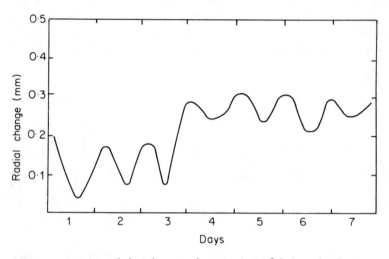

Figure 5.10 Diurnal shrinkage and expansion of the trunk of a *Pinus resinosa* tree in summer conditions. Data of Kozlowski (1965).

Klepper (1968), following the diurnal pattern of leaf and fruit water potential and fruit diameter in *Pyrus communis,* showed that leaf and fruit water potential could fall to −15 to −20 bars during the midday period. This fall was accompanied by a withdrawal of water from the fruit, causing a considerable reduction in diameter. At midday there was a slight, transient rise in leaf water potential which was attributed to stomatal closure; this was also observed in *Vitis vinifera.* Her experiments were conducted in conditions of intense atmospheric water deficit (30°C, r.h. 20%) which caused water loss to exceed supply during the day but recovery was rapid as the evening advanced.

WATER DEFICIT AND PHYSIOLOGICAL RESPONSE

Plant water deficits originating from low soil water potential or high transpiration demand have consequences which involve all of the physiological functions ranging from primary biochemical processes to the overall reduction of growth and development. Recent reviews by Crafts (1968), Gates (1968) and Zahner (1968) give a great deal of information in this field. Within the context of plant ecology it is essential that responses to water deficit are considered at all stages of the life-cycle and that the degree and duration of the effect should be related to the continued survival and competitive status of both individual and species.

Seed germination

Germination may be slowed or prevented by soil water deficit but the degree of sensitivity is specifically variable; for example, Harper and Benton (1966) used sintered glass tension plates on which to germinate the seeds of various species, and found that spiny and reticulate seeds were very sensitive, smooth seeds showed a graded response, the larger ones being less sensitive, while seeds which exude mucilage on wetting were least sensitive. These results suggest that sensitivity is related to the degree of contact between the seed and the surrounding soil particles. Osmotic effects appear to reduce germination less than matric effects; this would be expected if degree of contact plays a part and, over the period required for germination, the seed testa will be more or less 'leaky' to some osmotica so that water may be taken-up by an imbibitional process.

Lazenby (1955) attempted to study the effect of water supply on the germination of *Juncus effusus* alone and in various mixtures of grass and clover, by supplying water from water tables at different depths. A depth of eight inches below the soil surface prevented germination and intermediate depths caused considerable reduction compared with a treatment which was flooded to the surface. As the eight-inch water table also prevented the germination of the grasses, it seems likely that evaporation had caused the soil to lose capillary contact with the water. If this had not happened, the treatment would represent an equilibrium water potential, at the surface, of only -0.02 bar; Harper and Benton found considerable germination of some species at -0.2 bar, and Owen (1952) showed that wheat could germinate in soil below the permanent wilting percentage.

Lang (1965) notes that some seeds may be injured by soaking in water prior to germination; this may be caused by the leaching of some essential metabolite. In cases such as this the incidence of rain before and during the germination process will have an effect as it also will when the germinating embryo shows differing phases of drought sensitivity. Milthorpe (1950) showed that wheat seedlings were totally resistant to desiccation until the coleoptiles were 3–4 mm long. Until leaf emergence, up to 98% of the tissue water could be lost and only the roots were killed, but after this, the seedlings became very sensitive to water loss. Some desert plants are sensitive to rain and its duration as their germination depends on the washing-out of an inhibitor. The Californian plant *Filago californica* has a complex response: it will not germinate in moist soil, but any rainfall up to 8 inches causes an increase. Beyond this rainfall the germination declines again, reaching a minimum at 35 inches, beyond which it is unaffected by further rain (Lang, 1965).

Water uptake and transpiration. Inorganic nutrient uptake

From the previous discussion it should be apparent that increasing soil water deficit is likely to reduce water uptake; the corresponding fall in plant water potential and the consequent stomatal closure will limit transpiration loss. In the past there has been some controversy concerning the magnitude of negative soil water potential which would influence transpiration; it was suggested that transpiration was not reduced until the soil reached permanent wilting point (Veihmeyer and Hendrickson, 1927; see p. 151). Most recent work, however, shows that small depressions of water potential cause a decline in transpiration, the stomatal function becoming homeostatic to water loss as leaf water deficit rises. Under conditions of lesser deficit the transpiration loss may be influenced by light, carbon dioxide and those other factors which influence stomatal aperture.

Soil water deficits which reduce water uptake may also decrease salt uptake (Brouwer, 1965) though there is some conflicting evidence in earlier work. Correlation between water and ion uptake does not necessarily imply a mass-flow ion absorption mechanism as a reduced rate of water movement will slow the rate of ionic arrival at the root surface. Internal movement of ionic solutes is less affected by water deficit as the transpiration stream is normally more than sufficient for this function.

Photosynthesis and growth

Evidence has gradually accumulated that photosynthesis is reduced even by relatively small water deficits and certainly by water potentials higher than the -15 bar permanent wilting point. Schneider and Childers (1941), for example, compared the photosynthetic rates of watered and unwatered apple trees and found a steady decline during the five days prior to the first observed wilting. At this time both photosynthesis and transpiration had fallen to c. 40% of the control values. More recently Ashton (1956) conducted similar experiments with sugar cane during five drying and reirrigation cycles but in this case the decline occurred more abruptly after the build-up of a considerable soil water deficit.

Clark (1961) compared the photosynthetic response of *Picea glauca* and *Abies balsaminae* to soil water deficit. During the early drying cycles in his experiments, the photosynthesis of *A. balsaminae* was reduced to zero by drying to the PWP, but that of *P. glauca* only fell to about one-third of the maximum value. In this case the decline in photosynthesis again occurred gradually. Similar results were obtained by Brix (1962) with *Lycopersicon esculentum* and *Pinus taeda,* the reduction commencing after the leaf water potential had fallen to -7 bar in the former and $-4 \cdot 5$ in the latter plant.

Etherington (1967) found that very slight water deficits reduced photosynthesis in *Alopecurus pratensis,* confirming previous field and glasshouse observations that soil water potentials of minus one bar or less were sufficient to reduce the growth of this grass. These more recent observations are in conflict with those presented by Viehmeyer and Hendrickson (1927) who advanced the theory that growth and transpiration were not reduced until the soil reached the permanent wilting point. Their papers initiated a long-standing controversy concerning the concept of 'equal availability' of water between the field capacity and permanent wilting points. They upheld this view for almost 25 years (Viehmeyer and Hendrickson, 1950) but in an extensive review Richards and Wadleigh (1952) produced a great deal of evidence against the equal availability hypothesis and Stanhill (1957) analysed data from 80 papers describing water regime experiments and reported 66 cases in which growth was reduced before the PWP was reached. Experiments with osmotica also support the concept of gradual reduction of growth by decreasing potential (e.g. Slatyer, 1961).

The justification for the Viehmeyer and Hendrickson view lies partially in the nature of their experiments with orchard trees, of which the deepest extent of the root systems was not known for certain, and partially in the absence of *in situ* recording of soil water potential. Under the dry S. Californian climatic conditions the duration of soil water potentials between −5 and −15 bars would have been short, the transition representing only a small amount of water loss. On reaching c. −15 atm the rate of loss would be reduced due to stomatal and soil conductance effects; hence the growth reduction might be incorrectly associated with the onset of wilting. Their experiments with container-grown trees also rarely included treatments involving drying to points in the available water range midway between the field capacity and permanent wilting percentages.

Growth response to water deficit varies with the stage of the life-cycle and also with the physiological mechanism through which it is mediated. The growth–analytical approach shows that both net assimilation rate and leaf area ratio components of relative growth rate may be reduced by water deficit. This was shown, for example, by Etherington (1967) for *Alopecurus pratensis* (Table 5.2). The reduction of net assimilation rate conforms with the lowered photosynthesis described above and the reduced leaf area ratio suggests that water stress has limited full tissue expansion. This is confirmed by the partitioning of the *A. pratensis* leaf area ratio into its two components, leaf area/leaf weight and leaf area/plant weight; the former shows a reduction in the dry treatment, but the latter is unaffected.

Slatyer (1967) cites many workers who have found reduced cell enlargement under water stress. For example, Ordin (1958, 1960) found a marked relationship between turgor pressure and cell enlargement in *Avena* coleoptiles by using osmotica of different permeating characteristics; mannitol which was relatively non-permeating, caused the greatest reduction in extension. Impaired leaf expansion caused by soil water stress has been shown by Wadleigh and Gauch (1948) for cotton. Their measurements of

Table 5.2 The influence of soil water deficit on growth of *Alopecurus pratensis*
(Etherington, 1967)

	Soil water regime	
	Maintained field capacity	Dried to −5 bar before watering to field capacity
Net assimilation rate g dm^2 week^{-1}	0·22	0·19
Leaf area ratio dm^2 g^{-1}	0·86	0·74
	Leaf area ratio data partitioned into its two components	
Leaf area/leaf weight dm^2 g^{-1}	2·98	2·57
Leaf weight/plant weight g g^{-1}	0·29	0·29

leaf length show a decreasing rate of elongation which first became apparent at a soil water potential of −5 to −6 bars and reached zero at −13 to −14 bars. Jarvis (1963) found complete cessation of leaf expansion in *Saxifraga hypnoides* at −0·5 bar and in *Filipendula vulgaris* at −2 bar. The above experiments were carried out in pot culture and the overall measurements of soil matric potential may be considered reliable but it must be realized that the water potentials at the root surfaces are only a matter of assumption and may differ from the bulk soil values.

Respiration

Most workers who have measured respiration in relation to water deficit have observed that it is not so rapidly suppressed as the synthetic processes of metabolism, and in some cases moderate water stress has increased respiratory rates. Brix (1962) in his work with *Pinus taeda* and *Lycopersicon esculentum* found that photosynthesis was reduced to zero by leaf water potentials of −10 to −15 bar, but respiration was not reduced below c. 60% of the maximum rate. In *P. taeda* increasing the deficit beyond this point caused a rise in respiratory rate reaching 140% of the original value at −34 bar. In most cases increasing deficit reduces respiration: Jarvis and Jarvis (1965) showed a nearly linear relationship between increasing solute potential (0 to −14 bar) of bathing solutions and decreasing respiration of root sections of several tree species. In photosynthetic plants water stress may induce a net weight loss due to the more rapid suppression of photosynthesis than of respiration, but in storage organs such as seeds, the net rate of loss will be reduced as respiration declines.

Translocation

Experiments with ^{14}C-labelled metabolites and various herbicides show that the phloem translocation rate is reduced by water deficit but not to the same extent as the reduction of photosynthesis caused by equivalent deficits. The phloem operates under a positive hydrostatic pressure and appears capable of fulfilling its function even under conditions which deprive it of a large part of its water supply.

Biochemical changes

Some attempt has been made to interpret observations of the physiological consequences of water deficit at the biochemical level, but great experimental problems exist as most techniques require the disruption of cell or tissue systems in aqueous media with consequent changes of localized water potentials. The greatest success has been achieved in following long-term gross changes of plant composition, for example, the early work of Petrie and Wood (1938a and b) showed reduced synthesis of proteins from amino acids under water stress conditions. Subsequently, Gates and Bonner (1959) found a decrease in both RNA and DNA after moisture stressing in tomatoes, which may be considered in conjunction with the work of Zohlkevich and Koretskaya (1959) who showed that drought interrupted phosphorylation and consequently decreased the production of ATP, sugar phosphates, RNA and DNA, leading to the supression of protein synthesis. Gates and Bonner interpreted their data slightly differently as they found ^{32}P to be readily incorporated into water-stressed leaves: their suggestion was that drought caused an increased rate of breakdown of RNA.

In many cases it has been observed that soluble carbohydrate levels increase after drought (Stocker, 1960); Woodhams and Koslowski (1954) noted that this was a consequence of the conversion of starch to sugar but total carbohydrate levels were, however, reduced. The authors suggested that this was caused by raised respiratory levels accompanying the water stress.

Few attempts have been made to follow changes in metabolic pathway patterns under conditions of water deficit, the experimental work again being complicated by the need for aqueous media in preparation of organelle isolates. Nir and Poljakoff-Mayber (1967) attempted to measure the photosynthetic activity of chloroplasts taken from water-stressed plants and showed a reduction which persisted after isolation, but measurements such as these are of comparatively large changes consequent on severe treatments. Investigations of the biophysical attributes of drought resistance and drought hardening have also encounted much difficulty when taken to the molecular level.

Chapter 6

Plants and water deficit: ecological aspects

XEROMORPHY AND DROUGHT TOLERANCE

The existence of the terms xerophyte, hydrophyte and mesophyte implies that the distribution of plants may be influenced by their physiological response to water. This is an uncritical viewpoint which neglects the fact that water availability may show a very wide range at any site, but the concept of a xeric habitat as one experiencing a high frequency of 'deficit-days' is useful.

It is perhaps unfortunate that the term xerophyte is so widely used as it has such a breadth of connotation as to be misleading. Maximov (1929), in his early classification of the xerophytes, was aware that the limits of the group are ill-defined. The term is more useful when expressed as relative xeromorphy but even so the criteria used for its evaluation are too numerous for convenience. Oppenheimer (1960), in a literature survey, favours a very broad interpretation embracing all of the anatomical, morphological and physiological modifications which may assist the plant to cope with environmental water deficit.

A long-standing problem in the discussion of xerophytism arose from Maximov's (1929) claim that some xerophytes transpire at a greater rate per unit of leaf area than mesophytes. More recent work suggests that this may be incorrect (Holmgren, Jarvis and Jarvis, 1965; Cowan and Milthorpe, 1968). Values cited by Cowan and Milthorpe from Altmann and Dittmer's (1966) compendium are summarized in Table 6.1, and support the intuitive conclusion that sclerophyllous plants with heavy cuticles are better fitted to avoid water loss than mesophytes.

Modifications such as leaf rolling in some grasses, stomatal pits and superficial coverings of epidermal hairs have all been considered to aid water conservation but the experimental evidence is conflicting. In some

Table 6.1 Ranges of cuticular and stomatal resistance values for xeromorphic and mesomorphic plants. (Derived from Cowan and Milthorpe, 1968)

| | Resistance (s cm^{-2}) | |
	Xeromorphs	Mesomorphs
Cuticular	50–400	20–50
Stomatal + Internal	10–400	1–50

grasses, for example, rolling does not take place until lethal water deficits are established (Parker, 1968). In the light of recent evidence that the resistance of the boundary layer external to the leaf may limit transpiration under still-air conditions (page 145), it seems that these modifications must increase the thickness of the boundary layer and consequently its resistance. Epidermal hair coverings may also increase the reflection of radiation, thus reducing the temperature of leaf tissues.

The ratio of root to shoot and the depth and extent of penetration of the root system plays an important part in the plant's response to water deficit. Many desert plants shed leaves or whole shoots during the dry season with a consequent reduction of transpiring area (Parker, 1968). Weaver (1926) studied root development in native and crop plants and drew attention to the profuse development and deep penetration of roots in sandhill species, while Salisbury (1952) made the same observation in many British dune species. Maximov (1929) suggested that extensive root development provided a mechanism for drought tolerance, and Parker (1968), in his review, stressed the critical role of rapid root penetration in the seedling stage of drought-tolerant plants. Experimentally, the reduction of water supply has been shown to reduce the shoot/root ratio; for example, Davies (1942), working with Cyperus rotundus, found that dry treatments reduced the top more than the tuber yield, and Etherington (1962) found, with two grass species, that an irrigation regime corresponding to a very dry summer (S.E. England) gave a root/shoot ratio of 1·9 compared with 4·5 in a treatment maintained near field capacity. These results could be interpreted as an induced trend towards xeromorphy.

Amongst normal mesophytes root penetration may be of significance in drought survival. Goode's (1956) work with Lolium perenne and Poa annua showed the importance of root systems which permit plants to exploit different layers of the soil, thus avoiding competition for water supplies. Lolium is normally deeper rooting than Poa but frequent cutting causes a reduction in its rooting depth, the two species then come into competition and the sward becomes less drought-tolerant. Many natural habitats show a similar root stratification.

The attributes described above permit the avoidance of tissue water deficit by preventing desiccation; in the same category must be included the water storage tissues, seen at their greatest development in the succulents. For example, Kilian (1956) refers to the eu-succulents in which a thick cuticle is only sparsely perforated by stomata which are normally shut during the day; these plants may lose up to 95% of their water without injury and are characteristic of frequently droughted habitats. It has also been suggested that water storage in stems and roots of normal, mesophytic woody plants may buffer the plant against short-term deficits (Reynolds, 1967), and that cell-wall water forms an easily mobilized reservoir for the cell protoplasts (Carr and Gaff, 1961).

Maximov (1929) realized in his early work that desiccation tolerance was a very different phenomenon from those other xeromorphic characteristics which *avoid* the establishment of tissue water deficits. He referred to 'true xerophytism' as the ability of the cell contents to resist the effects of drying; this is much more common in the lower plants, in particular lichens and mosses (Parker, 1968). Some Pteridophytes also show tolerance of this sort; Rouschal (1937–8), for example, showed *Ceterach officinarum* to be resistant to an extreme degree of atmospheric drying, finding a five-day survival period at zero relative humidity over concentrated sulphuric acid. The term 'poikilohydrous' has been used to separate such plants from the 'homoihydrous' types which have more control over water loss and perish at comparatively slight tissue water deficits.

Most higher plants are homoihydrous but it seems that there are specific and varietal differences which do derive from cytoplasmic differences at the molecular level. Recent reviews (Parker, 1968; Oppenheimer, 1960) suggest that very little satisfactory information is available concerning such differences.

COMPARATIVE ECOLOGICAL STUDIES

Strongly developed xeromorphic characteristics are generally associated with very dry habitats of climatic origin but local variation within a climatic zone or even microclimatic variation at a specific site may impose a smaller scale water balance pattern which will interact with the distribution of species or ecotypes. Under these conditions, minor variations in the physiological response to water deficit may govern the habitat tolerances of species or ecotypes; the need for comparative experimentation to locate such differences has been stressed by Grime (1965a) and much experimental ecology is now being undertaken on this basis.

Examples of this approach are to be found in the work of Jarvis and Jarvis (1963a, b, c, d), Jarvis (1963), Etherington and Rutter (1964) and Bannister (1964). Jarvis and Jarvis described comparative experiments with *Betula verrucosa, Populus tremula, Pinus sylvestris* and *Picea abies* in which growth and transpiration were studied in relation to soil water potential. Transpiration was also measured in liquid rooting media of varying osmotic potential and the water balance of the tissues related to specific drought tolerance.

The results of these experiments, summarized in Table 6.2, exemplified the difficulty of interpreting comparative physiological experiments in the ecological milieu as there were discrepancies between some of the physiological characteristics for which unifying explanations had to be found. For example, the order of sensitivity of the four species to drought was almost the reverse of that to moderate soil water potential. The authors suggested that desiccation resistance and ability to grow well at moderate water deficit might be inversely proportional. Furthermore, the depression of transpiration caused by substrate water potential bore no

Table 6.2 Some responses of tree seedlings (*Pinus sylvestris; Picea abies; Betula verrucosa; Populus tremula*) to water stress. (Summarized from Jarvis and Jarvis 1963, a, b, c, d)

	Order of sensitivity
Growth response (RGR in relation to moderate soil water deficit)	*Picea > Pinus > Betula > Populus*
Transpiration (degree of reduction by soil water deficit or osmotic potential of substrate)	*Pinus > Betula > Populus > Picea*
Drought resistance (assessed as combined desiccation tolerance and drought avoidance)	*Pinus > Picea > Betula > Populus*

relationship to the effects of water potential on growth. However, most of the findings supported the general ecological observations concerning the distribution of the four species and confirm the thesis that experimentally observed differential responses may be used in the interpretation of ecological problems.

Jarvis (1963) described the influence of water regime experiments on the growth of *Filipendula vulgaris, Saxigraga hypnoides, Thelycrania sanguinea* and *Prunus padus*. These were chosen as plants growing at the extreme limits of their geographical ranges, *F. vulgaris* and *T. sanguinea* having a south-eastern (low rainfall) distribution in the British Isles, and the other two species a contrasting north-western (high rainfall) distribution (Figure 6.1). Despite difficulties with low growth rates and high replicate variability, Jarvis was able to conclude that the two N.W. species were more sensitive to soil water deficit than the S.E. species, at least in respect of leaf expansion; *S. hypnoides* was particularly sensitive, growth ceasing at c. −0·5 bar soil matric potential. The boundary values for the other species ranged between −2 and −3 bar, values well above the controversial permanent wilting point equivalent of −15 bar.

Etherington (1962) and Etherington and Rutter (1964) encountered similar difficulties of variable growth response in comparative water regime and irrigation experiments with *Alopecurus pratensis* and *Agrostis tenuis* but showed that *A. pratensis* is slightly more sensitive to water deficit

158

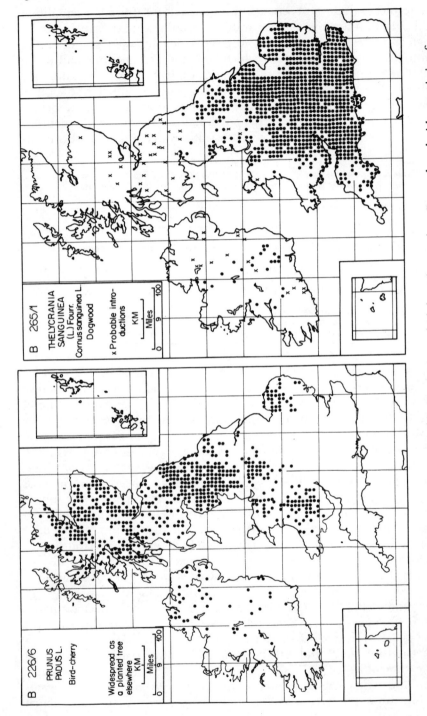

Figure 6.1a The distribution in the British Isles of *Prunus padus* and *Thelycrania sanguinea*. Reproduced with permission from F. H. Perring and S. M. Waters, *Atlas of the British Flora*, Botanical Soc. of the British Isles, 1962, Maps nos. 226/6 and 265/1.

159

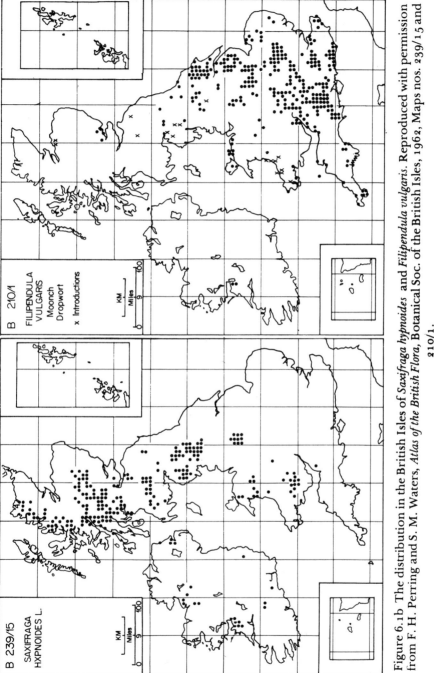

Figure 6.1b The distribution in the British Isles of *Saxifraga hypnoides* and *Filipendula vulgaris*. Reproduced with permission from F. H. Perring and S. M. Waters, *Atlas of the British Flora*, Botanical Soc. of the British Isles, 1962, Maps nos. 239/15 and 210/1.

than *A. tenuis*. This corresponds with the natural distribution of the two species, *A. pratensis* being a plant of moist meadow soils while *A. tenuis* is more characteristic of drought-prone sandy soils. It is likely in this case that the mineral nutrient status and pH of the soil are also important in determining the habitat range of these species, and it should be borne in mind that the experimental isolation of a single physiological factor in the ecological study of a plant species may be misleading if it is accepted uncritically.

Bannister's (1964a, b, c) work with *Erica cinerea* and *Erica tetralix* is a further example of the comparative, experimental approach. *E. cinerea* is a species of dry, heath soils while *E. tetralix* occupies much wetter, usually peaty soils. Bannister's observations showed that *E. tetralix* is very sensitive to soil water deficit compared with *E. cinerea,* the death of the plants being preceded by strong depression of relative water content and transpiration rate which did not occur so markedly in *E. cinerea* (Figure 6.2). Hence, at a whole-plant physiological level, there was a pronounced species difference which could account for the respective habitat tolerances. This difference was further reinforced at the wet end of the water availability scale by a

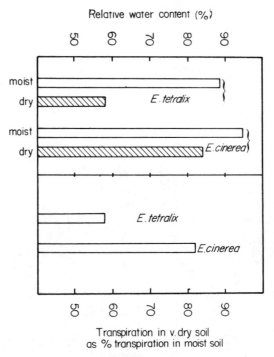

Figure 6.2 Relative water content and transpiration rate of *E. tetralix* and *E. cinerea* after two hours at 45% relative humidity; 20°C in a moist and in a very dry soil (below permanent wilting point). Data of Bannister (1964c).

differential response to waterlogging (page 200). In the same study *Calluna vulgaris* was compared with the two *Erica* species and, though showing a reduction of relative water content and transpiration rate on droughting, remained unharmed under the same conditions as those lethal to *E. tetralix*. This suggests that *C. vulgaris* may have an inherent desiccation tolerance and account for the observation that its habitat range overlaps that of *E. cinerea* and *E. tetralix*.

The foregoing experiments all share the common feature of interspecific physiological comparison which leads to the interpretation of an ecological situation. In some cases the experimental work was complicated by extensive morphological and anatomical differences between species which causes problems in the comparison of physiological attributes; for example, the external boundary layer characteristics of *Pinus* and *Picea* must be very different from those of *Betula* and *Populus* and there may be a risk of these differences confounding with experimental treatments. The choice of very similar species, as in Etherington's work, partially solves this problem, but even here there was the complication of potentially different requirements for other factors such as mineral nutrients. The comparison of two *Erica* species reduces the variation of morphology, anatomy and requisite soil environment to a level of similarity in which water availability and waterlogging are almost isolated as factors which may have a differential effect.

Comparison of varietal or ecotypic material probably furnishes the simplest experimental design. McKell, Perrier and Stebbins (1960) made this approach, comparing two subspecies of *Dactylis glomerata*: ssp *judaica* and ssp *lusitanica,* the first known to inhabit more xeric habitats than the second. Their results were, however, rather inconclusive. The mesic type showed a greater rate of water usage but leaf elongation ceased in both types between 3 and 5 bar. By contrast, the earlier work of Cook (1943), who compared 8 selections of *Bromus inermis*, showed significantly greater rooting depth and root axial length in the drought-resistant forms; in this case morphological and physiological differences interact to produce ecotypic differentials.

Agricultural experimentation to isolate drought-resistant varieties or species has stimulated a great deal of comparative work. Ashton (1948) has described the techniques and reviewed the findings of this work at length; there is much to interest the ecologist here as it has long been realized, in agricultural experimentation, that the isolation of specific factors may lead to results which are significant in the field. Even very simple screening methods may play a part; Carrol (1943) for example, exposed sod-samples of various pasture grasses to a slow drying process and assessed drought resistance on the basis of numbers of plants surviving after 21 days. Among the intolerant species were *Poa trivialis* and *P. nemoralis*, while *P. pratensis* and *Festuca rubra* were the most tolerant.

Though much can be done with simple techniques, increased sophistication of instruments has led to the introduction of physiological methods of some complexity. Comparative measurement of photosynthesis in relation to drought resistance of various cereal varieties has been undertaken by Todd and Webster (1965) using infrared gas analysis to monitor CO_2 exchange. Figure 6.3 shows the course of photosynthesis during five drought cycles in a tolerant variety (Kan-King) and an intolerant variety

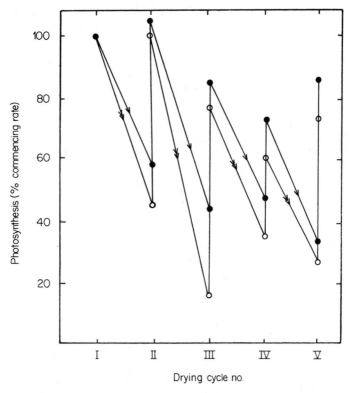

Figure 6.3 Decline in photosynthetic rate in two varieties of wheat caused by exposure to five drying : rewetting cycles. Var. Kan-King ●, Var. Ponca ○. Data of Todd and Webster (1965).

(Ponca) of wheat; the degree of drought tolerance was assessed by a conventional drought cycle/survival test. The photosynthetic measurements showed the sensitivity if the two varieties to drought very much more quickly than the conventional test and the authors proposed that the technique could be used in screening populations for drought resistance.

Klikoff (1965) found a strong correlation between the response of

photosynthesis to water deficit and the habitat distribution of some dominant species of the Sierra Nevada timberline. *Carex exserta* inhabits dry
montane meadows, *Potentilla breweri* moist meadows and *Calamagrostis
breweri* wet meadows in which the water potential never falls below −0·3
bar. By contrast the dry meadows develop water potentials of less than −15
bar by late summer. Figure 6.4 shows the relationship of photosynthesis to
leaf water potential determined by making measurements on pot-grown
plants at various times after watering; the response to water stress shows a
remarkable parallel with the distribution of the plants.

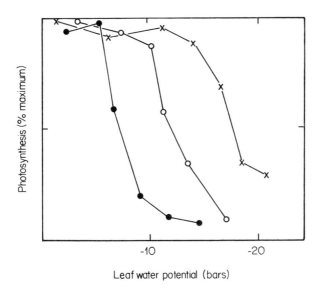

Figure 6.4 Relationship between photosynthesis and
leaf water potential of *Calamagrostis breweri* ●; *Potentilla
breweri* ○; and *Carex exserta* ×. Reproduced with
permission from L. G. Klikoff, Photosynthetic
responses. . . ., *Ecology*, **46**, 516–17, Figure 1 (1965).

Such variations in response to water deficit may be attributed to
differences in physiology and/or morphology: for example, Zelitch (1965)
cites a comparative experiment with the drought-tolerant grass *Cynodon
dactylon* and the more sensitive *Paspalum dilatatum*. Under his experimental
conditions the former grass was incapable of opening its stomata beyond
an aperture of one micron, whereas *P. dilatatum* showed full opening to six
microns. Zelitch considered that this attribute was of genetically controlled
origin and could be the cause of drought resistance in *C. dactylon*. Brown
and Pratt (1965), from a survey of many species, concluded that stomatal

inactivity was widespread in many grasses and may be related to survival in hot, semi-arid regions.

Hodges (1967) followed the diurnal course of photosynthesis in six conifer species and found that it was primarily controlled by changes in leaf water potential. The control mechanism, however, was variable: in *Pinus silvestris* and *Abies procera* it was stomatal movement, but in *A. grandis, Pseudotsuga menziesii, Tsuga heterophylla* and *Picea sitchensis* change of mesophyll resistance was the only satisfactory explanation. Amongst these changes Hodges noted a midday reduction in photosynthesis, which he attributed to falling leaf water potential, followed by an afternoon recovery.

Comparative experiments of this sort are ecologically useful but it must be remembered that some water deficiency situations are difficult to simulate under experimental conditions. Water availability varies widely in both space and time but extreme deficits are rare in the temperate zone, occurring possibly only once or twice each century. Nonetheless, they will cause far-reaching effects in the ecosystem, ranging from a temporary reduction in the population of plants with a short life-cycle to the total exclusion of other, longer life-cycle plants. Today's ecosystem is a reflection of yesterday's environment and its study requires a knowledge of the range of climatic variables; this is nowhere more apparent than in the distribution of precipitation and potential evaporation in which identical annual means may be accompanied by very different bio-climates of water balance.

The effects of occasional drought will vary not only with duration and intensity but also with the stage of the plant life-cycle. Milthorpe (1950) found three distinctly different degrees of susceptibility to desiccation injury during the early stages of germination and seedling growth in wheat. For example, until the emergence of the first leaf the loss of 98% of the tissue water did not cause serious damage. A single period of deficit may have a long-term effect; Gates (1955a, b; 1957), for example, showed that one wilting treatment in young tomato plants caused a persistent alteration of net assimilation rate and nutrient status (particularly phosphorus). It seems likely that a single drought at an early stage in the growth of a plant may alter its competitive ability and affect the ecosystem balance for a season or more (Figure 6.5).

In extreme cases the life-cycle of winter annuals and ephemerals may be terminated by drought but usually not before reproduction has occurred: examples of this are suggested by the work of Ratcliffe (1961) and Grime (1963b). Ratcliffe investigated adaptation to habitat in a group of winter annuals of shallow calcareous soils; the plants included *Erophila verna, Hornungia petraea, Saxifraga tridactylites* and *Arenaria serpyllifolia,* all of which lie at the northern extreme of a Mediterranean-centred distribution in Europe. The plants flower and set seed in the Spring but germination is delayed until the early autumn, thus avoiding the summer period of widely fluctuating water availability during which seedlings would have little

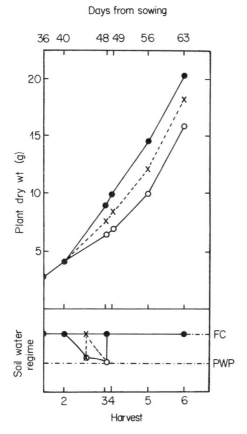

Figure 6.5 The influence of two repeated periods of wilting on the subsequent growth of Tomato plants. The lower diagram shows the soil moisture regime. FC; field capacity. PWP; permanent wilting point. Reproduced with permission from D. M. Gates, *Water Deficits and Plant Growth,* Vol. 2, © Academic Press, New York, 1968, p. 159, Figure 3.

chance of survival. The parent plants are probably killed by drought shortly after the seeds mature. Newman (1967) observed that in winter annuals such as these, any drought experienced during the flowering/seed-setting period hastened leaf senescence, and Grime found drought-killed individuals of such resistant species as *Festuca ovina* in identical habitats to those of Ratcliffe's work.

FIELD INVESTIGATION AND EXPERIMENT

The distribution of wild plants in relation to climatic wetness, their behaviour in response to short-term climatic fluctuations, the measurement of *in situ* soil water status and the elucidation of ecosystem water balance all provide a core of observational knowledge around which to build the experimental investigations suggested in the last section.

For many years botanists have attempted the subjective interpretation of plant distribution in relation to water; the habitat notes of most flora provide many examples. Critical distribution maps give a more definite picture of such relationships, for example, Figure 6.6 shows the distribu-

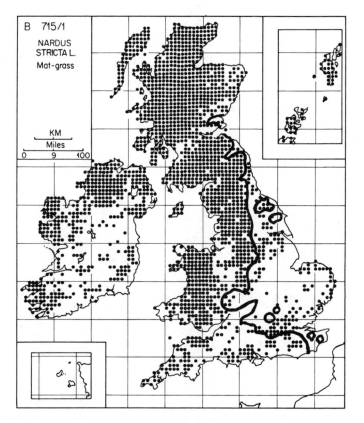

Figure 6.6 Distribution of *Nardus stricta* in Britain in relation to rainfall. Areas enclosed by or to the left of the heavy lines have a rainfall of above 40 inches per year. Reproduced with permission from F. H. Perring and S. M. Waters, *Atlas of the British Flora*, Botanical Soc. of the British Isles, 1962, Map 715/1.

tion of *Nardus stricta* in the British Isles in relation to precipitation. The apparently high positive correlation cannot, of course, be taken as proof of a causal relationship as the high rainfall of the northern and western regions is associated with areas of greater altitude and these in turn are correlated with older, weathering-resistant, nutrient-deficient rocks. However, a distribution of this sort suggests a relationship which could be investigated by experimental means.

The work of Woodell, Mooney and Hill (1969) is an example of a more critical attempt to correlate plant distribution with rainfall: Figure 6.7

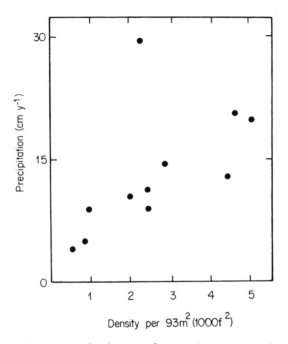

Figure 6.7 The density of *Larrea divaricata* stands in relation to precipitation at a number of sites. Reproduced with permission from S. R. J. Woodell, H. A. Mooney and A. J. Hill, Behaviour of *Larrea divaricata, J. Ecol.*, **57**, 37–44, Figure 2 (1969).

shows the density of *Larrea divaricata* stands in California plotted against mean rainfall. Except for one deviant stand which the authors explain by its differing degree of exposure, there is a markedly linear relationship of density to rainfall. Again causality cannot be proved but the authors suggest that the increased spacing in the drier sites is caused by competition for water. *L. divaricata* is a drought-tolerant plant found in some of the most arid regions of N. and S. America and for this reason rather extensive studies have been made of its ecological relationships and drought tolerance. Chew and Chew (1965) considered that the plant showed a true physiological tolerance of drought rather than avoidance, while Yong (1967) isolated at least two ecotypically differing populations of which the Californian type had a lower water requirement than that of more southerly States. This latter point must be taken as a warning against placing too much reliance on distributional studies alone without any check on the genetic constitution of the populations.

A similar use of climatic data may be found in Felgan and Low (1967) who studied the surface/volume ratio of a columnar cactus (*Lophocereus schotii*) at various sites in N.W. Mexico; the ratio was found to fall with reduced precipitation and this was taken as circumstantial evidence of a trend to increased xeromorphy. Again, experimental work is needed to prove the assumption.

Using shorter term rainfall data, Zahner and Donelly (1967) correlated the growth-ring width in *Pinus resinosa* with rainfall and calculated water deficits. Their multiple regressions of previous year's rain and deficit and current year's rain and deficit gave a correlation coefficient of 0·91. Relationships such as this, with short-term variables, may be taken generally as better evidence for causality than the long-term correlations described above.

Experimental work undertaken to test hypotheses founded on compilations of field data are well illustrated by a study of the three *Ranunculus* spp, *R. repens, R. acris* and *R. bulbosus,* described by Harper and Sagar (1953). They established a correlation between the ridge and furrow pattern of the meadow habitat and the relative distribution of the three species (Figure 6.8) and suggested that it might originate from differential sensitivities to soil wetness or winter flooding. Experiments with transplants and seedling establishment tests showed that seed germination and seedling survival followed a pattern in relation to water supply which could give rise to the field distribution (Figure 6.9).

Many plants respond to an environmental stress with a communal or individual plastic response; the increased spacing of *Larrea divaricata* with decreasing rainfall which is cited above is a good example of the former response. Karper (1929), working with grain *Sorghum* varieties, showed the influence of individual plasticity in response to water availability. The nontillering variety, *Kaffir*, showed an optimum row spacing of 36 inches (91 cm) during the driest two years of the study, but this was reduced to 3 inches (17·6 cm) in the wettest two years. *Milo*, which tillers freely, had an optimum spacing of approximately 24 inches (61 cm) in both driest and wettest groups of years. Karper suggested that the final crop density in *Kaffir* was dependent on the original row spacing but *Milo*, by its tillering habit, was able to 'self-adjust' to higher densities in wet conditions when there was no competition for water (see also Table 10.3).

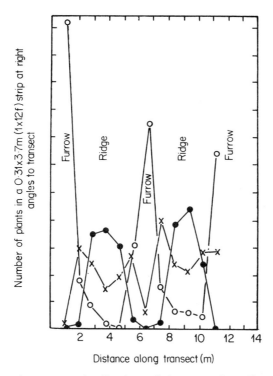

Figure 6.8 Distribution of three species of *Ranunculus* on ridge and furrow grassland. Reproduced with permission from J. L. Harper and G. A. Sagar, *Proc. 1st Brit. Weed Control Conf.*, 256–64 (1953).

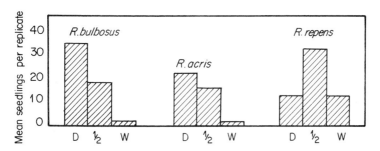

Figure 6.9 Seedling establishment of three species of *Ranunculus* in pots of differentially waterlogged soil. Reproduced with permission from J. L. Harper and G. A. Sagar, *Proc. 1st Brit. Weed Control Conf.*, 256–64 (1953).

EXPERIMENTS WITH SINGLE SPECIES

Most early investigations of plant response to water shortage were inspired by agricultural or horticultural needs and were often undertaken to provide a rationale for irrigation treatments. For this reason experiments were conducted on a 'one species at a time' basis and tended to be irreproducible unless undertaken in areas of very stable climate. The advent of controlled environment facilities has reduced this problem but water balance effects have not been studied extensively in this way as it is extremely difficult to simulate the natural energy balance which is the governing factor in water loss.

Richards and Wadleigh (1952), in a comprehensive review, quote many examples of single species experiments. Amongst these is Viehmeyer's (1927) report on work with French Prune trees grown in large, outdoor soil containers. A milestone in this type of work, it involved the recognition that soil water may be controlled only by returning the whole soil mass to field capacity. He concluded from his results that water-use was influenced only by leaf area and not by soil water content until the wilting percentage was reached. Wadleigh and Richards, however, disputed his suggestion that growth was not reduced except by the treatments which reached the wilting point; they furthermore discussed the whole range of publications in which Veihmeyer and his coworker, Hendrickson, formulated their 'equal availability' hypothesis which was presented in the last chapter. From this time onward it has become obvious that water is not equally available in the range from field capacity to the wilting percentage, and that any fall in soil water potential, however small, is accompanied by a reduction in growth and probably in transpiration.

One of the criticisms of Veihmeyer and Hendrickson's work was that they included insufficient treatments in the middle part of their 'availability' range; later workers paid more attention to this point. For example, Davies (1942) grew *Cyperus rotundus* under five different water regimes, the wettest of which was kept rather wetter than the moisture equivalent and the driest allowed to pass a little below the wilting percentage in each cycle. Each decrease in water content gave a significant reduction in top yield and also decreased the top/tuber ratio. Similarly, in earlier experiments with *Zea mays*, Davies (1940) had demonstrated, with the aid of porous pot tensiometers, that growth was reduced as soon as the soil water potential began to fall and stopped before the wilting percentage was reached.

More recently Wadleigh, Gauch and Magistad (1946) and Wadleigh and Gauch (1948) investigated the growth of *Parthenium argentatum* and *Gossypium barbadense* in response to the combined effect of soil matric and solute potentials. Figure 6.10 shows the growth response curve of *P. argentatum* and Figure 6.11 the leaf elongation of *G. barbadense* in relation to the imposed stress. These gradual reductions of growth with falling soil water potential are not unexpected in the light of the relationships which have

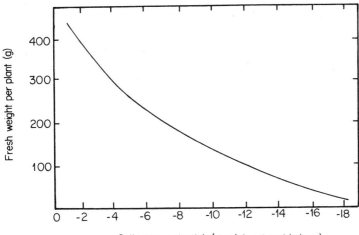

Figure 6.10 The growth of *Parthenium argentatum* in response to soil water potentials imposed by different drying and salt addition treatments. Reproduced with permission from C. H. Wadleigh and H. G. Gauch, Rate of leaf elongation as affected by TSMS, *Pl. Physiol.*, **23**, 485–95, Figure 3 (1948).

previously been described for photosynthesis and respiration. Recent trends in experimentation seem to have swung from the investigation of overall effects of water stress on growth to the more detailed analysis of response using physiological techniques to study the internal water balance of the plant, the accompanying changes of stomatal apertures and their effects on metabolic functions such as photosynthesis.

ECOSYSTEM WATER BALANCE

The increased study of plant–water relationships at the physiological level has been accompanied by a growing interest in the water economy of whole ecosystems. In the agricultural context studies of the water requirements of irrigated crops extend back to the late nineteenth century (Jensen, 1968) but more critical analysis of the ecosystem water budget has awaited recent developments in technique, instrumentation and recording facility.

Methods of measuring water balance

Rutter (1968), in a review of water consumption by forests, describes six basic methods which may be used in the measurement of water balance
(1) Micrometeorological estimation of the vapour flux from the canopy.
(2) Transpiration measurement.

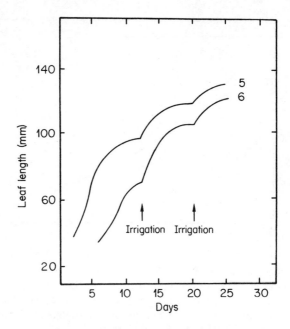

Figure 6.11 The growth of cotton leaves in relation to time from last irrigation. Sharp inflections in the curves show recovery from water
stresses which have approached the permanent
wilting point. 5 and 6–leaf insertion numbers.
After C. H. Wadleigh and H. G. Gauch, *Pl.
Physiol.*, **23**, 485–95 (1948).

(3) Measurement of interception.
(4) Measurement of soil water depletion.
(5) Lysimetry.
(6) Stream gauging on experimental watersheds.

Micrometeorological estimation of vapour flux
    Vapour flux away from the upper surface of the vegetation canopy may
be calculated either by the mass transport method from measurements of
the vertical gradient of humidity and wind speed or by estimation of an
energy balance based on the vertical gradient of humidity and
temperature. These techniques are of recent origin, having developed as
suitable sensors and recording instruments became available. Long (1968)
presented a critical discussion of the instrumentation which may be used,
and Tanner (1968) reviewed the whole subject of evaporation from plants

and soils, describing the requirements of measuring methods and the theoretical treatment of the profile and energy balance equations.

The work of Begg *et al.* (1964) furnished an example of the technique: working with Bullrush Millet (*Pennisetum typhoides*) they measured profiles of net radiation, water vapour and temperature (see Figure 2.5). At 20 cm intervals of height, latent heat fluxes were calculated from these profile values (Figure 6.12); they illustrate the gradual increase of transpiration during the day and its decline again towards dusk. The values of transpiration

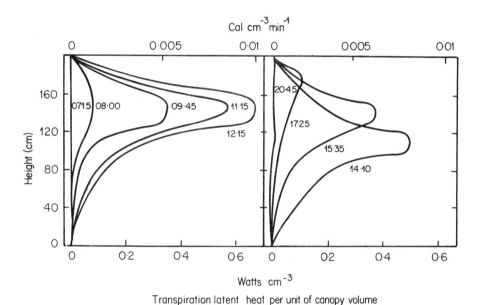

Figure 6.12 Transpiration latent heat profiles calculated from the micrometeorological profiles of a *Pennisetum typhoides* canopy. Reproduced with permission from J. E. Begg *et al.*, Diurnal energy and water exchanges in bullrush millet, *Agr. Meteorol.*, **1**, 294–312, Figure 6 (1964).

latent heat between 120–160 cm at 14.10 h are considerably depressed below the equivalent values at 12·15 and 15·35. This is probably due to the midday stomatal closure effect (p. 29) and the authors claim that this was the first 'whole ecosystem' demonstration of the phenomenon. The gradual descent of the peak transpiration into the canopy between 08.00 and 14.00 h and its subsequent rise provide evidence that the stomata in the upper layers of leaves close gradually during the morning, those of the highest leaves closing first; in the late afternoon they then re-open. At the time of the measurements the soil was near -15 bar water potential to a depth of 150 cm.

The problem with all micrometeorological estimates is that horizontal uniformity of physical conditions within the canopy is assumed. This not only seems unlikely but any attempt to correct for horizontal variations would make the technique unacceptably complex. However, provided that the profile samples are taken in an extensive and uniform stand of the vegetation permitting a large wind 'fetch', reliable results may be obtained. Micrometeorological measurement is more extensively surveyed in Chapter 2.

Transpiration

Most early field determinations of transpiration were made by the quick weighing, cut leaf or branch method, the assumption being that water loss continues unchanged for a short time after excision. With some plants, particularly branches of trees and shrubs, this may be valid, but in others there are rapid changes (Ivanov, 1928; Franco and Magelhaes, 1965). With suitable calibration the technique can be used, but it is better avoided.

A more recent development is the heat-pulse method, in which a localized area of the stem is warmed for a short period; the arrival of the heat pulse in the xylem contents is then noted at a higher level, using a thermocouple or thermistor sensor. This gives a direct measurement of the transpiration stream velocity and, if the cross-sectional area of the functional xylem is known, it may be converted to volumetric loss. Unfortunately, the functional extent of the xylem varies seasonally by gas embolism of vessels or blocking with tyloses, thus introducing some error into the estimate.

Various attempts have been made to measure transpiration loss by using 'transpiration tents', made of different types of plastic sheeting and supplied with air by large-capacity fans. The changing water content of the air passed through the tent may be measured with various psychometric sensors, either by condensation or absorption or by infrared gas analysis. The greatest problems with this method are the controlling of the enclosure microclimate so that it remains comparable with the exterior, and the expense of high volume-air-conditioning equipment.

Interception

Precipitation reaching the upper surface of the canopy is either re-evaporated directly, contributes to stem-flow, or may be absorbed. This is termed interception loss and is estimated by difference using standard rain gauges outside or above the canopy, ring funnels to measure stem-flow to the ground, and further gauges below the canopy to assess throughfall and canopy drip. Massive replication of such gauges is needed to cope with the horizontal variation of canopy distribution and the exposure of the gauges in clearings or above the canopy is fraught with difficulties caused by turbulence which may seriously influence the catch. Leyton *et al.* (1968) discussed these problems in detail.

Soil water depletion

This may be measured using any of the standard techniques, the rooting zone either being permanently instrumented with tensiometers and resistance units or periodic gravimetric samples or neutron scatter readings being taken. The neutron scatter technique has the advantage of giving a direct volumetric estimate of soil water, which is more accurate than the conversion, using a bulk density factor, needed for tensiometers or resistance units. Slatyer (1961) gives a detailed account of the methodology of a water balance study in which all of these techniques were used in an arid zone *Acacia aneura* woodland.

Lysimeters

The simplest, though most expensive, approach to the problem of crop water use is lysimetry. The lysimeter is a large soil container which is hydrologically isolated from the surrounding soil and permits the measurement of water income and loss. The derivation of the term implies the collection and measurement of drainage water but it has since been applied to large soil containers in which the water budget may be elucidated either by weighing or by soil water measurement. Hudson, in the discussion appended to a paper by Van Bavel and Reginato (1965), suggested the generic term 'Evapotranspiration gauge' and the specific epithets, lysimeter, weighable lysimeter and weighable container. It is also of considerable importance to specify whether the gauge contains repacked soil or encloses an undisturbed monolith. Hudson (1965) discussed the construction and specification of evapotranspiration gauges.

Penman (1963) reviewed the subject, commenting on the need for the crop cover on the gauge to be continuous with the canopy of the surrounding crop. Failure to ensure this leads to excessive exposure and overestimation of water loss by the 'oasis' or 'clothes-line' effect which is discussed by Stanhill (1965). The largest contributions to the oasis effect are the advection of sensible heat from the surroundings and evaporation from the 'sides' of a canopy arising from insufficient wind 'fetch' for typical profile development. Penman's comparison of results of Mather (1954) and McCloud and Dunavin (1954) provided an excellent example: Mather's gauges were exposed so that the contiguous canopies were level, but in McCloud and Donavin's work the crop projected about ten feet above its surroundings, thus giving an estimate of water loss which apparently exceeded the available radiant energy income by a large amount.

Plant cover of different types evaporates water at a remarkably similar rate once growth has closed the canopy, suggesting that the plant species is less important in regulating water loss than the energy income from the sun. If, however, the canopy is not closed or if it is not particularly dense, the water loss becomes dependent on plant height; this is probably due to the increased aerodynamic roughness of the canopy. For example, Stanhill (1965) showed that the energy exchange between free air and the evaporating surfaces of open water compared with a crop surface was nearly three times greater in the latter case, the aerodynamic roughness of the crop causing much more turbulence than the smooth water surface.

Etherington and Rutter (1964) used an array of 28 lysimeters to compare the response of two grasses, *Alopecurus pratensis* and *Agrostis tenuis* to water deficit and waterlogging. In this instance, water use was followed with tensiometers and resistance units and showed that water potential profiles similar to those occurring in the field could be simulated. The highest irrigation level was intended to maintain field capacity and was controlled by checking for drainage after each weekly irrigation: the subsequent irrigation could then be increased or decreased as necessary. This technique gave an almost exact correspondence between the applied water volume and the calculated weekly potential transpiration (see p. 179). Simulation of a natural environment, in addition to avoidance of the oasis effect, requires the establishment of soil, root and water profiles similar to those of the surroundings. Van Bavel and Reginato (1965) discussed these problems and also suggested that water should be removed from the base of the soil block to avoid impeded drainage at the soil/air interface. This effect has been investigated by Coleman (1946), who showed that it might have a considerable effect on the behaviour of water in lysimeters.

Experimental watersheds

Watersheds which are geologically sealed by an impermeable layer may be considered as gigantic natural lysimeters, the drainage of which may be observed in stream-flow. Reinhart (1967) has described techniques for calibrating such areas using covariance analysis of 'before' and 'after' treatment regressions. Treatments may include clear felling, selective felling or even extensive anti-transpirant application.

The most interesting findings of studies such as these have been the effects of variation in canopy structure and species composition on water balance and the relationship between water consumption and annual net radiation. Canopy structure and species influence the shortwave reflection coefficients (albedo) and the longwave reradiation characteristics of the vegetation surface. Resistance to vapour transfer from the canopy is also affected by surface roughness and its consequent effects on aerodynamic characteristics: surface roughness usually increasing with vegetation height. A further specific variable is the influence of stomatal resistance, which appears to be greater for tree leaves than for herbaceous leaves.

The combined influence of these variables causes the overall water loss of a tree cover to exceed that of herbaceous cover; for example, Hibbert (1967) showed that clear felling a watershed increased stream flow by the equivalent of 373 mm/ann which had decreased again by the equivalent of 300 mm/ann after 24 years of regrowth.

EFFECTS OF LIFE-FORM ON WATER LOSS

Early work in this field involved the comparison of agricultural crop plants of similar life-form and resulted in the suggestion, mentioned on page 175, that once the canopy was closed, the species composition of the cover did not markedly affect water loss. Further work has, however, established that this is an oversimplification; for example, the watershed studies which were referred to above show considerable differences in total transpiration of forest and herbaceous ecosystems.

The reasons for the differences are manifold: forest cover generally has a lower albedo than herbaceous cover, much reflected light being absorbed by the undersurface of the leaves in the upper parts of the deeper canopy (Monteith, 1968). This difference in reflectance is very noticeable in aerial photographs of adjacent woodland and agricultural areas; coniferous trees with their needle leaves and dark colour have an albedo which is even lower than that of deciduous cover. The canopy structure will affect aerodynamic roughness (see page 31) and hence the resistance to vapour transfer from the canopy to the external atmosphere. The roughness parameter, and consequently turbulence, generally increase with the height of the vegetation and also with more 'open' canopy structure. In most cases the surface roughness is a function of wind speed, since streamlining and 'surface sealing' may occur in grass and cereal cover while, in other cases, leaf flutter and branch movement at critical wind speeds may induce turbulence.

Water loss will ultimately depend on the radiation and temperature balance of the leaves in the canopy, the saturation deficit of the air in the canopy atmosphere and the physiological status of the leaves. Factors tending to increase leaf temperature and turbulence are specifically variable; for example, Gates (1965) has shown that small leaves are maintained nearer to the ambient temperature by convection than are large leaves. Rutter (1968) suggests that integrated stomatal resistance is also specifically variable and seems to be higher for tree than for herbaceous leaves. Figure 6.13 is an attempt to summarize the concepts of ecosystem water balance under conditions of both free and limited water supply.

EVAPOTRANSPIRATION

The previous sections should make it clear that plant water loss is governed by the physical attributes of the environment in such a way that it should be possible to derive estimates of evaporation and transpira-

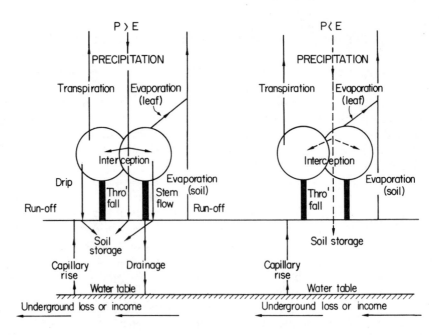

Figure 6.13 Ecosystem water balance. The left-hand diagram shows the path of water through the system when precipitation exceeds transpiration and evaporation. Rain may pass the canopy as throughfall or be intercepted by foliage from which it may either evaporate directly or be channelled into stemflow or leaf-drip and increase the input to the soil surface. Infiltration replenishes the soil storage reservoir or leads to drainage. If the infiltration capacity is exceeded, lateral run-off may occur, reducing the input to some parts of the soil and increasing it to others. Lateral movement below the water table may also be a source of income or loss. Water uptake by plants and its loss, by transpiration, represents the main return pathway to the atmosphere, reinforced by some evaporation from the soil surface. Between periods of rain the soil storage capacity buffers these changes and may be replenished by capillary rise from the water table.

The right-hand diagram shows the inversion of the dominant direction of water movement when evaporation exceeds precipitation: much less water enters the soil and capillary rise may be all important both in supplying the plant with water and in determining pedogenetic processes. Surface enrichment with solutes may occur compared with the strong leaching processes which are encountered when downward drainage of water occurs.

tion from meteorological data. Potential transpiration may be defined as the rate of evaporation from the extended surface of a plant cover which completely shades the soil and is not short of water. It should further be specified that the albedo of the surface should be standard; that no advection of energy water vapour or dry air from adjacent dissimilar areas should occur, and that the ratio of energy used in evaporation and in heating the air should remain constant. Within the limits of this definition it has proved possible to use meteorological measurements to assess crop water loss for irrigation control and hydrological purposes.

The source of energy for evaporation is solar radiation, which varies with season and latitude according to site and with cloudiness, slope and aspect according to local weather and topography (Penman, 1963). The income is independent of the nature of the surface but the first expenditure, in reflection, varies from a few per cent for clean water through 15% or more for forests up to about 25% for most green crops. Superimposed upon this shortwave balance is a longwave exchange between the surface and the atmosphere which may be shown in summated form as a heat budget:

$$H = R_I (1 - r) - R_B$$

where R_I is incoming shortwave radiation (Wm^{-2});

R_B is net outgoing longwave radiation (Wm^{-2});

r is the reflection coefficient.

The value of the heat budget (H) has its main environmental influence in evaporation and sensible heat transfer:

$$H = E + Q$$

where E is evaporation and Q (Wm^{-2}) is sensible heat transfer. E is expressed as Wm^{-2} evaporation equivalent. If Q/E remains more or less constant then it is apparent that evaporation will be almost entirely a function of a composite meteorological term based on radiation balance.

Following this line of thought Thornthwaite (1948; 1954) put forward a simple, empirical function of temperature and day length as an estimator of potential transpiration:

$$E_T = 1 \cdot 6 \, (10 T/I)^a$$

where E_T is the potential transpiration over a 30-day month;

T is the mean air temperature (°C);

I is a heat index derived from the sum of 12 monthly values of i;

a is a cubic function of I;

$i = (T/5)^{1 \cdot 514}$.

This equation has been widely used for irrigation control and in hydrological work, but another much more complex function, due to Penman (1948; 1956), has also found use in these fields. Penman's empirical equation requires more meteorological data, containing terms for wind speed, saturation deficit and energy budget (derived from net radiation flux). It also contains constants correcting for albedo, surface roughness and functions of the saturated vapour pressure curve and the wet and dry bulb psychometric relationship:

$$E_T = \left(\frac{\Delta}{\gamma}H_T + E_{aT}\right)\Big/\left(\frac{\Delta}{\gamma} + 1\right)$$

where Δ is the slope of saturated vapour pressure curve at the mean air temperature; is the psychometric constant;

γ is the psychometric constant;

$H_T = 0 \cdot 75 R_I - R_B$ (Heat budget corrected for reflection);

$E_{aT} = 0 \cdot 35 (1 + u/100) (e_a + e_d)$;

u is the wind speed at 2 m (miles/day);

e_a is the saturated vapour pressure at the mean air temperature (mm Hg);

e_d is the main vapour pressure of the atmosphere (mm Hg);

H_T is expressed in evaporation equivalents (1 mm = 59 cals/cm²/day) as are all other energy terms.

In a recent review Stanhill (1965) examined the concept of potential evapotranspiration and assessed its agricultural usefulness, particularly under arid zone conditions. His conclusion was that the calculated values may be considered reliable only if the areas receiving daily irrigation are very large. Failure to observe this condition introduces errors caused by the oasis effect (advection of heat or dry air). He also commented on the great influence of crop height in modifying the transpirational behaviour of vegetation surfaces. Despite this criticism, the technique has been widely used in irrigation control and, in areas where water costs are a major economic consideration, it is probably one of the least costly methods of optimizing irrigation treatments. As crops appear to differ widely in their growth and transpirational response to irrigation, there is an immediate need for research into these aspects of plant behaviour.

Chapter 7

Waterlogged soils

Contributed by Dr. W. Armstrong

INTRODUCTION

High rainfall, topogenic water accumulation of poor surface drainage may permit the development of waterlogged conditions and the pedogenesis of the hydromorphic soil types discussed in Chapter 4.

Waterlogged soils are of worldwide distribution. They range from subaquatic sediments, estuarine marshes and swamp through to ombrogenous organic peats. Flooding may be frequent, seasonal or permanent but even among non-waterlogged soil types there may be occasional temporary flood periods. Rice, which is the staple diet for so large a proportion of the earth's population, is cultivated chiefly as a wetland crop.

Waterlogging and the ability of plants to cope with such conditions is thus of wide ecological and agricultural importance, and as much of our knowledge in this area is still confined mainly to specialist literature, a full chapter dealing with a single soil condition seems to be justified.

PHYSICOCHEMICAL CHARACTERISTICS OF WET SOILS

This section gives an outline of the more fundamental characteristics of wet soils. A much more detailed account has been given by Ponnamperuma (1972).

Gas exchange

In the absence of rapid temperature and pressure fluctuations gas movement in soils occurs chiefly by diffusion (Grable, 1966). Fick's first law, published in 1885, states that the quantity of substance dn (moles) which passes in time dt (s) through a diffusion cylinder of cross-section A(cm^2), i.e. the rate of diffusion, of flux, is governed by the following expression:

$$\frac{1}{A}\frac{dn}{dt} = -\frac{D(C_2 - C_1)}{x} \tag{7.1}$$

where $C_2 - C_1/x$ is the concentration gradient of diffusate between source and sink (moles cm^{-3}/cm), and D is the proportionality constant and takes the units cm^{-2} s^{-1}. Diffusion is thus analogous to electricity flow as expressed by Ohm's law, viz. $I = V/R$ where I (current) $\equiv$ the diffusive flux, n/At; V (the potential difference or driving force) $\equiv$ the concentration difference between source and sink, and R (resistance) $\equiv x/D$.

Diffusion through well-drained soils occurs rapidly because of the extensive nature of the gas-filled pore space. However, the rate of respiratory gas exchange will depend ultimately upon the thickness of the water films which surround the soil organisms.

In a fully waterlogged soil the situation is somewhat different since the greater part of the pore space is preferentially occupied by water. Gas exchange is therefore entirely limited to diffusion through the soil solution. Since the diffusion coefficient of oxygen in water (0.26×10^{-4} cm^2 s^{-1} at $25°$) is about 10,000 times smaller than the coefficient in air (0.23 cm^2 s^{-1}) then, from equation (7.1) it follows that diffusive movement of oxygen in waterlogged soil will be very slow. Greenwood and Goodman (1967) have calculated a coefficient for wet soil which is about 20,000 times less than that for air. Consequently, in a freshly flooded soil respiring aerobic microorganisms will reduce the oxygen concentration to zero within a few hours (Scott and Evans, 1955).

Once the soil is depleted of oxygen, the diffusion rate from the external atmosphere is insufficient to maintain the supply for aerobic organisms and a new population of anaerobic microorganisms builds up, the already lowered redox potential continues to fall and chemically reducing conditions are established.

In a completely flooded soil, anaerobic conditions will prevail within a few centimetres from the surface (Figure 7.5). With a deeper water table there will be a fringe of anaerobiosis above the water level, the depth of which will be related to capillary rise (Boggie, 1972). The soil will be totally anaerobic below the water table unless it is being laterally flushed with oxygen-containing water (Armstrong and Boatman, 1967).

Anaerobic conditions are not always confined to fully saturated soils or soil horizons. Anaerobiosis may occur within the crumbs of wet but unsaturated crumb-structured soils. The extent to which this will occur can be theoretically predicted for different levels of soil microorganism respiratory oxgen demand (Currie, 1962) or measured experimentally (Greenwood and Goodman, 1967). Figure 7.1 shows Currie's calculated relationships of crumb radius to internal oxygen concentration for three different ratios of

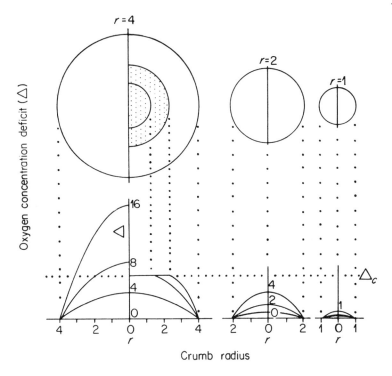

Figure 7.1 Calculated oxygen concentration profiles in respiring soil crumbs of three different radii. To the left of the vertical axis are plotted three profiles for three different ratios of soil oxygen demand to oxygen diffusion coefficient. On the right-hand side are the same profiles modified by the introduction of a critical oxygen deficit (Δ_c) below which respiration becomes anaerobic. The resultant anaerobic zones are stippled. Reproduced with permission from J. A. Currie, *J. Sci. Food Agric.*, **13**, 380–385, Figure 5 (1962).

respiratory oxygen demand to oxygen diffusion coefficient. On the left-hand side of each graph, the radial change of oxygen concentration within the crumb is plotted and, on the right-hand side, the radial change of concentration if a critical oxygen concentration (Δ_c), which inhibits microbial respiration, is introduced into the calculation. In the latter case it may be seen that central anaerobic zones may develop in crumbs above a critical radius. The total deficit Δ between the surface and the centre is proportional to the square of the radius. If a crumb of a particular radius is just wholly aerobic, a crumb of twice the radius will have 30% of its volume anaerobic and this may consequently restrict root exploitation. Similarly, a crumb with 10 times the critical radius would be 84% anaerobic. Greenwood and Goodman (1967) both predicted and experimentally measured a critical radius of about 0.35 cm for water-saturated spheres of a clay loam soil. Figure 7.2 shows the profile of oxygen concentration in such a sphere.

The commonest technique for the assessment of soil oxygen status is based upon polarography and uses a platinum microelectrode in place of the usual dropping mercury electrode. The method was originally described by Lemon and Erickson (1952, 1955) and has recently been

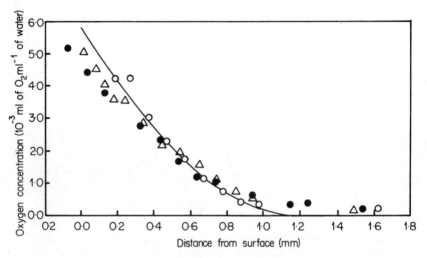

Figure 7.2 Oxygen profiles in a saturated soil sphere. 1st electrode insertion, Δ; 2nd, electrode insertion, ○; 3rd, electrode insertion, ●; Predicted relationship, ——. Reproduced with permission from D. J. Greenwood and D. Goodman, *J. Soil Sci.*, **18**, 190, Figure 2 (1967).

excellently reviewed by McIntyre (1970). Electro-oxidizable or electro-reducible substances have unique current-voltage curves which permit their identification and assay. At pH values above 3.5 Oden (1962) considers that the electrolytic reduction of oxygen at the platinum microelectrode will follow the equation

$$O_2 + 2H_2O + 4e^- = 4OH^-$$

and for each molecule of oxygen reduced there is a current transfer of $4e^-$. Figure 7.3 shows a typical electro-reduction curve for oxygen at a platinum electrode in both air-saturated and oxygen-deficient waterlogged loam. The measuring circuit is illustrated in Figure 7.4. The current reading is taken at a standard time after closing the switch A at a particular voltage setting. With a low applied voltage the current is low; no oxygen reduction is

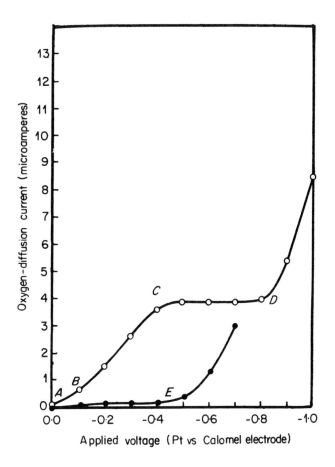

Figure 7.3 Current-voltage curves obtained with platinum versus calomel electrodes used in the polarographic circuit of Figure 7.4. 1. The curve for air-saturated, waterlogged loam. 2. The same soil approaching oxygen deficiency. Adapted, with permission, from W. Armstrong, *J. Soil Sci.*, **18**, 29–31, Figures 1 and 3 (1967).

taking place. In the absence of oxygen this 'residual current' extends out to point E of the current-voltage curve. In the presence of oxygen, at about point B, the curve steepens, rising sharply with voltage between B and C as the rate of oxygen reduction increases. Between C and D the curve forms a plateau, the rate of reaction being independent of applied voltage and governed entirely by the rate at which oxygen can diffuse to the electrode. The steady-state value of this plateau current (minus the residual current if

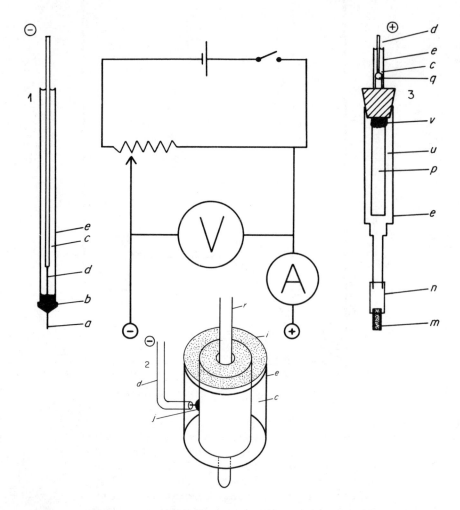

Figure 7.4 Polarographic circuit, with electrodes for assessment of oxygen diffusion rate in soil (1 and 3), and radial oxygen loss from roots (2 and 3). 1. Pt field electrode. 2. Pt cylindrical electrode. 3. Ag/AgCl anode; *a* Pt-wire (18 swg); *b*. Pt/Cu-wire junction jointed with conducting Araldite and set in household Araldite; *c*. epon-epoxy resin; *d*. sleeved Cu-wire; *e*. perspex tube; *f*. Pt-cylinder; *i*. celluloid guide (one at each end); *j*. solder joint; *r*. root; *m*. glass-tube with ceramic or fibre plug; *n*. polythene extension; *p*. Ag/AgCl core; *q*. Ag/AgCl/Cu-wire junction; *u*. saturated KCl solution; *v*. household Araldite.

known) is used to calculate the oxygen diffusion rate (ODR) to the electrode using the equation:

$$i_t = nFAf_{x = |0, t}$$ (7.2)

where i_t = diffusion current (amperes) at a time t after closing the electrode circuit; n = number of electrons required for the reduction of one molecule of oxygen, assumed to be 4; F = the Faraday (96,500 coulombs); A = surface area of the platinum electrode (cm²); $f_{x = 0, t}$ = oxygen flux at zero distance (x from the platinum surface at time t).

Because the dimensions and oxygen affinity of the platinum-wire electrode resemble those of a respiring root it has been suggested that the oxygen diffusion rate (weight/unit area/time) to the electrode simulates the potential respiratory oxygen supply to the root. This idea has been questioned for well-drained (three-phase) soils, but the analogy probably does apply to a two-phase system such as a saturated soil. The technique has been widely used to specify soil oxygenation status and the results are likely to be trustworthy in wet soils. It must be recognized, however, that certain soil factors may cause complications, for example, plateau shifts accompany the lowering of oxygen status (Figure 7.3) or increase in soil acidity (Armstrong, 1967a, Black and West, 1969). For this reason, great care must be taken to ensure that oxygen diffusion determinations are made at a realistic applied potential which may vary from soil to soil or even within a profile. This caution has not always been observed and the literature abounds with data in which the diffusion values may be erroneously inflated.

As oxygen depletion occurs in waterlogged or wet soils so carbon dioxide concentration increases. Recent work by Greenwood (1970) and Greenwood and Nye (unpublished) seems to confirm the general opinion that CO_2 concentrations in aerobic soils never become sufficiently high to inhibit plant growth. Despite the fact that the diffusion coefficient of CO_2 in water is less than that of oxygen, its solubility is more than 30 times greater so that dispersion from sites of production is more rapid than the opposite movement of oxygen. In waterlogged soils, high CO_2 concentrations may be encountered, particularly after addition of fresh organic matter as happens under rice cultivation. However, soil CO_2 levels rarely, if ever, reach lethal values (15–20%), even in rice cultivation conditions. Dilution by methane, produced from decomposing organic matter, is probably responsible for keeping CO_2 concentration down under these circumstances (see also p. 196).

Oxidation–reduction potential

When the oxygen supply is limited, a proportion of soil microorganisms make use of electron acceptors other than oxygen for their respiratory oxidations. This results in the conversion of numerous compounds into a state of chemical reduction and is reflected in lowering of the oxidation–reduction (redox) potential, a physicochemical property of the soil (Pearsall, 1938).

The redox potential (E_h) of a system is a measure of its tendency to receive or supply electrons and is governed by the nature and proportions of the oxidizing and reducing substances which it contains. For example, a common redox couple in soil is the reversible ferrous: ferric system $Fe^{2+} \rightleftharpoons Fe^{3+} + e^-$. In pure solutions, when the ferrous and ferric ions are at equal concentration, this system has an E_h of -771 mV relative to the standard hydrogen electrode of $E_h = 0 \cdot 00$ mV ($\frac{1}{2}H_2 \rightleftharpoons H^+ + e^-$).

The tendency of systems to gain or supply electrons can be observed quantitatively by measuring the potential at an unattackable electrode, e.g. platinum, which when immersed in a system takes on the electrical potential of that system. If the half-cell formed by the immersed platinum is electrically coupled to a standard half-cell, e.g. the hydrogen electrode, also in contact with the redox system, a cell is thus formed. The potential of the redox system is effectively the electromotive force of this cell and can readily be measured as the imposed potential required to prevent the passage of current through the circuit. (This potential will be equal and opposite to the electromotive potential of the system.) In practice, however, the redox potential is generally measured using a Pt standard half-cell electrode pair in conjunction with a high-resistance amplifying millivoltmeter. Most modern pH meters have a millivoltage scale which can be used for this purpose. Because of its very high input impedance an electrometer of this type will directly measure the 'cell' e.m.f. with no significant passage of electricity which could change the potential of the system under investigation. In normal usage the standard hydrogen electrode is replaced by other standard half-cells with their own potentials relative to the hydrogen electrode (e.g. the saturated calomel electrode $= + 250$ mV) the saturated Ag/AgCl electrode $= +220$ mV) and consequently $E_h = E + 250$ (or 220) mV where E is the meter reading.

Redox measurement is slightly affected by temperature (Hill, 1956), but more important is its dependence on pH which complicates comparisons between soils. To overcome this problem, redox potentials have been converted to standard pH values, for example E_6 or E_7, on the assumption that E_h increases by 59 mV for each unit of pH decrease. This may introduce error as the correction value is variable according to the nature of the redox couples involved (see Ponnamperuma et al., 1966). It is more desirable to specify both E_h and pH separately.

Redox measurements are chiefly of value in characterizing negative soil aeration, since, in spite of the chemical heterogeneity of soils, it has been possible, particularly for wet soils, to define the potentials at which a number of important chemical changes in equilibria occur. Nevertheless, it is important to realize that redox potential is a measure of intensity level and not capacity. In this it resembles temperature and pH, and just as temperature and pH give no information as to heat capacity and buffering power respectively, so redox potential is independent of the poising effect, the capacity term in oxidation–reduction potentials (Hewitt, 1948).

Figure 7.6 illustrates, in relation to redox potential, the sequence of events which follows the flooding of a mineral soil within a closed system. The concentration of oxygen declined first and was then accompanied by nitrate reduction. (Oxygen and nitrates are usually undetectable in soils at E_7 values below $+250$ mV.) Reduction of insoluble but easily reducible manganese compounds caused an increase in exchangeable manganese which persisted for the whole of the seven days period. By contrast, the reduction of iron did not begin until the 4–5th day when redox potential (E_7) had fallen to about 150 mV. This delay was probably due to the reserves of reducible manganese present in the soil. The iron concentration in solution then increased very rapidly to 300 μg/ml on the seventh day when redox potential had fallen to -150 mV.

Reactions which appear as a time sequence in the above examples usually occur sequentially with increasing depth in permanently flooded soils and this is reflected in both oxygen and redox potential measurements in Figure 7.5. In this case, however, the redox potential corresponding with absence of oxygen is lower than in Figure 7.6, probably because the lower end of the electrode was in soil at a potential below the oxygen extinction point.

Soil nitrogen

Certain facultative anaerobic microorganisms can use nitrate as an oxygen source in respiration causing denitrification by the liberation of gaseous nitrogen or nitrous oxide (N_2O). Ammonia may also be produced after submergence although, contrary to popular belief, little of this is formed from nitrate but originates from anaerobic breakdown of organic matter. At high pH, high temperature and low cation exchange capacity ammonia may volatilize from waterlogged soils.

Losses by denitrification may be very rapid if an energy source is available. Bremner and Shaw (1958) found that addition of organic matter greatly increased NO_3^- loss and various workers have recorded rates between 15 and 55 μg/g/day from soils to which no additional organic matter was added. In rice cultivation, serious losses from ammonium fertilizers may be caused by rapid nitrification followed by leaching of the nitrate into the underlying anaerobic layers where it is denitrified. Nitrogen usually reaches plants in waterlogged soils as the ammonium ion derived by slow release from decaying organic matter. Under these circumstances it is prone to leaching loss due to displacement by ferrous or manganous ions.

Denitrification is generally associated with oxygen deficient soils. When it does occur in apparently aerobic soils it is likely to be associated with

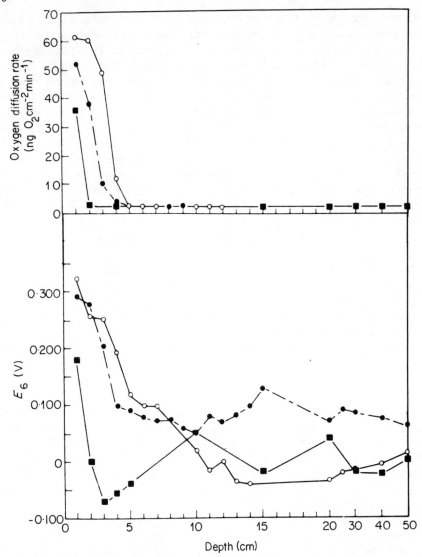

Figure 7.5 Profiles of oxygen diffusion rate and corresponding redox potentials in a valley bog peat with a surface water table. Reproduced with permission from W. Armstrong and D. J. Boatman, *J. Ecol.*, **55**, 103–105, Figures 1 and 2 (1967).

oxygen depletion in the centres of structural aggregates. Greenwood (1962) has shown that nitrification may proceed at half of its aerobic rate in oxygen concentrations as low as 1/80 of air saturation.

Nitrogen fixation by free living microorganisms is important in some

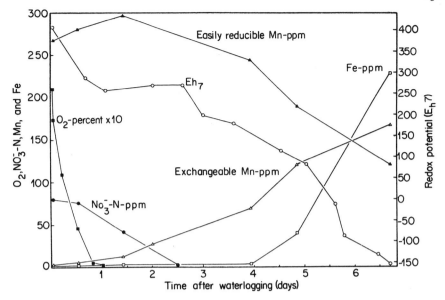

Figure 7.6 Changes in oxygen, nitrate, manganese, iron and redox potential of a silty clay after waterlogging. Reproduced with permission from W. H. Patrick and F. T. Turner, *9th. Int. Congr. Soil Sci. Trans.*, Vol IV, paper 6, Figure 4 (1968).

waterlogged fertile soils such as rice paddy where blue-green algae form an important nitrogen source but McRae and Castro (1967) also showed significant fixation in dark-incubated samples, suggesting that bacteria are equally important in these conditions. Ponnamperuma (1972) suggested that nitrogen transport through the internal atmosphere of the rice plant (p. 206) and loss to the rhizosphere may provide ideal conditions for fixation in a situation where soil resistance to diffusive movement otherwise limits fixation.

Manganese and iron

Because soils contain more iron than manganese, the dominant redox system is usually that of the iron hydroxides (Figure 7.7) rather than the manganese system of oxides and carbonate (Figure 7.8). However, just as manganese reduction does not occur until all free nitrate has disappeared (Takai and Kamura, 1966), so the presence of manganese dioxide or manganic compounds may delay or prevent iron reduction to the ferrous state. Additions of manganese dioxide have been used to buffer rice soils against the development of the extremely reducing conditions injurious to crops. These two Figures indicate the wide range of soil pH and E_h in which aqueous solutions of divalent iron or manganese may occur. More detail of

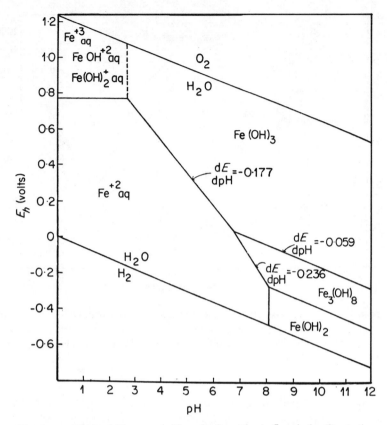

Figure 7.7 The stability areas of iron hydroxides in flooded soils relative to E_h, pH and an aqueous Fe^{2+} activity of one millimole at 25°C. Reproduced with permission from F. N. Ponnamperuma, T. A. Loy and E. M. Tianco, *Soil Sci.*, **103**, 380, Figure 1 (1967).

the behaviour of iron in waterlogged soil is given in Figure 8.2. High concentrations of iron and manganese are toxic to plants and some protective mechanism is required by plants which inhabit waterlogged soils (p. 197 and following). With organic matter addition, paddy soils may become extremely reducing and even rice, which normally has a high iron requirement, may develop symptoms of iron toxicity (Ponnamperuma *et al.*, 1955).

High concentrations of divalent iron and manganese are not confined only to agriculturally manured paddy soils; for example Jones (1967) measured the seasonal fluctuation of exchangeable iron and manganese in wet sand dune slacks, finding very high values during the late spring and early summer months. These high concentrations coincided with the

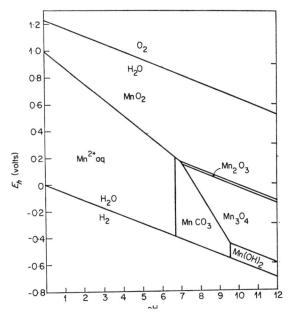

Figure 7.8 The pH/redox boundaries of manganese compounds assumed to be present in manganiferous soils. The E_h of the solutions of flooded soils high in manganese is usually 0.1 to 0.2 volt at pH 6.5 to 7.0, and P_{CO_2} is 0.05 to 0.1 atm. In this region, the stable solid phases are MnO_2 (at E_h higher than 0.17 volt), Mn_2O_3, and $MnCO_3$. If the activity of Mn^{2+}_{aq} and/or P_{CO_2} increases, the $MnCO_3$–Mn^{2+}_{aq} boundary will move left, enlarging the $MnCO_3$ area and excluding Mn_2O_3 and Mn_3O_4. The situation will be reversed if Mn^{2+}_{aq} or P_{CO_2} decreases. For soils low in manganese, in which the less reactive forms of the manganese oxides are involved, the MnO_2 – Mn^{2+}_{aq} boundary will be depressed 0.1 to 0.2 volt, narrowing the Mn^{2+}_{aq} area. Reproduced with permission from F. N. Ponnamperuma, T. A. Loy and W. M. Tianco, *Soil Sci.*, **108**, 55, Figure 1 (1969).

period of maximum growth and were presumably associated with increasing temperature and enhanced microbiological activity (Figure 7.9).

Sulphate: sulphide

As redox potential falls, the next inorganic reduction following the formation of ferrous iron is the reduction of sulphate. This is generally caused

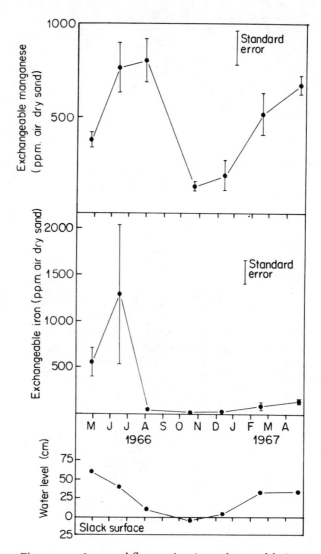

Figure 7.9 Seasonal fluctuation in exchangeable iron and manganese in a dune slack dominated by *Agrostis stolonifera*. Jones (1967).

by a group of obligate, anaerobic bacteria of the genus *Desulphovibrio* which use SO_4^{2-} as an electron acceptor in respiration. These bacteria are largely responsible for the accumulation of sulphide as iron sulphide in waterlogged soils, bogs and marine sediments. Further detail is given in Chapter 9, p. 257.

Sulphate reducers are more exacting in their requirement of anaerobic conditions than most anaerobes and, consequently, require very low ambient redox potentials. They function best near neutrality and have an ecological range from pH 5·5–9·0 (Starkey, 1966). They have a high sulphide tolerance, some species producing up to 2000 μg/ml of hydrogen sulphide in culture solutions. Many bacteria and other microorganisms can produce sulphides from other reduced sulphur compounds but these are normally insignificant compared with the sulphate reducers. Alexander (1961), Postgate (1959), Starkey (1966) and Zobell (1958) have reviewed the physiology and ecology of sulphate reduction.

Various workers have found that sulphate reduction starts at redox potentials between zero and -190 mV (Postgate, 1959; Takai and Kamura, 1966) while Connell and Patrick (1968) demonstrated very strict pH limits for the accumulation of sulphide in soil, reflecting the normal pH tolerance limits of the bacteria concerned. They also found that sulphate became unstable below $E_h = -150$ mV suggesting that sulphide may accumulate abiotically in very reducing soils.

The soluble sulphides (S^{2-}, HS^- and H_2S) are considered highly toxic to plants and other soil organisms. Allam (1971) has reviewed the influence of soluble sulphide on such enzyme systems as catalase, peroxidase, ascorbic acid oxidase, polyphenol oxidase and cytochrome terminal oxidase which affect root oxidative capacity in rice. However, Connell and Patrick (1968) point out that ferric iron is reduced to ferrous at higher potentials than that of sulphate reduction. In iron-containing soils, any sulphide formed by bacteria is likely to be precipitated by the ferrous iron already present. Ponnamperuma (1965) suggests that in normal soils the presence of Fe^{2+} will keep the concentration of hydrogen sulphide below 1×10^{-8} M ($0·0034$ μg/ml) in solution. Allam (1971) also suggests that the clay fraction may reduce soluble sulphide concentration by sorption.

The concentration of dissolved hydrogen sulphide which can exist in equilibrium with a particular concentration of ferrous iron increases as pH decreases. However, at the same time, the solubility of ferrous iron increases with the decrease of pH. Thus, if the soil has a high iron content the sulphide in solution will remain low but if the soil is iron-deficient it is possible for dissolved hydrogen sulphide and low concentrations of ferrous iron to coexist.

Phosphorus

The status of soil phosphorus is discussed in Chapter 8, p. 227, where it is noted that it may occur either as iron phosphate or as occluded phosphate surrounded by a sheath of iron oxide. In both forms the solubility product of the phosphate is very low but iron reduction on waterlogging may result in its release to the soil solution. The transformations of phosphorus in waterlogged soil have been reviewed by Patrick and Mahapatra (1968).

Silica

Increased levels of dissolved silica are characteristic of waterlogged conditions, apparently due to the reduction of iron in ferrisilica complexes. Evidence from rice cultivation suggests that increased availability of silicon in waterlogged soils is beneficial, apparently decreasing the uptake of iron and manganese from solution. Okudu and Takahashi (1964) have suggested that it promotes the oxidizing power of the root (p. 201), probably by improving internal diffusive oxygen transport rather than by any biochemical mechanism.

Organic products of anaerobic metabolism

Among the organic products of anaerobic microbial metabolism are methane, ethane, ethylene, propylene, fatty acids, hydroxy and dicarboxylic acids, unsaturated acids, aldehydes, ketones, alcohols, monoamines, diamines, mercaptans and heterocyclic compounds. Organic materials may, in fact, dominate the redox conditions of many waterlogged soils. For example, the soil solution of many rice soils may contain up to 20 me/l of permanganate oxidizable material of which more than half is organic in nature.

Little is know concerning the effects of organic materials on plant growth beyond Takijima's (1963) suggestion that the harmful consequences might be due either to utilization of oxygen in and around roots or to direct toxicity (see p. 200). It is possible that some organic products of waterlogging may have a beneficial effect by acting as growth promoters (Wang et al., 1967).

Soil pH

Redman and Patrick (1965) studied the pH behaviour of a large number of soils on submergence and noted that those originally above pH 7·4 decreased in value while those below pH 7·4 increased. The reasons for this are discussed in Chapter 4, p. 107.

Specific conductance

With the onset of reducing conditions the specific conductance of the soil solution usually rises. This is partially due to the increase in ferrous iron concentration and also due to the displacement of other cations such as magnesium and calcium from the exchange complex into the solution. Soils initially high in nitrates or sulphates may show a reduction in conductance. In calcareous soils, calcium and magnesium will be mobilized by the increase in carbon dioxide in solution. These cation displacements and CO_2 solvent effects usually cause increases in cation concentration following waterlogging even though they are not directly involved in the soil reduction process. Cations such as potassium may become prone to leaching or lateral drainage loss under these conditions, particularly on agricultural, coarse-textured soils with drains.

HIGHER PLANTS AND THE WETLAND ENVIRONMENT

Many of the characteristics of wet soils which have just been described create problems for plant growth and survival. Soil oxygen deficiency *per se* appears to offer little restraint to wetland species (see later); it may, however, seriously restrict growth in non-wetland plants. Field data published by Williamson (1964) show that for a number of non-wetland crop species optimum growth occurs at a soil ODR of 150 ng/cm^2/min. Between 0 and 30 ng/cm^2/min the growth reduction ranged from 25 to 75% according to species.

Although unable to survive in many situations where wetland roots remain healthy, most non-wetland species probably have some capacity for internal oxygenation (Greenwood, 1967a, b, 1971; Armstrong and Read, 1972; Healy and Armstrong 1972). This transport of oxygen from the atmosphere through the intercellular spaces to the roots may be expected to provide some degree of buffering against soil oxygen deficiency. Troughton (1972) showed that the growth of *Lolium perenne* in culture solution was related to the amount of intercellular air-space in the roots even where the solutions were artifically aerated.

The degree to which internal aeration renders non-wetland roots independent of soil oxygen largely remains to be discovered. Recently, however, Healy and Armstrong (1972) have evaluated the effectiveness of internal transport in Pea. In aseptic, anaerobic 3% agar jelly roots grew to 20 cm or more, but in anaerobic liquid culture periodically gassed with pure nitrogen, growth ceased at 8–9 cm. Mesophyte tissues have low internal air porosities and it was found that oxygen loss by diffusion from the root surface into the liquid medium was sufficiently rapid to deplete the limited internal supply and cause oxygen starvation at a root length of 8–9 cm. By contrast, diffusion into 3% agar is less rapid and in the absence of degassing permits the build-up of an oxygenated zone around the roots which further reduces the rate of oxygen loss from the root surface. A larger proportion of internally diffusing oxygen is thus available for root respiration and accounts for the prolonged root growth. It follows that because of the high oxygen demand in waterlogged soils, oxygen 'drainage' from the root's limited internal supplies should be a more immediate factor in root failure in mesophytes than the lack of soil oxygen. This may be sufficient to prevent many mesophyte species from occupying a wetland niche, although other soil factors can also be shown to play a part.

Tolerance of, or susceptibility to, high iron and manganese concentrations affects the natural distribution of many wetland species and may be responsible for the exclusion of some mesophytic species from waterlogged soils. Martin (1968) found that *Mercurialis perennis*, a woodland plant of well-drained soils, was very sensitive to ferrous iron toxicity in culture solution, compared with *Deschampsia caespitosa* which inhabits much wetter, heavier woodland soils. He also found other woodland species to have a graded tolerance of ferrous iron which corresponds with the wetness of their natural habitat soils (Table 7.1).

Bannister (1964) found that the distribution of the two heath plants *Erica cinerea* and *E. tetralix* was governed by water availability (p. 160) and waterlogging. Fairly short periods of waterlogging were lethal to *E. cinerea* but left *E. tetralix* unharmed. Jones and Etherington (1970) and Jones (1971a, b) studied the problem further and suggested that iron toxicity was

Table 7.1 Tolerance of species to ferrous iron in water cultures. (Reproduced by permission from M. H. Martin, *J. Ecol.*, **56**, 786 (1968))

Species	p.p.m. Fe^{2+} (observations on root system)	
	Survival	Death
Mercurialis perennis	2	4
Endymion non-scriptus	5	10
Brachypodium sylvaticum	–	15[a]
Geum urbanum	10	10–20
Circaea lutetiana	10	15
Primula vulgaris	10–20	20+
P. elatior	20	30
Carex sylvatica	30	30–40
Deschampsia caespitosa	50	80–100

[a] Not tested below 15 p.p.m. Fe^{2+}.

a key factor governing this differential response. On waterlogging pot-cultured plants, *E. cinerea* died quickly following the development of a characteristic waterlogging syndrome which included leaf discoloration and massive leaf water loss. On analysis, the *E. cinerea* plants were found to have taken up more iron than the *E. tetralix* plants which were unharmed by the waterlogging.

A second experiment showed that cut shoots of the two species differed in their sensitivity to iron, supplied in solution. *E. cinerea* was quickly killed by concentrations which did not harm *E. tetralix*. The appearance and water status of the *E. cinerea* shoots before death almost exactly simulated the waterlogging syndrome. In a final experiment *E. cinerea* was grown in two peat soils, one of high and one of low iron content. On waterlogging, the plants in the high iron peat developed the waterlogging syndrome but those in the low iron peat survived, confirming the iron toxicity hypothesis.

The influence of chemical processes in waterlogging may be studied, as above, by contrasting various plant species or by comparing waterlogged and well-drained habitats. The sand dune and dune slack catenary system (Jones and Etherington, 1971) is an ideal habitat for this purpose. Willis *et al.* (1959) had already demonstrated the marked relationship between the duration of winter flooding and the distribution of many dune and slack species. Jones and Etherington extended this type of study by comparing the waterlogging response of a number of dune and slack species. The slack species, *Carex nigra*, recorded by Willis *et al.* as requiring a long period of flooding was slightly stimulated by, or insensitive to, waterlogging compared with the dry dune species, *Festuca rubra*. The slack species, *Agrostis stolonifera*, was only slightly impeded by waterlogging and *Carex flacca*, which inhabits both slacks and dune slopes, responded best to a partial waterlogging regime.

Jones (1972) compared the iron and manganese contents of the plants in this experiment and found that uptake of both metals was greatly enhanced by waterlogging in most cases. Under waterlogged conditions the iron content of roots was much greater than that of shoots. Jones suggested that this might be caused by internal precipitation of iron hydroxides (see p. 201). *C nigra*, the species requiring the wettest slack habitat, showed a smaller increase in root iron, with waterlogging, than the remaining species. In a further series of experiments Jones (1972b) showed that the dune grass. *F. rubra* was harmed by high concentrations of manganese (200 μg/ml) in culture solutions whereas the slack plants *C. nigra* and *A. stolonifera* showed increased or unchanged growth compared with controls containing 2 μg/ml of manganese. Figure 7.9 gives some indication of the natural levels of iron and manganese to which these plants may be exposed.

Iron and manganese toxicity effects have also been documented for agricultural conditions and particularly for rice culture where soils are not only flooded for long periods but organic matter additions result in very low redox potentials. Ferrous iron is, for example, involved in the 'bronzing' disease of rice described by Baba, Inada and Tajima (1964) and Ponnamperuma (1965) has discussed the role of divalent manganese in relation to rice nutrition and toxicity effects. In the case of rice there are fewer instances of proven toxicity for manganese than for iron.

In extremely reducing soil, hydrogen sulphide is also responsible for a number of toxicity effects which have been most widely observed in rice culture. Sulphide or ferrous iron toxicity are suspected of causing 'suffocation disease' in Taiwan (Takahashi, 1960), 'straighthead' in the U.S.A. (Atkins, 1958), 'akagare' and 'akiochi' in Japan (Baba *et al.,* 1963; Park and Tanaka, 1968) and 'brusone' disease in Hungary (Vamos, 1959). Hollis (1967) has reviewed toxicity diseases in rice and also drawn attention to the occurrence of symptomless disease caused by soil reduction but manifested only as yield reduction. Tanaka *et al.* (1968) have suggested that the simultaneous occurrence of dissolved hydrogen sulphide and ferrous iron under low pH conditions may cause 'bronzing' disease as high sulphide concentrations interfere with the oxidizing activity of normal roots and allow them to take up excessive amounts of ferrous iron.

Root rot, often accompanied by blackening, is of fairly common occurrence in wet soils and is suggestive of sulphide damage. The 'die-back' disease of *Spartina townsendii* involves rotting of the rhizome apex and

Goodman and Williams (1961) reproduced the symptoms by addition of sulphide to water cultures. 'Die-back' soils have a very low redox potential and high sulphide content. Armstrong and Boatman (1967) suggested that *Molinia caerulea* was excluded from some valley-bog sites in N. England by the presence of up to 8 μg/ml of sulphide in soil solution. At the same site, *Menyanthes trifoliata,* which has a greater root oxidizing activity (p. 203) was unaffected by 2 μg/ml dissolved sulphide but a proportion of its roots which had entered deeper horizons with 7–8 μg/ml of sulphide were stunted or dead.

Among the organic waterlogging products which have been considered phytotoxic, butyric acid is the most commonly described, particularly with reference to rice (Mitsui *et al.*, 1954; Baba *et al.*, 1964). Wang *et al.* (1967) investigated waterlogging injury to sugar cane and found the monocarboxylic acids to be toxic but hydroxy- and di-carboxylic acids were growth-promoting at low concentrations.

Ethylene has recently been implicated in waterlogging damage as it is generated by soil microorganisms under these circumstances. Smith and Scott-Russell (1969) found that barley root elongation was strongly inhibited by 10 μg/ml and reduced to 50% by 1 μg/ml. Soil solutions from waterlogged fields gave air equilibria of 0·1 to 8 μg/ml but, because of the variable effects of ethylene on plant tissues, according to carbon dioxide and oxygen concentration (Burg and Burg, 1965), Smith and Scott-Russell were cautious in interpreting waterlogging damage as a consequence of ethylene toxicity (see also Smith and Restall, 1971). Waterlogging symptoms such as the leaf epinasty and abscission, observed by Kramer and Jackson (1954) within about four days of flooding, could well have been a response to ethylene generation. He also found that flooding reduced transpiration and caused wilting; similar effects to those observed by Bannister (1964a) and Jones and Etherington (1970). Plants potted in soil were more injured than those in sand, an observation compatible with either ethylene or ferrous iron toxicity.

ADAPTATIONS TO THE WETLAND ENVIRONMENT

Although the wetland environment is unusually hostile to plant life there is nevertheless a vast assemblage of species either endemic to wet sites or tolerant of some degree of soil anaerobiosis. Many plants can produce extensive and healthy root systems in all but the most reducing of soils.

The degree to which tolerance is achieved may depend upon one or more of a number of recognizable features characteristic of endemic wetland species. These will be discussed in the remaining sections and include: (i) the ability to exclude or tolerate soil-borne toxins, (ii) the provision of air-space tissue, (iii) the ability to metabolize anaerobically and tolerate an accumulation of anaerobic metabolites and (iv) the ability to respond successfully to periodic soil flooding.

Exclusion of soil toxins

Extraction of roots from waterlogged soil or observation of such roots through a glass screen almost always reveals red-brown deposits of ferric compounds associated with part of their length (Figure 7.10). Sectioning of these roots may even show oxidized iron deposits on the walls of the cortical intercellular space (Armstrong and Boatman, 1967). The role of iron as a toxin in wet soils has already been mentioned and these deposits indicate that the plant may have the ability to exclude significant quantities of iron from its roots by re-oxidation processes. As early as 1888, Molisch showed root oxidizing activity toward substances such as pyrogallol, and Raciborski (1905) stated that while no phanerogams lack such powers, there are specific differences in effectiveness.

Such differences have since been correlated with tolerance of waterlogged soil, for example Fukui (1953) related the ability of roots to oxidize a-naphthylamine with their ability to penetrate reduced paddy soil. Bartlett (1961) investigated the oxidation of iron compounds adjacent to plant roots in soils and culture solution and showed that high oxidizing capacity conferred the ability to limit iron uptake from a reducing environment. Rice, which is of such economic importance and is associated with very reducing soils, was claimed by Doi (1952) to have the highest root-oxidizing capacity of any known plant. Goto and Tai (1957) found that resistance to rice root rot was correlated with the ability to oxidize the redox dye, aesculin.

The exclusion of soil-borne toxins by oxidation can be related to radial oxygen loss (ROL) from the roots causing direct oxidation of the rhizosphere, to enzymatic oxidation at the root surface, and to microorganism-dependent oxidations adjacent to the root surface.

Radial oxygen loss

Cannon and Free (1925) commenting on the uptake of oxygen from soil by roots suggested: 'It is conceivable that . . . in appropriate conditions movement of oxygen may be in the opposite direction, that is to say, from the plant into . . . the atmosphere of the soil'. Van Raalte (1941, 1944) showed, by redox measurement, that rice roots were able to oxidize anaerobic media. He suggested that oxygen, which he detected in the cortical air space of the roots at concentrations of c. 8%, was transported from the aerial parts and that root oxidation ultimately depended on this supply. He speculated that oxygen was probably, itself, the oxidizing medium diffusing from the roots, but noted that some other organic redox-couple in its oxidized state might diffuse from the root surface. He called this oxidized substance a 'bio-indicator' by analogy with redox indicators.

Van Raalte's work established not only the concept of radial oxygen loss

202

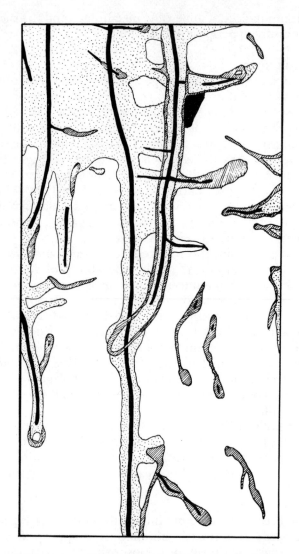

Figure 7.10 Ferric iron deposition around the
roots of *Eriophorum angustifolium* in waterlogged
soil – pH 6·2/6·4. (Reproduced to scale from a
colour transparency.) The diameters of the main
roots are in the range 0·06 – 0·10 cm. Reproduced
with permission from W. Armstrong, *Physiol Pl.*,
20, 924, Figure 1 (1967).
Key: Roots, ■ ; Hydrated ferric oxide light
deposit ⋮⋮; heavy deposit ▨; Soil remaining in a
reduced state □

from roots but also the need for internal diffusion from the aerial parts to replenish the supply. Isotopic techniques have subsequently established that the movement of oxygen in the stem and root intercellular space can be modelled as diffusion in a hollow tube. Evans and Ebert (1960) established this point using ^{15}O in *Vicia faba* seedlings and Barber, Ebert and Evans (1962) found that the build-up of ^{15}O activity in the roots of rice far exceeded that in barley, a finding which correlated well with the larger intercellular air space of rice. These workers suggested that the large volume of internal atmosphere was one factor determining the ability of rice to withstand waterlogging.

Radial oxygen loss from roots was first demonstrated using polarographic techniques (Heide *et al.*, 1963; Armstrong, 1964; Vartapetian, 1964). Armstrong (1964 and 1967b) measured root oxygen losses using a hollow cylindrical platinum electrode insulated on its outer surface (Figure 7.4). The root is threaded through the platinum cathode and the oxygen diffusion current measured using the circuit and technique previously described (Figure 7.4). If the root is in a deoxygenated medium the oxygen diffusing from it to the platinum cathode is the only source of an oxygen diffusion current and the radial oxygen loss from the root may be calculated from equation (7.2) where $A =$ the surface area of root within the electrode, and the oxygen flux is given as ng cm^{-2} root surface, min^{-1}.

Measurements of radial oxygen loss have shown that the lowest values are associated with non-wetland species such as *Mercurialis perennis* (Martin, 1968) and Pea (Healy and Armstrong, 1972). Oxygen flux from the roots of wetland species is usually higher and, corresponding with general oxidizing activity, is highest near the apex. Table 7.2 shows some of the interspecific or intervarietal differences which have been found. In the

Table 7.2 Radial oxygen losses from the roots of some wetland species (ng cm^{-2} min). All values refer to the apical centimetre of root. (Reproduced with permission from W. Armstrong, *Physiol. Pl.*, **20**, 540–53 (1967); **25**, 192–7 (1971))

Menyanthes trifoliata	163
Eriophorum angustifolium	128
Schoenus nigricans	128[a]
Juncus effusus	71
Spartina townsendii (s.l.)	67
Molinia caerulea	14
Rice cv Norin 36	
— waterlogged	150
— non-waterlogged	75
cv Norin 37	
— waterlogged	183
— non-waterlogged	71

[a] Over the $\frac{1}{2}$ cm segment behind the meristem the value rose to 190.

cases examined, increasing oxygen diffusion rate correlates with improved ability to root in wetland soil (Martin, 1968; Armstrong and Boatman, 1967; Boatman and Armstrong, 1968). Teal and Kanwisher (1966) concluded that the dominance of *Spartina alterniflora* in intensely reducing salt-marsh muds is related to internal aeration which is sufficient to supply both root respiration and also the oxygen demand of the reduced mud.

Protection by radial oxygen loss implies the formation of an oxygenated zone in the rhizosphere which forms a buffer between the cells of the root and the hostile soil environment and the pattern of iron deposition adjacent to roots (Figure 7.10) appears to support this view. Microorganism activity may be aerobic in this zone (Read and Armstrong, 1972). Armstrong (1970) has predicted the size of the oxygenation zone for particular species and known soil oxygen consumption using a mathematical model. The model is based on Fick's law of diffusion and relates the diffusion of oxygen from a cylindrical source (root) to consumption by the soil oxygen sink. It defines the thickness of the oxygenated root sheath in terms of two soil parameters, the oxygen consumption rate and the oxygen diffusion coefficient, and two plant parameters, the root radius and the oxygen concentration at the root surface. This concentration is assumed to be the solution equilibrium which corresponds with the oxygen concentration in the cortical gas space, when the root is in a steady state of oxygen loss to a reducing soil.

In Figure 7.11 the thickness of the oxygenated sheath, predicted from the model, is plotted against experimentally obtained value of radial oxygen loss for three British bog plants and two rice varieties. At higher values of ROL a higher root surface oxygen concentration can be maintained and the diameter predicted for the sheath becomes greater. This is important as many reduced inorganic toxins such as ferrous iron, manganous ions and sulphide are only slowly oxidized. Increased thickness of the oxygenated zone increases their residence time as they move in with the transpiration stream, and makes it more likely that they will be made harmless before reaching the root surface.

Figure 7.12 shows the predicted thickness of the oxygenated sheath in relation to root radius and for specified values of soil oxygen consumption. With soils of average oxygen consumption rate, narrow roots, e.g. laterals, may maintain an oxygenated rhizosphere which is many times their own diameter. The model still predicts an oxygenated zone for narrower roots in soils of high oxygen consumption. However, narrow lateral roots may lack the gas space provision necessary to provide oxygen concentrations within the root at a sufficient level to bring about the predicted rhizosphere oxygenation.

Enzymatic oxidations

It seems certain that direct oxidation by diffusing molecular oxygen only accounts in part for observed root oxidizing activity. Yamada and Ota (1958) found that extracts prepared from rice roots actively oxidized iron

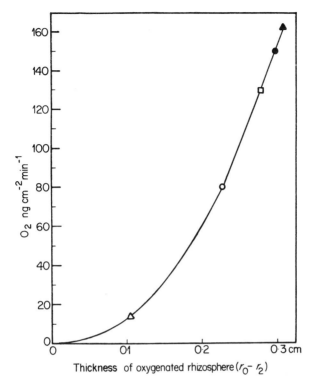

Figure 7.11 The oxygenated rhizosphere dimensions, $r_0 - r_2$, predicted for rice, cv. Yubae (●), cv. Norin 36 (○), *Menyanthes trifoliata* (▲), *Eriophorum angustifolium* (□), and *Molinia caerulea* (△), plotted against the values of ROL previously measured for these species. Reproduced with permission from W. Armstrong, *Physiol. Pl.*, **23**, 623–30 (1970).

in the presence of molecular oxygen. Intact root systems performed similarly. They attributed the activity to a new kind of peroxidase or new iron enzyme closely related to peroxidase, and the ratio of Fe^{2+} oxidized to oxygen consumed indicated that in the reaction $Fe^{2+} = Fe^{3+} + e^-$, $\frac{1}{2} O$ was used as an electron acceptor. Concentrations of $FeSO_4$ which proved to be very toxic to the rice plant retarded oxidation by the extract to the level of a boiled control.

Armstrong (1967c) attempted to compare the contribution made by oxygen *per se* and other oxidants to the oxidizing activity of *Molinia caerulea* and *Menyanthes trifoliata* roots. By comparing reduced dye-oxidation by living and artificial roots with assessments of radial oxygen loss he concluded that enzyme-mediated oxidation might account for up to 90% of the total.

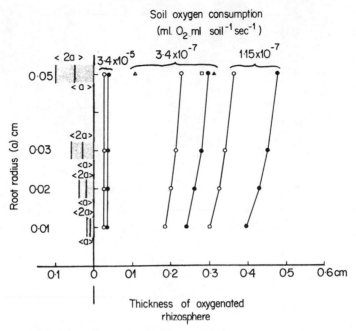

Figure 7.12 The predicted relationship between root diameter and the thickness of the oxygenated rhizosphere sheath for the two rice cultivars Yubae (●) and Norin 36 (○). Sets of curves are plotted for three different levels of soil oxygen consumption. Reproduced with permission from W. Armstrong, *Physiol. Pl.*, **23**, 623–30 (1970).

Microorganism induced oxidations

Pitts (1969) and Hollis (1967) have shown that the bacterium *Beggiatoa* may have a complex ecological relationship with rice and, perhaps, other marsh plants, which gives the plant protection from sulphide toxicity. *Beggiatoa* oxidizes hydrogen sulphide to sulphur, intracellularly. It is also known to need an external supply of catalase to prevent auto-intoxication by the hydrogen peroxide which it produces. Pitts demonstrated the release of catalase from rice roots and also found that *Beggiatoa* isolates, from a paddy soil, grew well in company with rice roots. There is thus a symbiosis in which *Beggiatoa* protects the rice roots from high sulphide concentrations and rice provides the catalase which maintains the bacterial system.

The provision of air-space tissue and its functional significance

Under normal soil water conditions, the oxygen entering the root from the soil will probably satisfy the oxygen demand of the cortex. Whether it will also supply the stele requirement is less clear (Fiscus and Kramer,

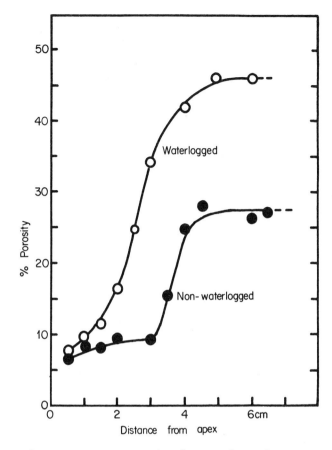

Figure 7.13 Root porosity changes along rice roots. Total porosity changes along waterlogged and non-waterlogged roots. Reproduced with permission from W. Armstrong, *Physiol. Pl.*, **25**, 192–7 (1971).

1970). As soil water content increases, the liquid path to the root will increase and a point reached where the external supply will be insufficient to meet the root respiratory demand. Under fully waterlogged conditions there will be no external oxygen supply at all. In nearly all plants there is, however, the alternative route of gas exchange in the ground tissue intercellular space which connects, ultimately with the atmosphere, through the leaves and stomata. In woody plants the lenticels usually provide the nearest contact with the atmosphere.

In normal mesophytes the porosity of the ground tissue is so low (2–7% by volume) that it is unlikely to be an adequate diffusion path for oxygen transport to the underground parts (see also pp. 203 and 210). In wetland

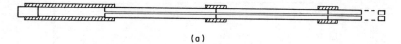

(a)

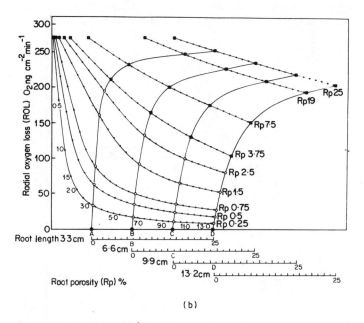

(b)

Figure 7.14 (a) An artificial silicone root. Glass microcap capillaries joined in series form the sub-apical parts of the root; the apex is 2 cm of silicone tube (hatched), sealed by a small glass bung at the free end. The open end of the root remains above the liquid to retain atmospheric connexion; the wall of the silicone apex is highly permeable to oxygen and thus simulates in size and diffusion properties the apex of a non-respiring living root. The length of the root and hence the internal diffusion path length, may be varied by altering the length of glass capillary. Gas-filled porosity is a function of the bore diameter of the capillaries used.

(b) The relationship found between radial oxygen loss (ROL), root length, and % root porosity using silicone roots. ROL was assessed at all times from the apical half-centimetre of root. The figure has been constructed as follows: four experimental ROL vs root length curves at the fixed porosities, 3·75, 7·5, 19 and 25% respectively, have been drawn separated from one another along the plane of the x-axis, in proportion to their respective porosities. (Instead of using individual abscissae scales for length, length has been indicated on each curve by dots. The interval between the two dots represents a 0·5 cm change in root length.) The separation of the curves has allowed the four experimental ROL vs porosity curves at the fixed root lengths 3·3, 6·6, 9·9 and 13·2 cm (curves A, B, C and D) to be drawn on the same figure. Experimental points are indicated by the symbol (■). The species up to 60% of the plant body is pore space and consequently diffusion resistance will be much lower and internal ventilation much improved. The increased porosity is a result of exceptionally loose packing of cortical parenchyma or the occurrence of much larger air spaces or lacunae. These air spaces are formed either by cell separation (schizogeny) during maturation of the organs, or by cell breakdown (lysigeny) (Arber, 1920; Sculthorpe, 1967). Cell collapse frequently accompanies schizogeny (e.g. in grasses and sedges).

Air-space formation is not confined to primary tissue and may be conspicuous in secondary tissues arising from a phellogen or from a normal cambial layer. Schrenk (1889) in fact originally proposed the use of the term aerenchyma to describe air-space tissue arising from a phellogen. It has subsequently become customary to describe all air-space tissues as aerenchyma.

Although more porous than normal parenchyma, lacunate tissues often contain sites of much lower porosity. For example, in a honeycomb type of structure the gaseous exchange between individual chambers is confined to the normal intercellular spaces of the bounding walls. Similarly, the gas-filled cavities in many hydrophyte stems are interrupted, at the nodes, by watertight cellular diaphragms of comparatively low porosity. Tissue of restricted porosity also occurs in the compacted zone at the root and stem cortical junction. These zones must act to some extent as rate-limiting barriers in gaseous transfer.

Plants vary in their ability to form aerenchyma depending on species and degree of soil anaerobiosis (Arikado and Aduchi, 1955; Arikado, 1959; Martin, 1968; Armstrong, 1971a, b). Figure 7.13 illustrates this point. Plants which cannot form aerenchyma appear to be intolerant of wet habitats, give low ROL values and have poor ability to oxidize inorganic toxins in the rhizosphere.

Williams and Barber (1961) reassessed the role of aerenchyma and concluded that its function was not primarily one of oxygen transport, or one

---

Figure 7.14—continued

spatial arrangement of the experimental curves has enabled a further series of ROL vs root length curves to be derived. Apart from the ROL at zero length, the points on these curves (°) were derived by extrapolating from the porosity values 0·25, 0·5, 0·75, 1·5 and 2·5% on each of the abscissae A, B, C and D to meet the four experimental ROL vs porosity curves A, B, C and D. NB: The ROL at zero length is the same for all ROL vs root length curves being the ROL from the 2 cm-apical Si portion only, which is common to all roots. This value has for the sake of convenience been taken as the ROL at zero internal resistance. Ordinate: O_2 ng per cm^2 root surface and min. Reproduced with permission from W. Armstrong, *Physiol. Pl.*, **27**, 176, Figure 4 (1972).

of storage, but that it fulfilled the necessary requirements for a mechanical-cum-metabolic compromise in the wetland plant body; a compromise involving a considerable reduction of respiratory oxygen demand while at the same time retaining adequate mechanical strength. The formation of lacunae they regarded as necessary primarily to reduce the respiratory demand of the plant body. The size of the lacunae, was unnecessarily large for an oxygen pathway. The honeycomb structure of aerenchyma provides maximum strength with the deployment of the minimum of tissue, consequently reducing the ratio of respiratory oxygen uptake to volume; obviously of survival value in an oxygen-deficient environment.

Williams and Barber's interpretation was not derived from experimental evidence and no account was taken of the oxygen transport requirement to maintain an oxygenated rhizosphere. Evidence is now available which relates radial oxygen loss to root porosity and length. Figure 7.14 shows such data, obtained with 'artificial roots' of silicone rubber tubing in conjunction with the cylindrical platinum polarographic electrode. At a root radius of 0·05 cm, oxygen loss continues to be increased by porosity increments up to and even beyond 25%. It can be calculated that smaller roots will be even more sensitive to porosity and, at a radius of 0·025 cm, the shape of the ROL vs porosity curves will remain as shown but will lie over the range 0–100%.

These results suggest that, in terms of rhizosphere oxygenation alone, the provision of aerenchyma to a maximum porosity of 60% may not be excessive. It has also been predicted from the above data that, because of their thinness and low frequency, cellular partitions within aerenchyma will be unlikely to impose substantial restraints on internal gas transport.

In the living plant the assessment of the gas space requirements for oxygen transport is further complicated by such variables as respiratory activity and root wall permeability. Model building provides the best approach to this multivariable system and Luxmore et al. (1970, 1972) have produced a computer model of the soil–plant oxygen diffusion system. More simply, an electrical analogue may be built in which resistors simulate root porosity, root wall permeability and diffusion path length. Voltmeters and milliammeters indicate, respectively, oxygen concentration and oxygen flux while a variable resistance leak to 'earth' simulates respiratory oxygen consumption (Figure 7.15).

A bank of 10 analogue units is used in this model to simulate the oxygen diffusion activity of roots 10 cm long. The units may be programmed to predict the oxygen relations of either aerenchymatous or non-aerenchymatous species in conditions of soil oxygen deficiency (Figure 7.16).

Alternatively, mixtures of characteristics may be introduced. Figure 7.17 clearly shows that root porosity is the overriding factor governing internal oxygen concentration and radial oxygen loss. Waterlogged rice, for example, has an apical oxygen concentration of 12·3% in the model. Changing the porosity structure to that of non-waterlogged rice reduces this concen-

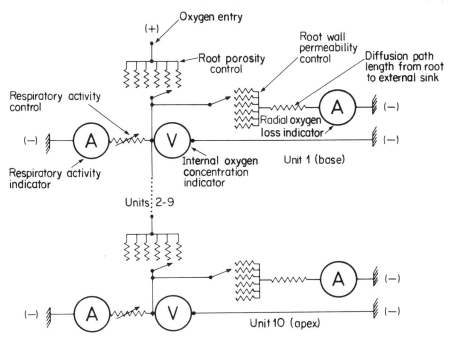

Figure 7.15 Electrical analogue model of the root oxygen diffusion system. The model is based on ten analogue units each of which represents a one cm segment of the root. Each unit may be programmed to representative values of simulated porosity, wall permeability, respiratory activity and diffusion path length to external oxygen sink. The equilibrium oxygen concentrations, respiratory oxygen consumption and radial oxygen diffusion loss may then be read from the three meters of each unit. It should be noted that the internal oxygen concentration indicator (meter V) has a very high impedance compared with the meter A. The current drawn by meters V has negligible effect on the system. Armstrong and Wright (unpublished).

tration to 7·7% and with a maize structure it falls to 2·6%. These modelled data correspond well with recent measurements on waterlogged and non-waterlogged rice (Armstrong, 1971b).

The foregoing discussion suggests that the oxygen transport requirement is probably the overriding factor governing the ecological need for aerenchyma development but that the structure also accords with the requirements of a mechanical-metabolic compromise. What little evidence there is does not seem to favour a reservoir hypothesis and it has been found recently (Gaynard and Armstrong, unpublished work) that the internal oxygen supply in the aerenchymatous bog species *E. angustifolium* is almost exhausted after one hour's submergence in the dark.

Although the functional significance of aerenchyma is now better understood, many questions are unanswered concerning its origin. It seems

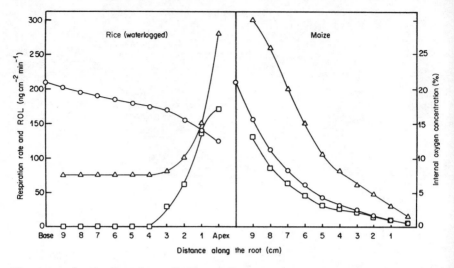

Figure 7.16 Analogue predictions of ROL, oxygen concentration, and respiratory activity profiles for roots of waterlogged rice and maize. (A) Internal oxygen concentration. (B) Respiratory activity. (C) Radial oxygen loss. Both B and C are expressed as oxygen used or lost per unit area of root surface. Armstrong and Wright (unpublished).

likely that low oxygen concentration is the factor which, either directly or indirectly, triggers air space formation, but the site of triggering is not known. It is possible that high respiratory rate near the apex of the root causes low oxygen concentrations which initiate air space development. In rice there is evidence that planes of middle lamella discontinuity arise at an early stage of root growth. Lack of cell adherence along these planes could lead to tearing, possible cell collapse and lacuna formation (Boeke, 1940). Similarly van der Heide *et al.* (1963), using barley, have found an inhibition of polysaccharide formation from sugars at low oxygen tensions.

Anaerobic metabolism

Occasions may frequently arise when tissues in the underground organs of wetland species will be subject to anoxia. For example, if the root systems develop in an aerated interflood period they may contain little aerenchyma and therefore will easily suffer oxygen stress during a subsequent flood period. The seasonal dying-back of the leaves of rhizomatous species provides a further example, for in high water table conditions this can lead to prolonged periods of oxygen stress.

Anaerobic respiration will naturally predominate under these circumstances and in some species ethanol accumulates to quite a high concentration, higher perhaps than the levels reached in non-wetland plants (Hook *et al.*, 1971). In other cases lactic acid may be a principal endproduct of anaerobic respiration (Chirkova, 1968; Hook *et al.*, 1971; Brown *et al.*, 1969). However, wetland species also exhibit forms of

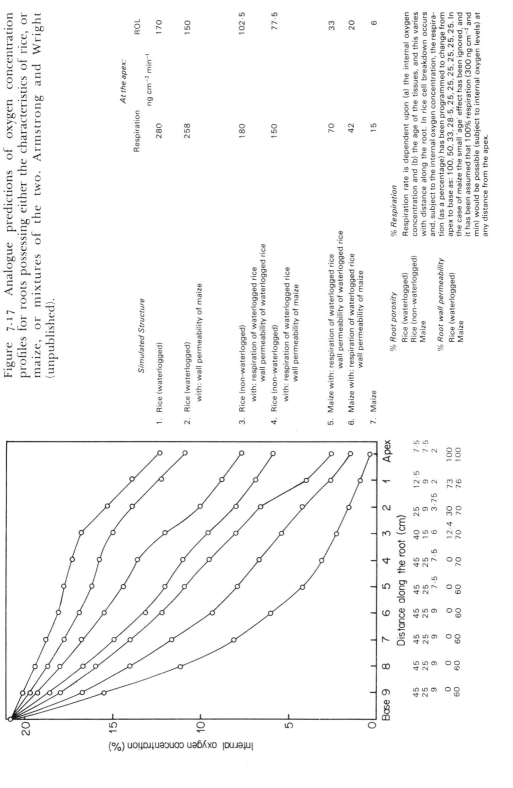

Figure 7.17 Analogue predictions of oxygen concentration profiles for roots possessing either the characteristics of rice, or maize, or mixtures of the two. Armstrong and Wright (unpublished).

At the apex:

Simulated Structure	Respiration	ROL
	ng cm^{-2} min^{-1}	
1. Rice (waterlogged)	280	170
2. Rice (waterlogged) with: wall permeability of maize	258	150
3. Rice (non-waterlogged) with: respiration of waterlogged rice wall permeability of waterlogged rice	180	102·5
4. Rice (non-waterlogged) with: respiration of waterlogged rice wall permeability of maize	150	77·5
5. Maize with: respiration of waterlogged rice wall permeability of waterlogged rice	70	33
6. Maize with: respiration of waterlogged rice wall permeability of maize	42	20
7. Maize	15	6

% Respiration

Respiration rate is dependent upon (a) the internal oxygen concentration and (b) the age of the tissues, and this varies with distance along the root. In rice cell breakdown occurs and, subject to the internal oxygen concentration, the respiration (as a percentage) has been programmed to change from apex to base as: 100, 50, 33, 28·5, 25, 25, 25, 25, 25. In the case of maize the small 'age' effect has been ignored, and it has been assumed that 100% respiration (300 ng cm^{-2} and min) would be possible (subject to internal oxygen levels) at any distance from the apex.

% Root porosity
Rice (waterlogged)
Rice (non-waterlogged)
Maize

% Root wall permeability
Rice (waterlogged)
Maize

anaerobic metabolism in which the accumulated end-products are much less toxic than ethanol, and this may better enable them to withstand prolonged periods of oxygen stress without injury. An example is the formation of shikimic acid in the rhizomatous species *Iris pseudacorus* and *Nuphar lutea* (Boulter *et al.*, 1963; Tyler and Crawford, 1970). This acid is relatively non-toxic and reaches high concentrations during the winter when loss of the aerial parts, coupled with raised water tables, severs gaseous connection with the atmosphere.

Many species, both wetland and non-wetland, are capable of the anaerobic dark fixation of carbon dioxide by carboxylation of phosphoenolpyruvate to oxaloacetate. In the presence of the malic dehydrogenase enzyme the oxaloacetate may subsequently be converted into malic acid. Mazelis and Vennesland (1957) were of the opinion that malic acid should be considered as a principal end-product of anaerobic respiration in many, if not all, plant tissues; of greater importance perhaps than ethanol, lactate and carbon dioxide. However, Crawford and coworkers (1966, 1968, 1969) have recently suggested that malic acid accumulation may be a characteristic feature of wetland species only. They consider that in non-wetland species accumulation will be prevented by the presence of 'malic enzyme' which will recycle excess malic acid to pyruvate and thence to ethanol and carbon dioxide. McManmon and Crawford (1971) have outlined these ideas as follows, and the pathways are summarized in Figure 7.18.

When intolerant plants are flooded their respiration is blocked and glycolysis proceeds to the formation of acetaldehyde and then ethanol. Acetaldehyde induces alcohol dehydrogenase activity which, together with a reduction in the apparent Michaelis constant of the enzyme–substrate reaction, accelerates glycolysis. Carboxylation of phosphoenolpyruvate can form malate via oxaloacetate but the malate is rapidly decarboxylated by 'malic' enzyme to form pyruvate and further contribute to acetaldehyde and ethanol accumulation which ultimately reaches toxic levels.

The flooding of tolerant plants partially blocks respiration to cause acetaldehyde and ethanol accumulation but the former fails to induce alcohol dehydrogenase, the Michaelis constant of the reaction is not changed and glycolysis does not accelerate. The malate which is formed via carboxylation of phosphoenolpyruvate cannot be decarboxylated again as these plants lack the 'malic' enzyme and consequently malate accumulates. It is non-toxic and remains in the plant until aerobic conditions are restored.

It is very much to be expected that these forms of anaerobic metabolism will be beneficial to submerged (non-absorbing) overwintering roots and rhizomes. Whether they can enable morphologically ill-adapted roots (non-submerged, non-aerenchymatous) to survive the rigours of a reduced soil in which phytotoxins abound remains to be shown. Active growth, however, seems always to depend upon a readily available source of oxygen.

(a)

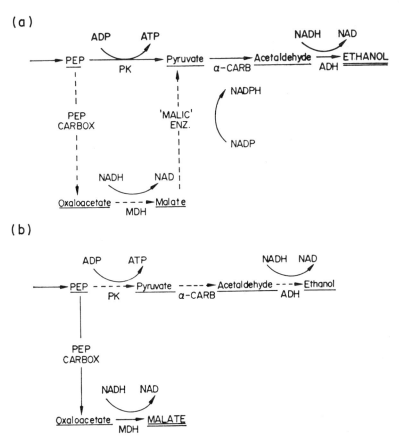

(b)

Figure 7.18 Metabolic schemes of events following the flooding of (a) flood intolerant and (b) flood tolerant species. PEP phosphoenol pyruvate; PK pyruvate kinase; α-CARB α-carboxylase; ADH alcohol dehydrogenase; PEP CARBOX phosphoenol pyruvate carboxylase; MDH malate dehydrogenase. Reproduced with permission from M. MacManmon and R. M. M. Crawford, *New Phytol.*, **70**, 299–306 (1971).

Other adaptations to waterlogging

A feature common to many wetland plants is the xeromorphic nature of their leaves and stems, despite the fact that they may be rooted in flooded soil. Good examples may be found in the genus *Juncus* and in wetland species of the Ericaceae such as *E. tetralix*. It seems likely that the primary function of xeromorphism in these plants is not the usual one of water conservation *per se* but is rather to reduce the velocity of water movement to the root surface. In this way, the time available for the oxidation of phytotoxins in the rhizosphere zone of oxygenation will be increased. Jones (1971a) showed that the use of antitranspirants on *Erica cinerea* could improve survival of waterlogging and also limit iron uptake. She suggested that the more xeromorphic nature of *E. tetralix* probably improves its ability to tolerate high levels of ferrous iron in the soil solution.

A change in root orientation in poorly aerated or anaerobic soil horizons may be frequently observed and the dwarfed plants of *Molinia caerulea* found in the surface water channels of valley bogs provide excellent examples of this obviously beneficial phenomenon (Armstrong and Boatman, 1967). In these places the adventitious roots become orientated into a horizontal growth pattern along a tolerable redox plane. In this way they avoid the deeper, more intensely reducing and hostile horizons. Under these circumstances the lateral roots of this species invariably grow upwards towards more oxidizing conditions.

The development of the 'shaving brush' effect in root systems exposed to a fluctuating water table appears to be an outward sign of yet another survival mechanism. The root apices are killed in the rising water table but the root bases remain healthy and new laterals rapidly grow from these and adjust to the new level. If the water table oscillates frequently, the cycles of death and regrowth produce a mass of brush-like roots. The roots are obviously harassed by these conditions but the mechanism which confines death to the apices, and allows rapid basal regrowth, must assist the plant to tolerate these marginal conditions. The 'shaving-brush' effect is common in conifers and other species not fully tolerant to the extremes of soil anaerobiosis, e.g. *Molinia caerulea*.

Various other biochemical effects of waterlogging have been noted, for example Mizrahi *et al.* (1972) found an increased abscisic acid content in the leaves of *Nicotiana rustica*, after salination, which was also associated with improved resistance to poor aeration of the rooting medium. They concluded that this might be due to an abscisic acid-induced lowering of the transpiration rate. Current research implicates ethylene in many aspects of plant response to stress, such as disturbed water balance, and attributes to it a causal role in promoting leaf abscission. Ethylene production from waterlogged soils or within waterlogging stressed plants may be associated with the development of the normal waterlogging syndrome.

WETLAND TOLERANCE IN WOODY SPECIES

Few tree species or other woody plants adapt well for growth in permanently flooded soil. Dormant trees may survive weeks of winter flooding, but a single day of flooding during the growing season may be very harmful (Kramer, 1969). The response of most trees in wet soils is probably similar to that noted for Lodge-pole pine (*P. contorta*) in some recent field experiments (Boggie, 1972). Young trees remained alive for little more than one season when the water table was permanently maintained at the soil surface. Survival increased enormously with a slight lowering of the water table while growth continued to improve with progressively lower water levels.

Because of their scarcity those species which are successful in the wetland habitat are relatively well known. They include the mangroves such as *Rhizophora, Avicennia* and *Sonneratia*, the swamp cypress (*Taxodium distichum*), as well as species of *Tupelo, Salix, Alnus* and *Myrica*.

A study of the swamp cypress and the tupelo gum (*Nyssa aquatica*) (Dickson and Broyer, 1972) shows that they grow best in waterlogged conditions, as do many wetland herbaceous species. Again, as with herbaceous species the provision of an adequate ventilating system appears to be essential for the successful wetland growth of woody species. Aeration in trees has recently been reviewed by Hook *et al.* (1972).

Special biochemical pathways may be available to cope with periods of anoxia brought about by any sudden raising of water tables, an example being the production of glycerol in *Alnus* (Crawford, 1971).

Perhaps the best known study of the ventilation of woody species is that of Scholander *et al.* (1955) who examined the gas exchange of *Rhizophora mangle* and *Avicennia nitida* in the tidal swamps of Florida. *Rhizophora* perches on arched stilt roots, which are richly provided with lenticels above ground and in the mud terminate in bunches of long, spongy, air-filled roots about finger-thick. The stilt roots are above the tide level so that submerged roots remain in gaseous contact with the atmosphere through the intercellular space system and the lenticels. The oxygen concentration in roots submerged in the reducing mud remained continuously high, 15–18%, but if the lenticels were blocked this concentration fell to 2%, or less, in two days. *A. nitida* (the black mangrove) is a bush or tree up to 20 m tall which produces prolific numbers of air roots (pneumatophores) protruding from the mud. A single tree may produce several thousand of these which are usually 20–30 cm high, one centimetre thick, soft and spongy, and studded with numerous white lenticels. In the mud they arise from radially-running main roots, which are also soft and spongy, and contain large amounts of air. Root aeration is influenced by tidal rhythm as the air roots and their lenticels are covered by each tide. Following submergence the root oxygen concentration falls, to rise again as the tide retreats. The authors noted that there was a progressive pressure drop in the internal gas space while the air-roots were covered and that air would be drawn in when the lenticels were again exposed. As the internal oxygen diffusion mechanism is so efficient in highly porous organs, it may be that this pressure change is not a necessary mechanism.

A number of other woody species produce structures which behave like air-roots, for example Bond (1952) observed that the nitrogen-fixing nodulated roots of *Myrica gale* grew upwards and he suggested these might improve the oxygen supply to the nodule tissue. The well known 'knees' of the swamp cypress, produced by upward proliferation of root xylem, have long been thought to improve gas exchange though Kramer *et al.* (1952) have cast some doubt on this.

Oxidation of reduced dye solutions has been used to show root-oxidizing activity in the woody species *Betula pubescens* (Huikari, 1954), *Salix atrocinerea* (Leyton and Rousseau, 1957) and *Nyssa sylvatica, N. aquatica* and *Fraxinus pennsylvanica* (Hook et al., 1972). Hook was unable to detect root-oxidizing activity in the non-swamp species *Liquidambar styraciflua, Liriodendron tulipifera* and *Platanus occidentalis.* Hook et al. (1971) concluded that tolerance of flooding in swamp tupelo is achieved by the combined adaptations of rhizosphere oxidation (ultimately dependent upon oxygen entering through the lenticels), anaerobic respiration at low oxygen concentrations and the tolerance to high carbon dioxide concentrations of the new roots produced after flooding.

Armstrong (1968) has demonstrated internal oxygen transport and radial oxygen loss with rooted cuttings of *Salix* species and the heath shrub *Myrica*. In these cases oxygen entered the stems through the lenticels, and if these were progressively blocked, from the base of the stem upwards, radial oxygen loss gradually declined. Radial oxygen loss ceased in *S. atrocinerea* and *S. fragilis* when only 3 cm of stem had been treated in this way.

If the lower stem lenticels form the main point of oxygen entry, sudden floods may severely limit the root oxygen supply, thus causing death, but in those species prone to flooding the rise in the water table is usually accompanied by a rapid proliferation of adventitious roots. These new roots arise from hypertrophied lenticels just below the new water surface.

The path of oxygen movement through woody species is not yet established with certainty. In young shoots and seedlings it is almost certainly the primary cortex while in a few plants, such as the legume *Aeschymomene aspera*, xylem aerenchyma is produced. The difficulty is with the remainder which in the adult condition have no cortical parenchyma and the most obvious route for gas transfer is through the empty conducting elements of the stele.

Although McVean (1965) concluded that a considerable quantity of the oxygen in Alder roots was present in the xylem the cambium has long been considered to act as a barrier to gas exchange with the xylem. However, there is now evidence that in wetland trees the cambium may be adequately pervious to oxygen. Hook and Brown (1972) have reported the presence of intercellular spaces and free gas exchange through the cambia of both swamp tupelo and green ash. In sycamore, tulip poplar and sweet gum, species which can withstand only short periods of inundation, cambial intercellular spaces were found to be either absent or so small that their continuity could not be established. Gas could only be drawn through these cambia under a tension of between 30 and 80 mm Hg.

Acknowledgements

I would like to thank Mr. R. Wheeler-Osman for drawing the illustrations, Miss E. M. Sharpe for typing the manuscript and my wife, Jean, for reading and criticizing this chapter during its preparation.

Chapter 8

Mineral nutrition

Plant physiologists have collected an enormous amount of data concerning the uptake and influence of the various mineral nutrient elements but, because of the complexity of the soil–plant system, the majority of these studies have been made in solution culture. Many investigations of nutrient uptake have, indeed, been made with excised roots. Despite these artificial approaches, much of the accumulated information has had implications for the ecologist, as well as defining the physiological problems of nutritional relationships. Of greater interest for the ecologist are the early field experiments which established, for many crop plants, the gross effects of nutrient elements on plant growth and physiological function. Work of this type was originally stimulated by Liebig's ideas concerning the mineral nutrition of plants and is directly responsible for the development of the modern artificial fertilizer industry. Russel (1961) provides a useful historical account of the agricultural development of plant nutritional studies.

The major advances during the second half of the nineteenth century were the realizations that plants require phosphates and salts of the alkali metals, that non-leguminous plants have a high requirement for nitrogen compounds and that artificial fertilizers incorporating these materials could maintain crop yield, for many years, in the absence of organic additions. Further developments, with even stronger ecological implications, were the discoveries that different crops had different inorganic nutrient requirements and that nitrogen might be 'fixed' from the atmosphere by the organisms in leguminous plant root nodules and by free living organisms in some soils.

The ecologist's interests in plant nutrition are threefold, comprising: (i) behavioural studies of the soil–plant system in the context of naturally occurring nutrient ranges; (ii) investigation of differential responses between and within species; (iii) investigation of competitive relationships between and within species (Clapham, 1969). Most early nutritional physiology has not helped in this sphere as crop plants or convenient 'laboratory plants' were used. However, the techniques of whole-plant physiology and the methods of soil analysis, originally developed with agriculture in mind, have provided an extensive background of experimental method with which to tackle these new questions.

Bradshaw (1969) stresses the importance of nutrition as an ecological factor by noting that there are virtually no British grasslands which do not respond to the addition of one or more of the plant macronutrient elements. This observation may be extended to the majority of the soils and vegetation types of the world; if water is not a limiting factor, nutrient additions produce a response, either in yield or in species composition. In each case this suggests that nutrient deficiency is a limiting factor in either photosynthetic productivity or competitive ability.

Recent ecological investigations suggest that variation in soil nutrient levels, at all scales, influences species composition and growth of vegetation. At the smallest scale, vertical variation in the soil profile and horizontal variations, within a few centimetres, occur, and are of great theoretical importance in plant competition and the maintenance of species diversity by niche differentiation (see Chapter 10). On a larger scale the differing plant associations of various soil types partially owe their existence to differences in nutrient status.

Responses to local or regional nutrient variation may span the range from deficiency to toxicity and result in differential behaviour of plant species or ecotypes. The consequent relationships of nutritional physiology and competitive efficiency, cause complex interactions between cohabiting species.

NUTRIENT REQUIREMENT

Plants require sixteen essential elements and the absence of any one of these will cause failure. Three of the elements, C, H and O, which are the constituents of the structural and primary energy storage compounds, are not usually described as nutrients. The remaining thirteen elements are subdivided into the macronutrients, N, P, K, S, Ca and Mg which are required in comparatively large amounts, and the micronutrients, Cu, Zn, B, Cl, Mo, Mn and Fe which are required in smaller amounts.

All of the nutrient elements, with the exception of N, are usually derived from the weathering products of the soil minerals, though it should be noted that rainfall may contain significant amounts of nutrients, particularly S and Cl and, under extreme conditions, may be the sole source of plant nutrients in ombrogenous soils. Nitrogen differs in its ubiquitous presence as atmospheric N_2, from which soil-dwelling, or root nodule bacteria and other microorganisms, may fix it as organic N.

MINERAL NUTRIENTS IN THE SOIL

Nitrogen

Most natural soils contain between 0·01 and 0·25% N in the surface layers (Bear, 1964) but the content is less in the deeper horizons. Some peat soils contain larger amounts: up to 2–3%. It is usually assumed that most of the soil-N is organic though recent work has shown that ammonium-N may become bound between the lattice layers of expanding clays such as montmorillonite (Bremner, 1967). In some subsoils up to 50% of the nitrogen may be held in this way.

It is safe to generalize, for surface soils, that most of the N is organic and, on hydrolysis, up to half of it may be recovered as amino-N. This suggests that the soil-N exists as bound protein or bound amino acids, the binding sites being on the clay fraction. Alternatively, the amino units may be incorporated in the humic complex molecules; earlier workers often described the humic complex as ligno-protein (Quastel, 1963). Attempts to isolate proteins from soil organic matter have, however, always failed, except for showing the ratios of hydrolytically produced amino acids, to be compatible with the constitution of an 'average' protein (Fellbeck, 1971).

The remainder of the organic-N is of varied composition, comprising amino sugars, purines and pyrimidines at very low levels, traces of amino acids and numerous other N-compounds. As all of these constituents occur at low concentrations, there remains considerable doubt concerning the composition of at least half of the soil organic-N which does not hydrolyse to amino-N.

Nitrogen is absorbed by roots as nitrate in aerobic soils and, consequently the nitrogen-supplying status of the soil depends more on the rate of mineralization of organic-N than on the total N content. Most of the nitrate-N is in soil solution, very little being adsorbed by any form of anion-binding (Fried and Broeshart, 1967). Biological mineralization follows the pathway: organic-N→ammonium→nitrite→nitrate; producing soil solution concentrations of nitrate ranging from 5–50 me/l in natural soils. The relative concentrations of ammonium, nitrite and nitrate-N in the soil are variable in relation to the populations of the various microorganisms involved in the transformations. Ammonia may be liberated by a whole range of soil microorganisms and its release to the soil depends on the nature of the organic material which is being decomposed. If it is rich in non-nitrogenous material such as carbohydrate, most of its nitrogen content will be re-utilized in the synthesis of bacterial or fungal matter, but if proteins are being metabolized considerable amounts of surplus nitrogen may be released to the soil in the ammonium form (Quastel, 1963).

Nitrification and denitrification

The oxidation of ammonium cations to nitrite is caused by bacteria such as *Nitrosomonas* spp and the oxidation of nitrite to nitrate by *Nitrobacter* spp. Experiments using perfusion techniques have shown that this bacterial nitrification is a relatively slow process (Quastel, 1963), the rate of which is related to the degree of adsorption of the ammonium cations on the exchange complex. The nitrifying bacteria appear to inhabit the surfaces of soil aggregates and to nitrify largely at the expense of exchangeable ammonium ions. The population of nitrifying bacteria in the soil solution is small and, for this reason, the maximum rates of nitrification are associated with full bacterial occupancy of the aggregate surfaces and a high level of ammonium saturation of the exchange complex.

Nitrite is rapidly biologically oxidized to nitrate, in aerobic soils, so that the overall factors likely to limit nutrient-N supply to plant roots will be the populations and metabolic rates of the ammonia-forming organisms, the nitrogen content of the organic matter, exchangeable ammonia supply and, finally, other factors influencing the metabolism of the nitrifiers. As might be expected, the levels of nitrate in the soil solution show great fluctuation in response to temperature, water supply and root activity. The rates of nitrification also vary from soil to soil in response to pH and nutritional differences as *Nitrosomonas* has a pH optimum of 8·5 and fails to nitrify below pH 4·0. As noted in Chapter 4 (Figure 4.6), nitrates are generally absent from acid soils and, because the nitrifying bacteria requires an oxygen supply, are also absent from waterlogged soils. Under these circumstances it is likely that ammonium nitrogen is absorbed by roots.

The dynamic aspects of soil nitrogen balance are further complicated by bacterial *denitrification*. Mulder *et al.* (1969) note that many bacteria are capable of reducing nitrate to nitrite but a smaller number may produce gaseous reduction products such as N_2O or N_2 which are lost from the soil. Bacteria such as *Thiobacillus denitrificans, Micrococcus denitrificans, Pseudomonus aeruginosa* and *Bacillus nitroxus* utilize nitrate as a source of oxygen in the oxidation of organic matter. Some of these bacteria are obligate anaerobes but others are facultative aerobes which can only develop under anaerobic conditions if there is a nitrate supply.

Denitrification is, therefore, a process which is most marked in oxygen-deficient, waterlogged soils. Quastel (1963) notes that even 1% of oxygen in the atmosphere is sufficient to suppress denitrification considerably but there is some evidence of gaseous nitrogen loss in natural aerobic soils. Under field conditions even quite freely drained soils may denitrify very efficiently as ammonium ions, which are protected from leaching by adsorption on the exchange complex, once nitrified to nitrite or nitrate anions, are no longer retained. As they leach downward they are exposed to lowered oxygen concentration and may be denitrified.

Nitrogen fixation

The original source of soil-N is *biological fixation* since very few geological parent materials are rich in nitrogen, being either laid down as anaerobic, denitrifying sediments or solidified from high temperature magma. The earliest stage of nitrogen fixation in soil development is carried out by free living microorganisms and by the blue-green algal component of pioneer lichens. During the last 25 years, N-fixation has been shown and measured in a large number of microorganisms by the ^{15}N mass isotope enrichment technique (Stewart, 1966) and more recently by the acetylene: ethylene method in which acetylene is specifically converted to ethylene by the nitrogenase enzyme of the organism. This test is 10^3 times more sensitive than the ^{15}N method and makes possible very wide-ranging screening for N-fixation (Hardy *et al.*, 1968).

Nitrogen fixation, by superficial crusts of blue-green algae, is widespread (Mulder *et al.*, 1969) and McGregor and Johnson (1971) suggest that it may be the only source of nitrogen in some desert-crust soils. Nitrogen fixing in the blue-green algae is by species which form the thick-walled, colourless *heterocyst* cells; common examples are *Nostoc* and *Anabaena* spp. The blue-green algae of the lichen symbiosis and some associated with higher plants (*Azolla*; *Gunnera*; *Cycas* etc.), as mentioned above, also fix nitrogen. Fixation of nitrogen by blue-green algae is probably of greatest importance in rice paddy in which waterlogging encourages strong bacterial denitrification.

The capacity of blue-green algae, as autotrophic nitrogen fixers, is important in raw-soils as they pioneer the sere of N-fixing microorganisms which become heterotrophically active as soon as dying organisms or organic exudates become available. Bacteria of the *Azotobacter* group are prominent nitrogen fixers and are most likely to be ecologically important in highly deficient habitats. Hassouna and Wareing (1964) suggested, for example, that *Azotobacter* might inhabit the rhizosphere zone of the dune pioneer *Ammophila arenaria*, thus being supplied with organic metabolites diffusing from the grass roots.

It has been stated that *Azotobacter* is generally unlikely to make a major contribution to the N-nutrition of vegetation. In the temperate zone it is absent from acid soils, having a pH optimum of about 7·0 and a range from pH 6·0 upwards. In the neutral to alkaline soils which form its habitat, the rates of fixation must be seriously limited by the availability of organic energy sources since, even in ideal culture conditions, *Azotobacter* needs some 50 g of carbohydrate to fix one gram of nitrogen (Campbell and Lees, 1967). However, discussion of this topic has largely concerned agricultural soils; in natural soils where competition for nitrogen is usually intense it is possible that the *Azotobacter* contribution may be significant in permitting some individuals, within populations, to receive survival levels of N. This may particularly be the case if the bacteria are associated with the plants in a rhizosphere sheath. There is some evidence that *Beijerinkia*, the tropical counterpart of *Azotobacter*, is present in greater numbers in the rhizosphere (Mulder *et al.*, 1969). *Beijerinkia* differs from *Azotobacter* in its broader pH spectrum (pH 3·0–7·0) and low calcium requirement (Virtanen and Meittinen, 1963).

By far the largest contribution to the soil nitrogen pool comes from the symbiotic fixation of N by bacteria to the genus *Rhizobium* which inhabit the root nodules of various leguminous plants. These are, per unit weight of bacterial cell, 100 to 200 times as efficient at N-fixing as the free-living organisms (Mulder *et al.*, 1969). Some other root nodule organisms, presumably *Actinomycetes*, occur in association with some woody, non-leguminous plants (e.g. *Alnus, Myrica, Hippophaea*).

The genus *Rhizobium* comprises a number of species and strains which form root nodules specific to different genera and species of the Leguminoseae. The initial infection of the plants takes place from soil-born rhizobia which penetrate a root hair cell of the piliferous layer. Even at this stage they appear to form a plant growth hormone which causes deformation of the infected root hairs. The infection rapidly penetrates to the inner cortex, leaving a visible infection thread through the outer cortical cells. According to species, the cells of either the inner cortex or of the pericycle begin to proliferate to produce a nodular swelling which pushes out the overlying cortex. The cells of the nodule are parenchymatous and rapidly become filled with pale, rod-shaped *Rhizobium cells*. An outer network of vascular tissue often develops around the nodule. As the nodule develops the bacterial cells become swollen and banded in appearance, in many plants they branch to form Y- or X-shaped, rod-shaped cells called bacteriods. At the same time they form a red pigment, leghaemoglobin, which colours the nodule, and nitrogen fixation commences (Virtanen and Meitinen, 1963). It has been shown that bacteroids are formed only in cells where synthesis of leghaemoglobin occurs and that, in the absence of leghaemoglobin, N-fixation never takes place. As the nodule cells age, their colour changes from red to green and the branches or swollen bacteroids are replaced by normal bacterial rods. Senescence of the nodule tissue liberates these into the soil solution where, by taking on a motile coccoid form, they may migrate for short distances to infect other roots.

N-Fixation and soil N

N-fixation is the major source of N for plant growth though some small contribution is made by photochemical and electrical discharge fixation in the atmosphere which, added to nitrogen compounds from industrial smoke, appears in precipitation as nitrates. Some alkaline soils and waters contribute gaseous ammonia to the atmosphere while denitrifying soils may volatilize nitrogen oxides which are ultimately recycled in rain (see Chapter 9).

The symbiotic fixation of N may be very large, reaching 2–400 kg/ha/ann in some leguminous crops and, in optimum greenhouse conditions, rising to the equivalent of 1000 kg/ha/ann (Virtanen and Miettinen, 1963). Free-living bacteria produce less with a maximum of, perhaps, 20 kg/ha/ann. In the case of the free-living organisms some excretion of organic-N compounds to the soil does occur but it is not large. This, in addition to limited organic substrate supply, probably accounts for the great difference in fixing efficiency between these and the symbiotic forms. In the latter case much of the organic-N is removed, as it is formed, into the host's tissues or may even be excreted to the soil though there is some controversy in the literature concerning the latter point. Excretion to the soil is demonstrated experimentally in Figure 8.1: pea plants, innoculated with rhizobia, were cultured in nitrogen-free sand. During the

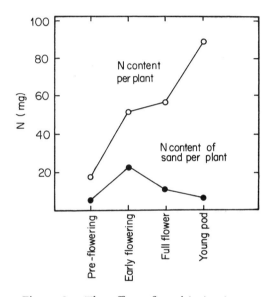

Figure 8.1 The effect of symbiotic nitrogen fixation on the nitrogen content of pea plants and the nitrogen content of the originally nitrogen-free sand culture medium. Data of Virtanen and Miettinen (1963).

experiment considerable amounts of nitrogen were fixed and, in the early stages of growth, about 30% of the fixed-N was excreted to the sand. As the plants continued growth to maturity and pod formation, the excreted nitrogen was taken up again. In similar experiments Virtanen and coworkers showed that a non-leguminous plant grown in the same container as a nodulated leguminous plant also benefited from the excreted nitrogen. The results of these experiments are apparently indisputable but, in the field, it is much more difficult to demonstrate that plants are obtaining their N supply in this way, rather than through microbiological breakdown of dead nodule and root tissue. This may seem to be an academic point as the fixed-N ultimately reaches the soil pool of organic-N, but, in relation to competition for N, the availability of even a minimal supply could be very important.

The relationship of symbiotic N-fixation to soil conditions has generally been investigated in the context of agricultural plants such as clovers (*Trifolium* spp.), alfalfa (*Medicago sativa*) and various peas and beans (*Pisum* spp., *Lathyrus* spp., *Vicia* spp. etc.) and, to this extent, attention has been divered from the ecological implications of the process.

Cultures of *Rhizobium* appear to have optima, for growth, between pH 6·5 and 7·5 while the same range is optimal for N-fixation in the field by species of *Trifolium* and *Medicago sativa*. Most agricultural legumes fix nitrogen in the range pH 5·0 to 8·0 but many wild members are active below this value, for example, *Ulex europaeus* may extensively and successfully colonize acid, nutrient-deficient soils. Other wild legumes occur on very nutrient-deficient soils: *Ononis repens* and *Lotus corniculatus* are common as sand dune plants in N. Europe while various *Lupinus* spp. have been utilized as pioneer N-fixers in the reclamation of poor quality and derelict soils. Non-leguminous N-fixers such as *Hippophaea rhamnoides* also occur as colonists of sand dune soils while *Myrica gale* may fix at pH values as low as 3·8 with an optimum of pH 5·4. Stewart (1966) notes that *M. gale*, supplied with nitrogen in its rooting medium, has a growth optimum of pH 3·3 though its optimum for nodulation is 5·4. If, however, the pH is lowered after nodulation it continues to fix N, suggesting that the plant may have an efficient internal pH regulating mechanism.

The rhizobia are aerobic organisms and, possibly for this reason, few N-fixing leguminous species seem to occur on heavily waterlogged soils but the non-leguminous *Alnus* spp. are often lakeside or stream margin dwellers, sometimes on quite acid soils. There is good evidence that they may contribute considerably to the nitrogen content of the soil during their growth: Stewart (1966) cites an annual fixation by *Alnus* spp. between 60–225 kg/ha/ann.

To summarize: the greater part of the global reserves of soil nitrogen has been produced by symbiotic fixation though there is some evidence that free-living organisms may play a more important role than formerly suspected, particularly in natural ecosystems. Walker (1965) goes so far as to say that: 'In most cases the important colonizing plants are legumes, nodulated non-legumes known to fix nitrogen, or other plants with known or suspected mechanisms for fixing atmospheric nitrogen'. Excellent accounts of these topics may be found in a number of publications: Waksman (1952) discusses the microbiology of nitrogen fixation and more general reviews are given in Virtanen and Mietinnen (1963) and Stewart (1966).

Phosphorus

Natural soils usually contain between 0.02% and 0.5% of phosphorus, most of which is derived from the parent minerals (Bear, 1964). The surface layers of the soil are often enriched with P which has been absorbed by roots and passed to the surface in litter fall. As a result of this biocycling, much of the soil-P is usually organic; Bould (1963) suggests that it may range from 3 to 75% of the total.

Inorganic-P is often present as apatite (calcium phosphate) of which the commonest form is fluorapatite $(Ca_{10}F_2(PO_4)_6)$. In alkaline soil, apatite may form 50% or more of the inorganic-P but, as weathering proceeds and pH falls, so the percentage of iron phosphate increases, to reach perhaps 40%. Other forms of inorganic-P are aluminium phosphates and occluded-P, the latter sealed into mineral grains with a coating of insoluble iron oxide. These two forms of soil-P usually represent some 10–20% each of the inorganic-P content. Occasionally, very high soil-P concentrations may be associated with unusual conditions, for example, vivianite (ferrous phosphate) may be precipitated under peat in bog soils while a few soils, derived from biogenic marine deposits or from guano, are immensely rich in calcium phosphate (Bear, 1964; Black, 1968).

The occurrence of other forms of inorganic-P is less well documented, but Jeffery (1964) has suggested that *Banksia ornata*, an Australian sclerophyll-heath plant, may synthesize and retain inorganic polyphosphate in its roots. Many bacteria also synthesize polyphosphate and Kuhl (1962) notes that some microorganisms may have a large percentage of their total-P in this form. Naturally occurring polyphosphates have many thousands of orthophosphate residues in their molecules and are relatively stable materials; their persistence in soil has not, however, been reported in the literature. Under waterlogged, anaerobic conditions inorganic-P may be reduced to phosphine (PH_3) and lost by volatilization.

The common forms of inorganic-P in soil all have very low solubility products and, consequently, dissolved-P rarely exceeds $0.01–1.0$ mg/ml in soil solution. Fried and Broeshart (1967) cite data for a number of N. American soils, some of which had received P-fertilization, showing the P concentration range between 0.04 and 0.08 μg/ml to be the most frequent. In the pH range of natural soils most of the dissolved-P is present as the ion species $H_2PO_4^-$ or, in very alkaline soils, HPO_4^{2-}. A further consequence of these low solution concentrations is that plant uptake rapidly depletes the dissolved-P so that the rate-limiting process is solution from adjacent soil particles. Larsen (1967) notes, as a further complication, that P forms soluble complexes with many metal ions and it is, perhaps, an oversimplification to assume the bulk of the soluble-P to be $H_2PO_4^-$ and HPO_4^{2-}.

The organic-P entering a soil is largely derived from plant litter; though animal remains and faeces may be important under heavy grazing. The known groups of organic-P compounds in this material are inositol phosphates, phospholipids, nucleic acids, nucleotides and sugar phosphates, but only inositol phosphate has been detected, in quantity, in soil and even this may be of secondary, microbiological origin. Cosgrove (1967) notes that organisms occur, in the soil and rhizosphere, which are capable of dephosphorylating all known organic-P compounds of plant origin and that plant roots may also have similar phosphatase activity at their surfaces. Woolhouse (1969) further investigated this possibility and found differential phosphatase activity between various ecotypes of *Agrostis tenuis*. Despite this, a large proportion of the soil-P remains in the organic form, presumably because any inorganic-P which is produced by phosphatase activity is very rapidly mopped up by microorganisms. The fractionation of soil-P is fraught with difficulty and it may be that much of the soil-P, formerly thought to be organic-P, is held in living microbial cells and cell debris (Larsen, 1967).

In concluding discussion of the phosphorus status of soil, it may be noted that it shares with nitrogen the characteristic of being immobilized in organic form and is strongly dependent on soil microorganisms for its natural cycling. Unlike nitrogen, however, its inorganic compounds are rather insoluble and impose a further rate limitation on biological transfer processes.

Potassium

Natural soils contain much more potassium than phosphorus or nitrogen. Black (1968) cites values between $0 \cdot 3$ and $2 \cdot 5\%$. Potassium in soils is generally derived from alumino-silicate minerals such as feldspars and micas and, consequently, finer textured soils with high silt and clay contents tend to have a higher percentage of potassium.

Chemically, K is a mobile element, its simple compounds being very soluble in water and, were it not for its incorporation in the fairly stable lattice structure of the alumino-silicates, it would rapidly be lost by leaching. In soil only a very small proportion of the K is soluble or exchangeable, the remainder is a non-exchangeable component of the soil matrix. Black (1968) cites a mean figure of $99 \cdot 6\%$ of the total $-$K in the latter form and, of the remaining $0 \cdot 4\%$, $0 \cdot 17 - 0 \cdot 87$ *me*/100 g were exchangeable and $0 \cdot 003 - 0 \cdot 02$ *me*/100 g were in solution. Fried and Broeshart (1967) record solution concentrations between 2 and 6 g/ml K ($0 \cdot 05 - 0 \cdot 15$ *me*/l) as being the most frequent in N. American soils.

The cation exchange complex of most soils carries only a small proportion of K as Ca and Mg are relatively more abundant in the exchangeable form. Cation exchange capacity ranges between about 5 and 50 *me*/100 g, in all but the coarsest textured soils, but it is rare to find more than about 1 *me*/100 g of K whereas Ca often ranges up to 30 *me*/100 g and Mg to 20 *me*/100 g in cation-saturated soils. In highly leached soils the exchangeable K is again low but the balance is made up in exchangeable hydrogen (data of Soil Surv. Staff, 1960). The K in solution is in equilibrium with that adsorbed on the exchange complex and its low concentration protects the soil against leaching loss. This protection is probably made more effective because rainfall contains a certain amount of K and more is leached from leaves in the canopy (see Chapter 9). The K content of the downward percolating water may thus be at a level which inhibits dissociation of K from the exchange complex and root uptake is likely to be from cyclic K in the soil solution, rather than at the direct expense of exchangeable K. In the context of K mobility, organic combination is not important as it is very easily leached from both dead organic matter and living plant parts such as leaves and roots.

Sulphur

Sulphur forms between 0·01% and 0·15% of temperate zone soils (Bould, 1963) and there is reason to believe that much of it is immobilized in organic combination (Freny, 1967), not more than 50–500 μg/g being soluble sulphate. This contrasts with arid zone soils in which calcium or magnesium sulphates may accumulate after upward transport in capillary water, and evaporative concentration at the soil surface.

Sulphur, in soil, originates from the mineral matrix in which it may occur as various metal sulphides (e.g. FeS_2; ZnS) or as crystalline sulphates. Generally, sulphides occur in igneous rocks and sedimentary rocks which were laid down under reducing conditions and sulphates in sedimentary rocks produced in oxidizing conditions—often as evaporites. In industrial countries a significant amount of sulphur is now added to soil in precipitation, industrial production of SO_2 causing an annual input of up to 36 kg/ha of S (Bear, 1964). There is a natural rainfall input of which, perhaps, 6 kg/ha/ann originate from H_2S released from waterlogged soils, lake muds and, probably most important, continental shelf sediments.

Like N and P, S may be immobilized in the organic form so that rates of microbiological mineralization of the element are significant factors in pedogenesis and plant nutrition. It differs from P in that sulphate, its common inorganic form in soil, is moderately soluble and rapidly leached in high P/E conditions. Sulphate ions are not held by anion exchange in alkaline soils but some binding may occur in acid conditions. Availability from free sulphate and rainfall input is generally greater than from organic-S mineralization.

A wide range of organic-S compounds is formed by living organisms and may enter the soil in litter. These include S-amino acids, sulphonium compounds, sulphate esters, sulphur-containing vitamins and antibiotics, organic sulphides, sulphoxides and isothiocyanates (Freny, 1967). The majority of these are rapidly decomposed and become indetectable in soil. The amino acids cystine and methionine are produced on acid hydrolysis of soil, suggesting that they may occur in polypeptides or proteins incorporated in the humic complex or bonded to clay particles. About half of the organic-S is in this form while the remainder is possibly present as sulphate esters though there is considerable doubt about its true composition.

Plants absorb S as sulphate though there is some evidence that S-amino acids may also be assimilated. The mineralization process is, consequently, of considerable interest but no definite pathways for conversion of S-amino acids to sulphate (aerobic) or sulphide (anaerobic) have been identified (Freny, 1967). Some knowledge of environmental influences on mineralization of S has accrued, for example increasing pH and/or temperature (between 10 and 35°C) increases SO_4^{2-} release, while drying of the soil slows the process. These observations suggest that mineralization is bacterially mediated.

Calcium

The calcium content of soils is widely variable, rendzinas and chernozems derived from Ca-rich parent material or enriched with evaporites, may have $CaCO_3$ contents of 40–50% (16–20% Ca) while acid sandy soils often contain less than 0·5% Ca and no free carbonate. In non-carbonate soils the Ca originates from the alumino-silicate and Ca-Mg silicate minerals with the consequence that Ca and clay content are often correlated.

In rich soils some 70–80% of the cation exchange complex may carry Ca ions and in non-carbonate soils this may be a relatively large proportion of the total soil-Ca. Black (1968) records a group of soils with a mean Ca content of 17·3 me/100 g as having 23% of this Ca on the exchange complex. The equilibrium concentration in the soil solution is also high compared with other cations such as K. Fried and Broeshart (1967) suggest a mean value of about 10 me/l for Ca compared with 0·1 me/l or less for K.

Calcium is, pedogenetically, a significant element in the sense that the presence or absence of calcium carbonate may be diagnostic of the P/E regime. This concept is applicable only to parent materials which are rich in free $CaCO_3$, but, as limestones are geographically ubiquitous and carbonate-rich loess deposits widespread in mid-continental areas, the concept of carbonate leaching is most useful in interpreting present-day soil conditions. Incoming rainwater is in solution equilibrium with the atmospheric CO_2 and, as it enters the surface layers of the soil, may dissolve more. The downward percolating water may be able to carry up to 250 μg/ml of $CaCO_3$ by solution as $Ca(HCO_3)_2$ (Perrin, 1965).

If the P/E is much greater than unity, the surface layers of the soil may be completely leached of free $CaCO_3$ and an advancing front of decalcification passes down the profile at a rate governed by the P/E ratio, the original carbonate content of the soil, temperature regime and seasonal distribution of the rainfall input. If the P/E ratio falls much below unity during the summer months then upward movement of capillary water in response to surface evaporation causes $CaCO_3$ (and other salt) enrichment of the A horizon. Even if the winter precipitation is high it may not reverse this summer effect as the leaching process is less efficient at low temperatures or, more commonly in midcontinental areas, the soil is frozen.

Biocycling of calcium is important in restricting leaching loss by returning Ca to the soil surface but there is no question of serious organic immobilization of Ca as it is quickly released by the decomposition process. Its more important biological implication is in relation to the composition of the soil biota: both earthworms and bacteria are much reduced in numbers by low Ca status. Though Ca is not immobilized in organic matter, it does show chelation reactions with a number of organic acids, for example, citric and gluconic (Stevenson, 1967). Formation of Ca-chelates may have important effects on both pedogenesis and plant nutrient uptake.

Before leaving the discussion of Ca in soil it should be noted that $CaCO_3$ is the commonest soil constituent responsible for high levels of soil alkalinity, many $CaCO_3$ soils having pH values between 8·0 and 8·4. The relationship between such high pH values and plant nutrient availability is discussed later (pp. 244 and 263).

Magnesium

Magnesium behaves similarly to Ca in soils being derived from alumino-silicate, silicate or sulphate minerals on non-carbonate parent materials or from dolomite $(MgCa(CO_3)_2)$ and magnesite $(MgCO_3)$. The total soil content of Mg, like Ca, is widely variable, ranging from 0·003% to 0·6% Mg in normal soils (Bould, 1963) to much higher values in dolomite soils where 1·2% Mg may be exceeded (Bear, 1964).

Exchangeable Mg in soils is usually less than exchangeable Ca despite the fact that the total Mg content of non-carbonate soils often exceeds total Ca. Bear (1968) gives a mean value of 1·6 me/100 g for exchangeable Mg and 40·7 me/100 g for total Mg in a number of N. American soils. The concentration of Mg in the soil solution is similar to that of Ca at about 10 me/l (Fried and Broeshart, 1967).

Iron

Though iron is a micronutrient for plants it has a much wider pedogenetic and microbiological significance than its quantitative uptake would suggest. Iron makes up some 0·7–4·2% of temperate zone soils and between 14–56% of many tropical latosols. Small percentages of iron occur as lattice components of the clay alumino-silicates but the greater part is present either as hydrous ferric sesquioxide in particulate form or as coatings on other mineral particles. Under anaerobic conditions iron may become soluble in the divalent (Fe^{2+}) form, may form organoferrous complexes or precipitate as FeS, $FeCO_3$ or $Fe(OH)_2$ as shown in Figure 8.2. In normal neutral or alkaline soils the Fe^{3+} concentration in the soil solution is extremely low, but falling redox potential may allow Fe^{2+} to be produced, sometimes rising to toxic concentrations of several hundred

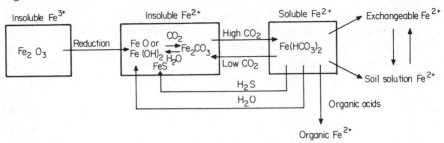

Figure 8.2 The reactions of iron in waterlogged soils. Reproduced with permission from L. N. Mandal, Fe and Mn in waterlogged soil, *Soil Sci.*, **91**, 124 (1961).

μg/ml (see p. 191). Similarly, exchangeable ferric iron is normally at rather low levels but reducing conditions permit Fe^{2-} to occupy some of the cation exchange sites. In these soils, iron availability is strongly dependent on redox status, but in acid soils ferric iron is more soluble, even under oxidizing conditions, so that iron deficiency is usually a characteristic of well-drained calcareous soils.

The distribution of ferric sesquioxide in soil is often a useful parameter of pedogenetic processes as its solubility, relative to that of other soil components, is widely variable. In alkaline to neutral conditions iron is not easily mobilized other than in waterlogged soils where low redox potentials permit its conversion to the more soluble ferrous form. In consequence, well-drained soils have silica/sesquioxide ratios close to those of the original parent material (Table 8.1). Acid, temperate zone soils may show podsolization with iron loss from the A horizon and redeposition in the B horizon. Though the silica/sesquioxide ratio of the A horizon is increased and that of the B horizon reduced, the mean ratio for the whole profile is not much altered unless soluble iron is flushed away to the ground water. The only case of massive Fe_2O_3 loss from temperate zone soils occurs, with waterlogging, if ferrous iron is flushed downward to a *laterally moving* water table.

Table 8.1 Silica: sesquioxide ratios ($SiO_2/(Fe_2O_3 + Al_2O_3)$) of various soils

Podsol	A_1 17·9	A_2 41·7	B_1 15·3	B_2 19·0	C 21·6	Mean $A_1A_2B_1B_2$ 23·5

Chernozem	A 4·6	B_1 4·9	B_2 3·8	C 4·1	Mean $A B_1 B_2$ 4·4

Latosol	1 Parent rock 3·2	2 Inner weathered crust 0·16	3 Outer weathered crust 0·14	4 Laterite 0·23	Mean 2,3,4 0·18

Under high temperature regimes and high rainfall in the tropics the situation differs as the silicate minerals are very unstable and a large loss of silica, probably as colloidal silicic acid, occurs. The ratio of silica/sesquioxide becomes very low, the Al and Fe sesquioxides of the parent minerals being resistant to decomposition and some synthesized *in situ* from ionic Al and Fe liberated during alumino-silicate breakdown. This process of *laterization* appears to be most common on base-rich parent materials and leaves a residue of *primary laterite* containing little else but iron and aluminium sesquioxides and a few resistant parent minerals (Mohr and van Bahren, 1959). Nearer to the water table, primary laterite may become resilicified, by deposition of silica from solution, to form *argillaceous laterite* which covers wide areas in the tropics.

The characteristic colours of soils are, in general, conferred by iron compounds. Sesquioxides give colours ranging from yellow-brown, through dark brown to rust red to bright red or pink according to degree of hydration. The dehydrated form, so characteristic of oxic soils, is *haematite* (Fe_2O_3) which is bright red while the hydrated form, *goethite* ($Fe_2O_3.H_2O$) is brown or reddish brown. Most reducing soils have a drab olive-grey, greenish or bluish tint which is probably attributable to organoferrous compounds and they are often mottled with ochreous spots of hydrated iron oxides or hydroxides. In the presence of free sulphide dark colorations from grey to black may be caused by the formation of FeS or FeS_2.

The migration of iron under podsolizing conditions has been the source of much controversy and various workers have considered it to be: (a) mobilization as inorganic ferrous iron in response to changing E_h–pH conditions, (b) movement of soluble organo-iron complexes formed from humic compounds or (c) movement of soluble complexes formed with organic leachates from fresh litter. It has also been suggested that iron may be eluviated in colloidal form, either alone or in association with organic materials.

Bloomfield (1965) maintained the view that humic compounds are not involved, except as an energy source for microorganisms, and postulated that iron is reduced to the ferrous condition by polyphenols leaching from fresh litter and then travelling down the profile as organo-iron complexes with polyphenols. Microbiological activity in the *B* horizon probably decomposes the organo-iron complex, initially liberating ferric hydroxide as coatings on mineral grains and intercalated between soil particles. The bacterium *Pedomicrobium* appears, commonly, to be associated with this process (Aristovskaya and Zavarzin, 1971).

The mobilization of ferrous iron in waterlogged soils is generally caused by a fall in redox potential and production of reducing organic compounds by a non-specific microbial population. The bacterium *Gallionella*, which oxidizes ferrous iron, is responsible, amongst other organisms, for the accumulation of ferric concretions and mottles in gley soils, for the formation of bog-iron desposits and for the precipitation of hydroxides in natural waters.

Manganese

In soil, manganese behaves similarly to iron and consequently shows similar patterns of distribution in profiles, related to $pH-E_h$ variations and valency–solubility change. The total manganese content may vary from a few $\mu g/g$ to about 1% though this is no index of its availability as it may occur in the di- tri- and tetravalent state of which only the first is particularly soluble. Manganese exists in soil as the higher valency oxides, as manganous carbonate and hydroxide or as exchangeable or soluble Mn^{2+} (Erlich, 1971). The manganous ion begins to appear in solution at a pH below 8·0 and a corresponding E_h of 600 mV.

Many microorganisms are involved in the oxidation and reduction of the soil Mn and are probably responsible for maintaining the availability of Mn^{2+} in well-oxidized, high pH soils. The reduction of Mn to the divalent form does not necessarily imply that it will appear in solution as $MnCO_3$ and $Mn(OH)_2$ may precipitate above neutrality. Microorganisms are also responsible for the oxidative production of manganese concretions in poorly aerated and gleyed soils. Manganese deficiency, like iron deficiency, is most commonly associated with high soil pH.

Copper

Copper, in soils, derives from various primary minerals, the commonest being chalacopyrite ($CuFeS_2$). Its concentration may range from 0·1–1000 $\mu g/g$ in mineral soils though less than 1 $\mu g/g$ is likely to be in solution. Its solubility is much reduced by high pH and by the presence of carbonates and sulphides (Erlich, 1971). In solution it may exist as the Cu^{2+} cation in equilibrium with the exchange complex and it may also appear in solution as an organic complex. Deficiency of Cu is usually associated with low concentration in the parent material.

Zinc

Zinc occurs in a wide range of primary minerals and is easily released by weathering either as the Zn^{2+} cation or as an organic complex. The normal soil content is between 10 and 300 $\mu g/g$. Above pH 5·0 its availability may be reduced by precipitation of $Zn(OH)_2$. In acid soils much of the Zn content is exchangeable but in calcium soils a large proportion may only be extracted by dilute acid treatment (Bould, 1963). Deficiency of Zn is associated with a low Zn content of parent minerals.

Boron

Boron in soils originates from borosilicates, calcium and magnesium borates and iron and aluminium boron complexes. The total content is usually between 2 and 100 $\mu g/g$ (Bould, 1963) and it may appear in solution as an equilibrium mixture of *ortho-, meta-* and tetraborate anions (Colwell and Cummings, 1944). Usually less than 3 μg/g is soluble as borate but much higher values occur in arid zone saline soils where boron toxicity may occur. Boron deficiency is associated with light, sandy, easily leached soils.

Chlorine

Until comparatively recently the essential role of chlorine as a plant nutrient was not recognized (Hewitt, 1963). It occurs ubiquitously in soils as the Cl^- anion and its concentration is often similar to that of NO_3^- and SO_4^{2-} in the soil solution. In the saline soils of salt deserts and salt marshes the concentration of Cl, associated with Na, may be very high, reaching toxic concentrations and sometimes appearing as a crystalline efflorescence at the soil surface or within soil pores.

Molybdenum

Molybdenum occurs in natural soils as the sulphide or the molybdate, usually entering soil solution as the molybdate anion. It may, like phosphorus, become immobilized by the production of insoluble aluminium or iron molybdates. The usual soil content is 1–10 μg/g but most of this is insoluble. Deficiencies may be associated with absence from the parent minerals or fixation, particularly in acid soils. A few calcareous, P-rich soils show excessive Mo availability which causes 'teart' disease in cattle grazing on such land. Molybdenum is also an essential element in bacterial N-fixation.

Other elements of ecological interest

Various non-essential elements play a part in soil formation and plant ecology. Quantitatively, the most significant of these is *Aluminium* which forms a part of the structural lattice of most clay minerals but also enters the exchange complex of some acid soils. The possible confusion between exchangeable hydrogen and exchangeable -H was discussed in Chapter 4, p. 102. In acid soils, aluminium may also be involved in the immobilization of P by production of very insoluble phosphates. At high concentrations Al^{3+} is a plant toxin and is one factor in the environmental differentiation of calcifuge species (see p. 264).

Sodium is a common constituent of many salt marsh and salt desert soils and its widespread occurrence is associated with the ecological differentiation of halophytic species which withstand soil salt concentrations which would be physically or chemically harmful to normal mesophytes. There is also some evidence that sodium may be an essential nutrient, at least for some species of plant (Hewitt, 1963) and it is, of course, essential for the animal component of the ecosystem.

A few other elements may reach toxic concentrations in soils: some, like *arsenic*, usually inhibit plant growth before they can become a threat to animals, but others, such as *selenium*, reach high concentrations in ac-cumulator species (e.g. *Astragalus* spp.) and cause toxicity diseases in grazing animals. Heavy metal toxicity effects are discussed on p. 272.

NUTRIENT AVAILABILITY

The foregoing discussion has shown that only a small proportion of most nutrients is present in a rapidly assimilable form. The soil solution is the primary source of inorganic nutrients for plant roots and, in the case of the more important cations, the concentration of this solution represents an equilibrium with the cations adsorbed on the exchange complex. Thus, for many cations, the soluble + exchangeable component is a good index of availability for plant growth, though the situation is rather different for calcium (and perhaps magnesium) in calcareous soils where free car-bonates are present.

Short-term availability of cationic nutrients will be governed by the solution concentration, and the rate of replenishment from the exchange complex, unless contact exchange plays any significant part. Overstreet and Jenny (1939) claimed that plants accumulated cations more rapidly from clay suspensions than from equilibrium dialysate solutions of the clays, because contact exchange of cations occurred due to overlapping of the ionic atmospheres surrounding the roots and the soil particles. This hypothesis is also supported by evidence from physical systems in which cation transfer between solid adsorbents is more rapid, if contact occurs, than if cation diffusion in liquid is the only possible pathway (Black, 1968). The evidence for contact exchange is, however, still tenuous and cannot be fully accepted without further experimentation.

The availability of some micronutrient cations is governed more by their solubility relationships than by exchange phenomena, thus iron may be very insoluble in well-aerated calcareous soil due to precipitation of ferric hydroxides and in a similar fashion manganese may disappear from solu-tion at high pH.

Anionic nutrient availability is more difficult to specify as the anion exchange activity of soils is a rather variable parameter, and also because several anionic nutrients such as sulphur, nitrogen and phosphorus, became involved in biological cycles which effectively remove them from the physicochemical equilibrium relationship.

Phosphorus shows a much stronger binding reaction with soil than does either nitrogen or sulphur and it may be removed almost quantitatively from solution if added as orthophosphate (Broeshart, 1967). This im-mobilization of P is due not so much to ion exchange as to formation of in-soluble phosphates at the surfaces of Fe-, Al- and Ca-containing minerals. In theory, if organic matter is mineralizing fast enough to supply adequate N, it should also be releasing sufficient P for plant growth but, unfortunately, much of the phosphate-P so produced is precipitated before it can be absorbed and also, because the solution-P is at a very low concentration, P tends to be rather immobile. Unless the mineralizing organic matter is in the immediate vicinity of absorbing roots the released inorganic-P will not reach root surfaces and, for this reason, organic-P cannot be considered to form an immediate source of nutrient-P.

Sutton and Gunary (1969) note that the immediate uptake by roots depletes the soil solution of dissolved P and further phosphate ions will then be drawn into solution from the soild phase. The quantity which may be liberated in this way is called the *labile* fraction. To specify the total P-supplying potential of the soil it is also necessary to know the P-adsorbing capacity as some inorganic-P will be temporarily trapped by anion exchange, at positively charged sites, before entering solution. Sutton and Gunary describe the P-supplying behaviour of a soil by using a well analogy (Figure 8.3): the water in the well represents the immediately available solution-P; the water in the porous matrix, the labile-P and the volume of the porous material is a measure of the adsorbing capacity. If two soils have the same adsorbing capacity then the labile-P content of each is sufficient to specify their relative P-supplying power but, if the absorbing capacities differ, then both these and the labile-P contents must be specified. Because P is so immobile, the P-supplying power of a soil must be considered as its ability to supply P to an individual absorbing root; the total length of absorbing root thus becomes very important in the context of whole-plant supply. A soil which may contain adequate P to support an established plant with an extensive root system may, at the same time, be quite unable to support a seedling with a single, short root (Sutton and Gunary, 1969).

ABSORPTION OF NUTRIENTS BY PLANTS

Nutrient absorption, like water absorption, is a function of the soil capacity-intensity factors, of the kinetics of release and movement in the soil, of surface area of root–soil contact and of internal transport characteristics of the plant. The root system of a plant is a much branched structure, the terminal parts of which intimately permeate the soil pore space and have their surface contact further increased by the development of root hairs which enter even very small pores and also become adpressed to the surfaces of soil particles. It is currently thought that the major pathway for water and solute mass flow to the root stele is through the microporous structure of the cortical cell walls (Weatherley, 1963; 1969). This pathway is interrupted by the suberized Casparian strips of the endodermis but, beyond this barrier, the mass flow pathway is again available in cell wall micropores and in the xylem lumina. At the root apex where the Casparian strips are not fully suberized, and further up the root, where

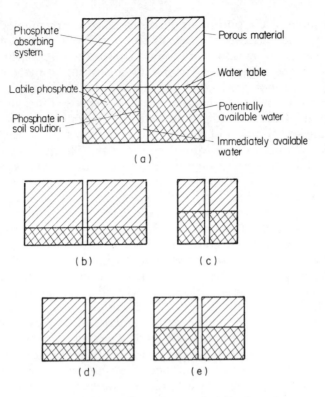

Figure 8.3 The well analogy of soil phosphorus behaviour. (a) depicts the analogy between the porous matrix surrounding a well and the phosphorus absorbing system. The amount of water in the porous medium represents the labile phosphorus supply. In comparing soils, if the absorbing system is of equal size as in (b) and (c), it is only necessary to specify the labile quantity but, if the adsorbing systems are unequal as in (d) and (e), both the size of the adsorbing system and the labile quantity must be specified. Reproduced with permission from C. D. Sutton and D. Gunary, *Ecological Aspects of the Mineral Nutrition of Plants*, Blackwell, Oxford, 1969, pp 127–34.

passage cells or endogenous root branches penetrate the endodermis, mass flow entry of the soil solution to the xylem can occur, perhaps up to 25% of the total intake (Weatherley, 1969).

The earliest view of nutrient absorption was that solutes entered the plant passively with the transpiration stream but this view was rapidly discredited by permeability studies which showed that water molecules and the various ionic species each have characteristic transport coefficients so that some solutes may be accumulated while others are excluded. Because decapitated plants show root exudation (root pressure), the root system is believed to act as an osmometer in which the external solution is separated from the exuding xylem fluid by a semipermeable barrier (see p. 146). If it is accepted that the cortical cell wall micropores form a mass flow pathway for soil solution, this barrier is likely to be the endodermis and, to its exterior, ions may move freely by diffusion or be swept in with the transpiration stream. If the plant is transpiring fast, in addition to osmotic transfer of water into the xylem there will be a transfer caused by the hydrostatic water potential gradient drawing water through the endodermis by 'ultrafiltration'.

The mass flow of water towards and into the outer tissues of the root, followed by transfer across the endodermal semipermeable barrier, may result in a dilution of the xylem fluid and an increase in the ionic concentration of the liquid external to the endodermis. Both of these effects would favour increased uptake and partially explain the enhancement of absorption at high transpiration rates which some workers have noted. There is, however, some conflict in the experimental evidence for this enhancement, for example, salt-starved plants show less transpiration dependence than salt-rich plants (Weatherley, 1969). The metabolic status of the plant is also likely to interfere with the enhancement relationship because active ion accumulation relies on metabolic energy sources. A further complication in passive ion uptake is the presence of the cell wall which is a weak acid anion exchanger. Some workers have noted the differing exchange capacities of the roots of various species and groups and suggest that the exchange activity plays a part in ion uptake. Care must be taken in interpreting this type of work as dead roots have a non-specific exchange activity which follows the normal valence and lyotropic laws (Broeshart, 1967) but the activity of living roots is specific and may represent the first stage of ion binding to carrier sites. It is unlikely that non-specific adsorption is important in nutrient absorption.

The protoplasts of the endodermal cells may be considered as an osmotic through-route, while those of the cortex, and most other living cells in the plant, are osmotic *cul-de-sacs* in dynamic equilibrium with the ionic status and water potential of the passing transpiration stream. The entry of solutes into these cells has been the source of endless experimentation and discussion and numerous models of the cation–anion uptake–exclusion system have been postulated. More recently these have centred around metabolically driven 'pumps', biochemically akin to the ATP-ase coupled pumping of potassium into, and sodium out of, animal cells (Jennings, 1969). Because of their experimental convenience, most of the work suggesting that ATP-ase pumps function in plants has been done with algal giant cells in such species as *Nitella translucens,* other Characeae and *Hydrodyction africanum.*

McRobbie (1971) suggests that a cation exchange pump, driven by ATP, causes K-influx and Na-efflux, in *N. translucens*, and is accompanied by an anion pump causing chloride influx. This differs from the cation exchange pump in being insensitive to uncoupling agents and must be driven by some other energy source than ATP. In an earlier review, Dainty (1962) expressed the opinion that plant cells probably differ from animal cells in possessing anion pumps, which are the favoured mechanism for producing a high internal osmotic water potential, and are an evolutionary consequence of the possession of a cell wall. Chloride is pumped in the algae and organic acid anions in the higher plants. Dainty commented that his views were speculative and, in 1969, Jennings was still able to write; '. . . almost everything which has been written about ion pumps in higher plants is almost all speculation'.

The preceding paragraphs are not intended to be a comprehensive survey of ion uptake by plant cells; rather, they indicate the current viewpoint in this branch of physiology and serve as a source of reference. If it is accepted, as now seems most likely, that higher plant cells have metabolically driven pumps which cause both anion and cation uptake, it follows that regulation of their activity may establish electrochemical potential gradients and bring about movements of other ions. Furthermore, the metabolic production of ionic, from non-ionic, substrates may cause ion accumulation. For example, synthesis of organic acid anions causes passive cation accumulation. Once these mechanisms are functional it is not necessary to postulate active pumping of each ion species; creation of a potential gradient by pumping of a single ion, in conjunction with differential permeability characteristics at the plasmalemma, will result in differential rates of passive ion accumulation. Such differential permeabilities could be related to the specific protein linings of membrane pores and would be under genetic control (Jennings, 1967). Other views have been expressed, for example, Epstein and Jefferies (1964) in discussing the genetic basis of selective ion transport in plants concluded that selectivity is usually carrier-mediated by the transient formation of ion-specific or group-specific complexes but Jenning's (1967; 1969) interpretation permits an alternative explanation.

The existence of differential permeability or ion-selective pumping mechanisms suggests that capacities for absorption and transport are biochemical phenomena subject to the normal genetic control and, in the context of the widely varying natural soil environment, may account for the various distribution patterns of plants in relation to nutrient supply and tolerance of unfavourable habitats which ecologists have described (Loughman, 1969). Differential effects of this type may extend beyond the living cell surfaces and out into the rhizosphere zone: Woolhouse (1969) has observed that the acid phosphatases of root surfaces may be ecotypically variable in relation to phosphate availability and the ionic composition of the soil solution. At the opposite extreme, in the absence of exclusion processes or transport limitation some ions may enter plants very freely and internal tolerance mechanisms must play some part in their ecological relationships.

Epstein (1969) noted that the metabolic machinery of plants has to cope with the chemical environment in three ways: (a) they must selectively acquire essential elements, (b) they must cope with those elements present in excess and (c) they must acquire water. The selective absorption of essential nutrients often involves considerable concentration from the external solution, for example, potassium may often have a concentration of less than 0·001 me/ml in the soil solution but reaches several hundred times this concentration in the vacuolar fluid. Despite its chemical similarity, and generally higher concentration in soil solution, sodium is not taken up in quantity, nor does its presence competitively restrict potassium uptake to a serious extent. In conflict with Jennings' (1967) suggestions, Epstein considered most nutrients to be taken up by carrier mechanisms (pumps). For most nutrients he postulated two distinct mechanisms, one operating at low external concentrations of the specific ion and the other at high concentrations. For potassium it is the first mechanism which is highly selective and discriminating against sodium.

Epstein discussed the halophytes, an extreme group of plants which have to cope, not only with the normal problems of selective uptake, but also with levels of salt concentration toxic to normal plants and with osmotically induced difficulties of water uptake. In this case he suggested that the second mechanism of potassium uptake is active and does not discriminate so strongly against sodium uptake as the low concentration mechanism. In consequence, a high internal ion concentration is built-up, protecting the plant against the water loss which would otherwise ensue because of the high osmotic concentration of the external medium. Thus halophytes are distinguished from other plants, not by their capability of withstanding 'physiological drought', but by their ability selectively to accumulate ions to very high internal concentrations and to tolerate the metabolic effects of such high concentrations. The general succulence of the halophytes may be teleologically interpreted as an increase in cell volume/dry weight with a resultant dilution of the cell solutes.

Alternative explanations are possible and Jennings (1968) has put forward a theory which accounts not only for halophyte succulence but also for the induced succulence caused by aridity and increasing light intensity. He postulates pumps, similar to those described above for various Characeae: an outwardly directed sodium pump at the plasmalemma and an inwardly directed anion pump. In addition, sodium is pumped from the cytoplasm, across the tonoplast, into the vacuole. The sodium pumps are ATP/ATP-ase driven and Jennings suggested that a high external sodium concentration may reverse the pump, causing synthesis of ATP. The tonoplast pump exports the sodium irreversibly into the vacuole as the tonoplast appears to have a very low passive permeability to solutes. High light intensities and aridity, by increasing transpiration rates, may cause an increase in ionic concentration external to the endodermis, but they also increase root permeability and decrease ion selectivity so that higher than normal ratios of Na/K enter in the xylem fluid. In each case the succulence may be interpreted as a dilution homeostasis effect, while the ATP synthesis by the reversal of the Na pump may increase wall plasticity or synthesis of wall material. High light may further enhance such a relationship by increasing photophosphorylation.

In conclusion, it may be said that plant cells are demonstrably capable of selectively accumulating cations and anions, against concentration gradients, both by active pumping and by passive diffusion. Selectivity is achieved through the specific configuration of biochemical carriers or by modification of the passive permeability of biological membranes to particular ions. Diffusive uptake must be related to the creation of electrochemical potential gradients by the activity of pumps, or to the production of organic ions from non-ionic substrata within the cell. A good example is the synthesis of organic acids which may then exchange hydrogen ions for metal cations through the cell membranes. Bicarbonate, formed by decarboxylation in roots, may exchange for nitrate ions from the external solution, the nitrate then being carried, in association with potassium, to the leaves via the mass flow system of the transpiration stream. Bicarbonate is formed during nitrate metabolism and, after conversion to carboxylic acids, is transported, with potassium, via the phloem to the roots. A cyclic absorption process is thus established in which one anion is exchanged for another.

ROOTS, SOIL AND NUTRIENTS

A great deal of our knowledge of ion absorption by roots is derived from experiments with excised roots in a state of nutrient and metabolite depletion, and from solution-cultured plants which may be physiologically abnormal and certainly have different kinetic relationships with the surrounding liquid by comparison with soil-grown plants. It is possibly for this reason that more extreme ecological situations, such as halophytism or tolerance of toxic elements, have been a more rewarding eco-physiological field of study than investigation of normal plants. The physiological differentials associated with these rather unusual situations are probably sufficiently marked to persist and be measurable even under artificial culture conditions.

The problem in the investigation of soil-rooted plants is twofold. Firstly, the roots are in contact with only small volumes of solution, in close proximity to the solid-state exchange surfaces of soil particles, so that the diffusion and mass flow characteristics of dissolved ions are very difficult to specify. Secondly, the biological relationships in soil are entirely different from those in nutrient solutions. The root is enveloped by a sheath of biological influence, the rhizosphere, which not only differs chemically and physically from the surrounding soil but also has a specialized microflora. Barber (1969) has reviewed this subject and concluded that the inorganic nutrition of plants cannot be understood without considering the role of rhizosphere and root surface microorganisms. For some nutrients, in particular phosphorus which is often in short supply, experiments with sterile and non-sterile roots often produce very different results. Generally, if the phosphorus is present in an insoluble form, the presence of microorganisms on the roots increases its availability.

The rhizosphere, in addition to having differential characteristics due to the passive leakage of metabolic biochemicals, may also contain materials which are specifically secreted to aid in nutrition and water uptake. The root surface phosphatases and other enzymes may fall into the former category and the gels which Floyd and Ohlrogge (1971) found at root surfaces might well stabilize the contact between roots and soil under water-deficient conditions. Chemical gradients may be established during absorption of nutrients, particularly when they are present at low concentration in the soil solution and are mobilized only slowly from the solid state. Barber et al. (1963) demonstrated gradients of cation depletion around roots, in soil, using ^{86}Rb as a tracer and Weavind and Hodgeson (1971) found similar gradients in the absorption of ^{59}Fe from an agar gel under iron deficiency conditions.

It should be remembered that most natural soils are fairly deficient in at least some of the essential elements and, under these circumstances, the translocation of nutrients from the large soil volume in which they were collected and their redeposition either by leaf fall or death of whole plants, must play an important part in establishing both horizontal and vertical gradients of nutrient distribution in ecosystems. Zinke (1962) demonstrated such gradients as a response to tree growth: nitrogen content of the surface mineral soil increased by a factor of about 10 under the canopy of *Pinus contorta* compared with the soil outside the canopy. Considerable reduction of pH under the canopy and differences in exchangeable cations were also shown. Goodman and Perkins (1959) investigated *Eriophorum vaginatum* growing in very nutrient-deficient blanket peat and concluded that the plant tussocks act as localized reservoirs of nutrients, particularly potassium and phosphorus, which have been gathered from a much larger area than that of the tussock base. These observations suggest that, at least in the earlier stages of succession, plants are likely to impose some degree of local patterning by their nutrient gathering characteristics and, for the less mobile nutrients, the pattern mosaic may persist for much longer than the life of the plant.

Soil solutions contain a mixture of nutrient ions and, often, some toxic elements. The relative concentrations of these ion species are difficult to determine and fluctuate with soil water content and plant activity. An ecologically relevant study of nutrient uptake is, consequently, more difficult than the physiological characterization of absorption from simple ionic mixtures. From the earliest years of experimentation it has been noted that mixtures of ion species may interact with each other, either antagonistically or synergistically; thus calcium has been recognized as a 'depoisoner' which antagonizes the uptake of some other toxic elements: sodium, for example, is tolerated at much higher concentrations than normal in the presence of calcium. This particular antagonism has been used to ameliorate the harmful effects of saline irrigation water, by adding large amounts of soluble calcium salts. Sodium seems to differ from the other alkali metal cations as its uptake is inhibited by equivalent concentrations of most other nutrient cations, whereas potassium and rubidium uptake are generally stimulated by bivalent and some trivalent cations. It is difficult to generalize about cation synergism/antagonism as there are many conflicting reports in the literature; Broeshart (1967), for example, cites references to both stimulation and inhibition of potassium and rubidium uptake by calcium. It is equally difficult to interpret the literature in relation to anions, but calcium, and some other divalent cations, generally stimulate the uptake of nutrient anions. The situation in the field must, then, be a complex function of plant species and the relative concentrations of ions in solution; direct competition for uptake sites, antagonistic effects and synergism all being involved.

MINERAL NUTRITION AND pH

A further complication in the nutritional relationships of plants arises from the wide range of pH values which may be encountered in the field. The common soil pH range from pH $8 \cdot 0$–$3 \cdot 0$ represents an increase in hydrogen ion concentration of $\times 10^5$ amd this is accompanied by extensive differences in the chemical characteristics of the soil. The majority of the effects of pH on plant growth are mediated directly, or indirectly, through mineral nutrition. Most plants show physiological optima between about pH $5 \cdot 0$ and $7 \cdot 0$ and, above this range, tend to have nutritional problems such as iron or manganese deficiency, due to the favouring of the less soluble valency form, or phosphorus deficiency caused by the formation of insoluble calcium phosphates. Iron deficiency in calcareous soils often appears as lime-induced chlorosis, for example Hutchinson (1968) showed that susceptibility to lime-induced chlorosis and iron-deficiency chlorosis were strongly correlated in *Teucrium scorodonia* ecotypes. Grime and Hodgeson (1969) consider lime-induced chlorosis to be a consequence of a physiology adapted to acid soils. The chelation mechanism responsible for preventing excessive Al^{3+} uptake also limits absorption of iron.

Generally, acid soils cause a wider range of plant stress symptoms than alkaline soils and the causes of the 'soil acidity complex' have been summarized by Hewitt (1952).

(i) Direct injury by hydrogen ions (low pH).

(ii) Indirect effects of low pH.

    (a) Physiologically impaired absorption of calcium, magnesium and phosphorus.

    (b) Increased solubility, to a toxic extent, of aluminium, manganese, and possibly iron and heavy metals.

    (c) Reduced availability of phosphorus partly by interaction with aluminium or iron, possibly after absorption.

    (d) Reduced availability of molybdenum.

(iii) Low base status.

    (a) Calcium deficiency.

    (b) Deficiencies of magnesium, potassium or possibly sodium.

(iv) Abnormal biotic factors.

    (a) Impaired nitrogen cycle and nitrogen fixation.

    (b) Impaired mycorrhizal activity.

    (c) Increased attack by certain soil pathogens.

(v) Accumulation of soil organic acids or other toxic compounds due to unfavourable oxidation–reduction conditions.

Direct injury by high H^+ ion concentration does not usually occur in soil as abnormally acid conditions are required for cell membranes to be damaged. Cation absorption may, however, be reduced by competitive inhibition if H^+ ions occupy too many of the primary binding sites of the uptake mechanism. The early work of Arnon and Johnson (1942) supports this contention as they found a number of test plants to grow equally well between pH 4·0 and 8·0 if an adequate nutrient supply was maintained. Phosphorus absorption may be impeded as rising pH, within the biologically tolerable range, shifts the ion form from $H_2PO_4^-$ to the bivalent HPO_4^{2-} and then, at very high pH values to PO_4^{3-}. In addition to this complication, the increasing solubility of iron and aluminium, at low pH values, may cause precipitation of insoluble phosphates in the soil or at root surfaces. The combined effect of these factors and the insolubility of calcium phosphates at high pH causes potential phosphorus deficiency at both extremes of the soil pH scale. Low pH soils are often coarse-textured, of low cation exchange capacity and easily leached, thus suffering losses of easily mobilized elements such as calcium, magnesium and potassium either by desaturation of the exchange complex or by solution of calcium and magnesium carbonates. In tropical oxisols the high degree of weathering causes a great reduction in cation exchange capacity and, again, deficiencies of the mobile cations are likely. Some oxic soils carry a net positive charge and have no cation-exchanging activity (Soil Survey Staff, 1960).

Variation in soil pH *per se* does not directly influence plant growth; the very marked differences between the natural vegetations of acid and

alkaline soils seem to arise either from mineral nutritional disturb-
ances or from nutritional effects on the soil microflora. Nitrogen is often
deficient, for example, in acid soils as nitrogen-fixing organisms require a
high calcium supply. The various plant species which respond as calcicoles
or calcifuges have become selected for the possession of physiological
mechanisms which permit them to cope with nutritionally abnormal environments, or tolerate toxic effects, thus allowing them to escape competition from otherwise more vigorous plants. The calcicole–calcifuge
problem is discussed more fully in the following chapter (p. 263).

THE FUNCTION OF THE ESSENTIAL NUTRIENTS

To understand the ecological implications of mineral nutrition it is
necessary to appreciate the physiological role of individual nutrients and
their interactions in so far as they influence either competition or survival
in extreme habitats. Extensive reviews may be found in Steward (1963) and
Ruhland (1958). The detailed knowledge of elemental deficiency symptoms
is almost all drawn from experiments with crop plants. For obvious
reasons, wild plant species do not normally occur under conditions which
impose extreme growth limitation but those wild species which have been
grown experimentally under nutrient stress appear to respond similarly to
cultivated plants though generally having a lower nutrient requirement.

INTERACTIONS

In the field a number of deficiencies may occur simultaneously and with
strong interactions. This is particularly apparent in the relationship of the
three major elements, nitrogen, phosphorus and potassium, as illustrated
by Figure 8.4 which shows the effects of factorial additions of N, P and K on
leaf number in *Ipomoea caerulea*. The plants were grown in a basal medium
which supplied sufficient NPK to prevent deficiency symptoms but caused a
considerable limitation of growth. Addition of N to the basal medium
caused a growth increment but P, K and PK, in the absence of N, had little
influence. N shows a slight interaction with P, K and PK, its response line
becoming steeper with each further interaction. Much stronger interactions are apparent when K is added in the presence of N and NP or P in the
presence of N and NK. This particular set of interactions is fairly simple
and may be interpreted as the result of N limiting the response to P and K.
Less markedly K or P act as limits in the NP and NK treatments, thus
reducing the response to N in these two cases. Much more complex interactions may occur and the involvement of synergism, antagonism or
chain-linked physiological effects makes interpretation and investigation
very difficult. In natural ecosystems a single nutrient element addition may

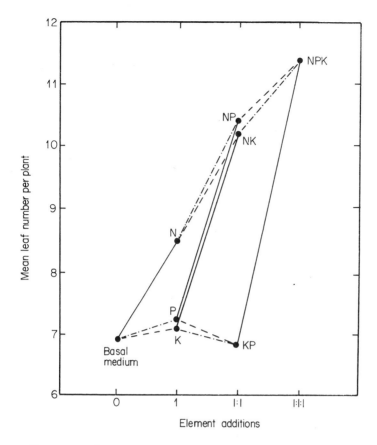

Figure 8.4 The interaction of leaf number with factorial NPK addition in *Ipomoea caerulea* (Data of Njoku, 1957). The method of plotting is that of Richards (1941) which permits interactions to be identified visually. If there are no interactions, the set of response lines for a single element addition will be parallel. In this case N shows a slight interaction with P, K and PK, but the K interaction with P, N and NP is very much stronger. ——— N response; –·—·—·— P response; – – – – – – K response.

often elicit a response but usually it also reveals the existence of other limiting deficiencies. Furthermore, in mixed species associations, differential responses often lead to changes in the vigour and representation of individual species. The very great differences in floristic composition and productivity which exist between eutrophic and obligotrophic habitats, and which often overrule climatic considerations, may be interpreted as the outcome of differential competitive effects in relation to nutrient supply.

Chapter 9

Biogeochemical cycling and the ecology of mineral nutrition

With the origin of life on earth there came a change in the physical and chemical processes which normally lead to the random dissipation of free energy. In a non-living system, with no external energy source, the general trend is for ordered arrangements of molecules to become disordered by thermal agitation, the loss of orderly structure being defined by the increasing *entropy* of the system. The first chemolithotrophs were able to harness the free energy of naturally occurring chemical compounds to create localized accumulations of organic matter with an extremely complex, ordered structure, while the evolution of the photosynthetic process further sophisticated the impact of life on the geochemical environment as the exploitation of solar energy allowed much more extensive synthesis of living material.

The utilization of an external source of energy, the sun, permitted the apparent reversal of the universal entropy stream at the earth's surface and thus, the evolution, firstly of the chemo- and photolithotrophs and then of more complex heterotrophic organisms, imposed a directional force on virtually all surface geochemical processes and caused an immense alteration in the history of the development of planet earth. The most dramatic change was the institution of photochemical oxygen production in an atmosphere which was, primevally, in the reducing condition. The first chlorophyll-bearing plants appeared some three to three and a half thousand million years ago in the Pre-Cambrian and the atmospheric oxygen content immediately began to rise, first reaching its present level at the beginning of the Carboniferous. During this rise, the carbon dioxide content of the air was reduced by immobilization of carbon in organic compounds and by biogenesis of insoluble carbonates caused by photosynthetic withdrawal of CO_2 from dissolved bicarbonates in seawater. A large proportion of the earth's limestone deposits were formed in this way by algal photosynthesis.

At present the $O_2:CO_2$ ratio of the atmosphere approximates a 'closed box equilibrium' in which the CO_2 concentration of 300–350 v.p.m. represents the CO_2 compensation point of a photosynthetic system which is illuminated for approximately half of the 24-hour diurnal cycle, and also contains respiring heterotrophs in balance with the photosynthetic energy fixation. It may be inferred that plants operate at an ambient CO_2 level which is suboptimal for photosynthesis (Nichiporovich, 1969) and the proof of this is found in the experimental increase of photosynthetic rates by CO_2 enhancement in full sunlight. The prevailing oxygen content of air (21% v/v) is supraoptimal for photosynthesis as shown by Bjorkman (1966), who found enhanced rates down to 2% O_2 or below (Warburg effect).

The photosynthetic plants of the earth's surface inhabit an atmosphere which they, themselves, have produced and a soil system which is a reflection of the dynamic balance between the input of plant materials and the catabolism of heterotrophs. The consequence of these relationships is the establishment of a number of cycles of essential elements, some of which are essentially local and some global in scale (Figure 9.1). The local cycles involve the less mobile elements in which there is no mechanism for long-distance transfer but the global cycles have a gaseous component which links all of the world's living organisms to form the giant ecosystem which we know as the biosphere. In relation to geological time scales the functioning of all the essential element cycles is extended by the various processes of continental denudation and sedimentation or other redeposition of materials. Within this reference framework, biological activity is responsible for the process of *biogeochemical cycling*.

These cycles may be divided into three global types.

(i) The gaseous cycles of carbon, oxygen and water, elements which are not normally regarded as nutrients, but are either utilized or produced in photosynthetic and respiratory processes.

(ii) The gaseous cycle of the nutrient element, nitrogen.

(iii) The non-gaseous, sedimentary cycles of the remaining nutrient elements. Sulphur is to some extent intermediate as H_2S or SO_2, formed under some circumstances, add a gaseous component to its normally sedimentary cycle.

The sedimentary cycles are, in the short term, localized within ecosystems and the transport of their components by sedimentary processes permits them to become fully closed cycles only with the passage of geological time. The localized portions of the cycles, which are directly involved with biological activity, are strongly influenced by ecosystem energy flow so that an increased flow usually causes more rapid cycling. The strongest nutrient fluxes become associated with the main energy flow pathways in the producer–consumer food web; Figure 9.2 shows this relationship and also indicates the biological concentration of nutrients which may accompany the growth of an energy flow pathway in an evolving

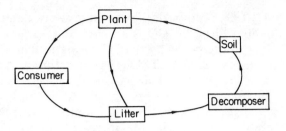

(a) Local cycles of P, K, Ca, Mg, Cu, Zn, B, Cl, Mo, Mn and Fe

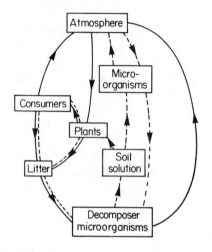

(b) Global cycles of C, N, O and H (pecked lines represent path of N)

Figure 9.1 The two contrasting types of essential element cycles. Sulphur is, to some extent, intermediate as it normally has a local type of circulation but may become global if it enters the atmosphere as hydrogen sulphide. (a) Local cycles of P, K, Ca, Mg, Cu, Zn B, Cl, Mo, Mn and Fe. (b) Global cycles of C, N, O and H.

ecosystem. It should be appreciated that the concentrations of nutrients at soil surfaces, caused by litter fall, and the accumulation of nutrients, in living organisms, from a low ambient concentration, are energy-demanding processes.

The relationship between energy flow and mineral cycling has been one of the more fruitful fields for the application of systems analysis method,

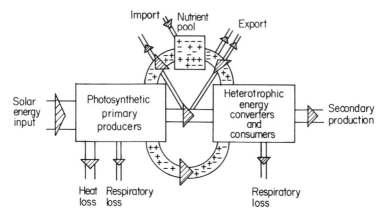

Figure 9.2 The relationship of the ecosystem nutrient cycle to energy flow. After E. P. Odum, *Fundamentals of Ecology*, 3rd. Edn., Saunders, Philadelphia, 1971, p. 87.

the classic Liebig limiting factor approach being too inflexible to cope with the complex interactions which are now known to occur in plant nutrition. Furthermore, as the flux of materials through an ecosystem gives some measure of its continuity and stability, and also follows the pathway of energy flow in food webs, Pomeroy (1970) suggests that the analysis of mineral cycling is a useful strategy with which to approach ecosystem analysis.

THE CARBON CYCLE

The primary consequence of the photosynthetic process is the storage of energy in reduced carbon compounds. These may then be exploited by the consumers and decomposers of the ecosystem so that the throughput of energy is accompanied by a cyclic exchange of CO_2 with the atmosphere. This exchange takes place over the whole land and ocean surface of the earth wherever temperature is high enough, or water supply adequate, to support life. It may be considered to be the most significant of the biogeochemical cycles as it is through the synthesis of carbon compounds that ecosystem energy flow is linked to the cycling of all other materials. Figure 9.3 illustrates the global carbon cycle and indicates the quantities involved in the various storages and annual fluxes. Though these estimates cannot be particularly accurate they do indicate relative magnitudes; on a geological time scale it may be seen that photosynthetic carbon consumption would rapidly exhaust easily available resources if it were not for the respiratory return of CO_2 to the atmosphere. This return is at present nearly equal to the photosynthetic production as the prevailing world climate is not favourable to the formation of extensive fossil organic matter deposits.

The atmospheric CO_2 concentration is globally rather constant, though a small seasonal variation over the earth's surface is correlated with vegetation cover and photosynthesis (Lieth, 1970) and local enrichment derives from urban-industrial burning of fossil fuel. The constancy of the atmospheric CO_2 concentration is due to the buffering effect of solubility in ocean water, coupled with the mixing caused by large- and small-scale air turbulence. Despite the large amount of fossil carbon which has been returned to the atmosphere since the industrial revolution, the ocean buffer has so far prevented an excessive increase in concentration, but it is

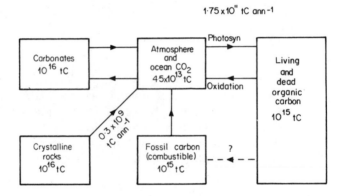

Figure 9.3 The global carbon cycle. Data from Nichiporovich (1969), Wassink (1968) and Revelle and Fairbridge (1957).

estimated that, by the year 2000, there will have been a 25% rise. The consequences of such an increase can only be guessed, but it is thought that the energy budget of the earth's surface may be altered by the 'greenhouse effect'; the trapping of longwave black-body reradiation of energy, from the earth's surface, into space. This effect is due to the transparency of CO_2 to the shortwave income from the sun and its opacity to the much longer wave, low temperature reradiation. The obvious consequence would be an increase in atmospheric temperature but further alterations such as an increase in global cloud cover might also be expected to have a homeostatic affect (Revelle et al., 1965).

THE OXYGEN CYCLE

For each molecule of CO_2 taken in during photosynthesis, a molecule of O_2 is released. Such is the scale of this process that the whole of the atmospheric oxygen content could be generated in about 2000 years, geologically an incredibly short time. Contrasting this with the immensity of geological time, it becomes easier to accept the fact that the present atmosphere has been biologically generated. The constancy of its composition reflects the efficiency of plants as regulatory organisms: because CO_2 is at a strongly limiting level, any increase in its concentration will immediately be offset by photosynthetic consumption. Over periods of geological time there have probably been some oscillations in atmospheric composition, for example, during the late Palaezoic there appears to have been a considerable decline in oxygen and increase in CO_2 concentration, which may have been responsible for triggering the 'autotrophic bloom' which created most of our fossil fuel reserve (Odum, 1971).

THE WATER CYCLE

Some $4·4 \times 10^{14}$ tonnes of water are precipitated and evaporated, annually, over the earth's surface. This is by far the largest global process in which plants are involved, most of the water evaporated from fully vegetated land surfaces being lost via the transpiration stream. Comparatively, only a very small proportion of the water taken up by plants is used in the photosynthetic process. The larger proportion is lost by transpiration from leaves and the latent heat exchange during its evaporation is a very significant contribution to the ecosystem energy budget.

The water cycle also plays a transporting role in geochemical cycles. More than 75% of global evaporation is from ocean surfaces, but less than 75% returns directly to the oceans as precipitation. There is a consequent run-off from land surfaces which removes soluble and particulate materials and deposits them in ocean basins, thus forming the transport mechanism for the sedimentary cycles of the less mobile elements. Further discussion of the water cycle may be found in Chapter 6, p. 177, and Hutchinson (1957) also gives an excellent account of the subject.

The continuing, worldwide study of the water and carbon cycles has taken on greater urgency with the realization that man is beginning to interfere with both, to an extent which is causing marked changes in fluxes and storage. To make objective decisions concerning the management of the world's water and CO_2 resources, it must quickly be established what direct damage may accrue from these changes and whether any of them have a trigger function in climatological processes. An example of an altered flux is the removal of large areas of natural vegetation, which increases surface run-off rates and reduces infiltration, thus preventing the full recharge of natural aquifers. At a time when man is already over-pumping many underground water sources this could promote serious ecological, agricultural and industrial-urban crises. In the context of triggering effects, change of atmospheric CO_2 concentration may not, in itself, be significant but the change of mean global temperature, by even less than 1°C, might have far-reaching consequences such as the melting of polar ice and an accompanying sea-level change. The answers to these problems are not known and highlight the need for a global systems ecology.

THE NITROGEN CYCLE

Gaseous nitrogen is the most abundant element in the atmosphere and its global circulation provides an inexhaustible reservoir for the nitrogen-fixing organisms which supply almost all of the nitrogen utilized by plants. The quantity of nitrogen combined in living or dead organic matter is small compared with the total capacity of the atmospheric reservoir, a characteristic in which nitrogen differs from carbon cycling. The nitrogen cycle may, thus, be divided into a simple gaseous reservoir of enormous capacity, and a complex, soil-based cycle which is localized and of small magnitude (Figure 9.4a and b). In most ecosystems the N-fixation step is rate-limiting and plants are in strong competition for soil-N. Under these circumstances the rate of N-recycling within the system is very important, for example, in some peat soils, a high percentage of the ecosystem-N is immobilized as organic matter in the soil, thus limiting primary production and energy flow. By contrast, many tropical rainforests have the majority of the available-N in the standing crop, recycling from litterfall is extremely fast and the ecosystem has a high energy throughput.

The nitrogen cycle is geochemically stable and not easily disturbed by loss to some inaccessible pool, such as the sedimentary deposits of insoluble compounds which disrupt the phosphorus cycle. The stability is mainly due to the buffering effect of the very large atmospheric reservoir but also the efficiency of the denitrifying process in most soil environments.

THE PHOSPHORUS CYCLE

Simpler than nitrogen, the P-cycle has fewer pathways and chemical conversions but, lacking a large and mobile reservoir, it is more easily interrupted by both natural processes and human interference. Figure 9.5 shows that the cycle has two main storage pools of insoluble organic and inorganic-P while a much smaller pool of dissolved-P permits movement between them. Mobilization of P from the organic form is entirely a

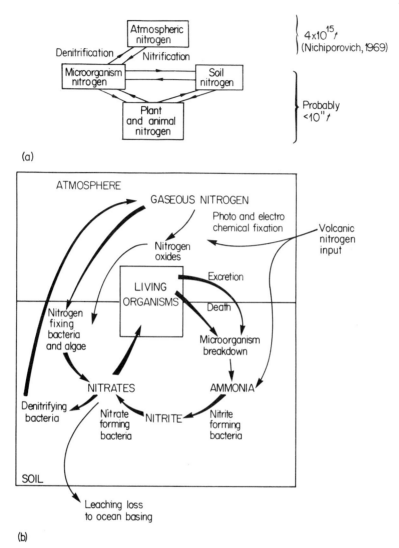

(a)

(b)

Figure 9.4 The nitrogen cycle. (a) Major relationships between the very large atmosphere pool of gaseous nitrogen and the biosphere. (b) The complex interrelationships of the soil-based portion of the cycle.

biological process while microorganisms and root surfaces probably play a large part in dissolving inorganic-P.

The removal of P to ocean basins by sedimentary processes enormously increases the time scale of the cycle which becomes complete only when geological processes expose sediments on newly formed land surfaces. The loss has now been accelerated by the mining of phosphate rock for fer-

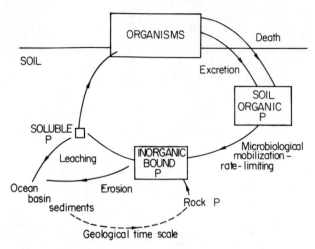

Figure 9.5 The local cycle of phosphorus. Both the organic and inorganic pools of soil phosphorus are essentially rather immobile with the result that the mobilization of organic P into the plant-available phosphate form is rate-limiting to the cycle, and also that the leaching loss of soil P is fairly slow. In many cases it is physical loss by soil erosion which removes P from natural ecosystems and agricultural land. Loss by these two mechanisms is only counteracted on a geological time-scale when ocean basin sediments become elevated to form new land surfaces. Soil P is normally very insoluble but the addition of fertilizer P causes a great increase in the rate of leaching to ground water or water courses.

tilizer, a form in which it is easily leached to the groundwater and thence to rivers. Though the present resources of phosphate rock are adequate, the irreversible loss may ultimately be disastrous for agriculture unless some means of recycling the surplus P can be found. At present the impact of this loss is felt, not in P shortage but in the superabundance of P which is entering natural waters, so forming part of the pollution–eutrophication syndrome.

THE SULPHUR CYCLE

The sulphur cycle has affinities with the gaseous nature of the N-cycle and with the sedimentary nature of the P-cycle because the loss to ocean basins may be partially reversed by the evolution of hydrogen sulphide which ultimately returns to land surfaces as SO_4^{2-} in precipitation (Figure 9.6). The gaseous phase does not, however, form a large reservoir as it does with nitrogen. The soil-based part of the cycle involves a microbiological oxidation–reduction interchange between sulphate, elemental sulphur and sulphide. Under aerobic conditions a variety of heterotrophic organisms decompose organic-S compounds and release SO_4^{2-} but under anaerobic conditions H_2S is produced. In addition, anaerobic sulphate reduction is caused by *Desulphovibrio desulphuricans* which also produces

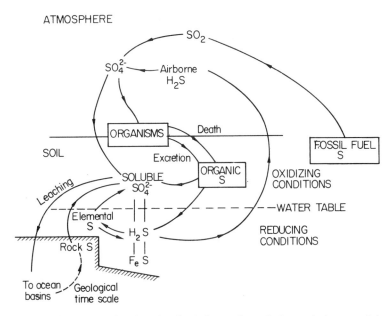

Figure 9.6 The hybrid cycle of sulphur. Though the cycle is essentially local it may be opened by the leaching of soluble sulphate into a reducing zone below the water table. If hydrogen sulphide is produced it may be liberated to the atmosphere where it rapidly oxidizes to sulphate. The conversion of organic sulphur to the plant-available sulphate and the various interconversions of sulphur in wet soil are all bacterially mediated processes.

H_2S, while various sulphur bacteria may oxidize H_2S to elemental-S and then to SO_4^{2-}. When a waterlogged soil dries, and becomes aerobic, bacteria such as *Thiobacillus thiooxidans* oxidize H_2S to SO_4^{2-} and, as a result, extreme soil acidity may arise. Problems have occurred where metal or concrete structures are exposed to the corrosive effects of such soils. Commonly, in soils, H_2S reacts with ferrous iron to precipitate black ferrous sulphide but, in continental shelf muds at rather high pH values, insufficient ferrous iron is available and H_2S is liberated to the atmosphere where it oxidizes to SO_4^{2-} and may be returned to terrestrial ecosystems in precipitation. Most fossil fuels contain a considerable amount of sulphide, or elemental-S, as they were deposited under anaerobic conditions conducive to sulphide accumulation. The burning of these fuels has added a sulphur load to the atmosphere which is, perhaps, larger than that deriving from continental shelf muds. Sulphur pollution enters the atmosphere as SO_2, which is fairly rapidly oxidized to SO_4^{2-}, but, near to the sites of emission, SO_2 phytotoxicity may occur, ranging from the complete destruction of surrounding vegetation and erosion of soil which occurred as a result of sulphide ore smelting, in Copper Basin, Tennessee, to the establishment of lichen and bryophyte deserts in industrial cities (Leblanc and Sloover, 1970). Gilbert (1970a, b) has suggested that lichens and bryophytes may be used to assess levels of SO_2 pollution because their sensitivity is so great.

OTHER ELEMENTAL CYCLES

The remainder of the nutrient elements have cycles of a sedimentary nature, of which phosphorus was cited as a typical example. In the case of manganese and iron, a number of interesting, microbiological redox changes are involved. Some of the changes are caused directly by microorganisms and others by the redox changes consequent on biological activity. The behaviour and mobility of iron and manganese, in soil and in geochemical cycles, is strongly influenced by redox and pH conditions because the higher valency forms are generally less soluble, and the solubility of all forms is reduced by raised pH (Figure 8.2). The very high concentrations of iron compounds which accumulated as iron ores in the geological past, and are still accumulating in bog, lake and marine basins, are probably all attributable to the biological mobilization of ferrous iron and its local reoxidation and precipitation. The manganese dioxide concentrations which may be found in many poorly aerated soils, and as marine basin nodules, are also the outcome of a biological reduction–oxidation process. Quastel (1963) suggests that the presence of free manganese dioxide in soil may be necessary to buffer the soil against accumulation of toxins under conditions of transient waterlogging. An example is the accumulation of thiol compounds under reducing conditions, which is prevented by the presence of MnO_2.

The cycling of elements through ecosystems results in local changes of concentration related to the deposition of organic litter on the soil surface. These changes commence when a freshly exposed substratum begins to be colonized by living organisms, and progress as the sere advances, to reach equilibrium when a climax vegetation is established. It was noted in Chapter 8 that biomass ceases to increase when climax is reached. This suggests that energy flow is maximal in climax conditions and, as nutrient cycling is a function of energy flow, it is likely that the cycling rate is also maximal under climax conditions. Whether a climax can be considered to have very long-term stability, in the absence of climatic change, may well depend on the balance between irreversible loss from the system by leaching or migration export, and the income from precipitation, blown dust and continued weathering of parent rock material.

SOILS AND NUTRIENT CYCLES

Commercial afforestation has made it possible to compare the effects of sudden changes in vegetation composition on nutrient distribution within profiles. Wright (1956) investigated dune soils, in Scotland, under different-aged plantations of *Pinus nigra* var. *calabrica*, and showed, for several elements, that absorption of nutrients reduced availability in the deeper parts of the profile, but the return in litterfall increased availability near the surface (Figure 9.7). The effect was particularly marked for magnesium and calcium but potassium initially decreased in the surface soil and the loss was never quite made up by the litter input. Leaching, superimposed on the biological cycle, may have been responsible for the overall lowering of potassium availability.

Conifers, and in particular pines, are less active in nutrient biocycling than deciduous species. Rodin and Bazilevich (1967) cite values of 50–100 kg/ha of ash elements annually returned in the litter of conifers and 200–270 kg/ha for deciduous species. This difference is particularly marked for calcium, which is much more strongly cycled by deciduous species than by conifers, and is probably responsible for maintaining the cation-saturated status of most Brown Forest Soils by counteracting the leaching process, returning nutrients from the deeper horizons to the surface (Figure 9.8).

In productive ecosystems a large quantity of nutrient may become immobilized in the standing crop and litter. In nutrient-rich systems, though the amounts immobilized are large, they form only a small proportion of the total soil resource but it has been postulated that the vegetation of oligotrophic soils may immobilize a considerable proportion of the ecosystem nutrient. Rennie (1955) suggested that afforestation of nutrient-poor soils and removal of timber may cause a nutrient loss which will degrade the habitat and prevent more than one timber crop from being grown. This conclusion is of great silvicultural significance but is still in dispute, for example Miller (1963) gave estimates of nutrient removal in the timber of 100-year old *Nothofagus truncata* which, though large, were of the same order as the annual additions from rainfall. Afforestation with conifers may, however, be unwise in some marginal soil areas as their more limited activity in biocycling of nutrient elements may increase rates of leaching loss; a subject which is in need of further research.

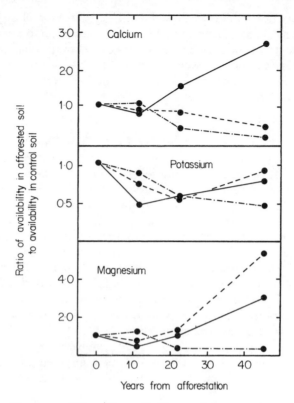

Figure 9.7 The effect of afforestation on nutrient availability in the upper 60 cm of a soil profile. The data for 0 years are from control sites adjacent to the afforested plots.

Soil depth ——— 0 cm; — — — — — — 15 cm;
— · — · — · — 60 cm. Data of Wright (1956).

Despite the large quantities of nutrients held in the living plant cover, the annual turnover is considerable. It mainly derives from litterfall but there is leaching of some elements from leaves by precipitation throughfall and stemflow. The turnover is a closed biological cycle and is responsible for establishing the nutrient distribution patterns described above. The cycle is not completely closed, a geochemically open component arising from the leaching of nutrients to the groundwater, soil erosion and input of nutrients in precipitation. As noted earlier in this chapter, the geochemical part of the nutrient cycle becomes closed only with the passage of extended periods of geological time.

Estimates of input in precipitation vary widely as sample collection and the measurement of low concentrations of nutrients in precipitation are

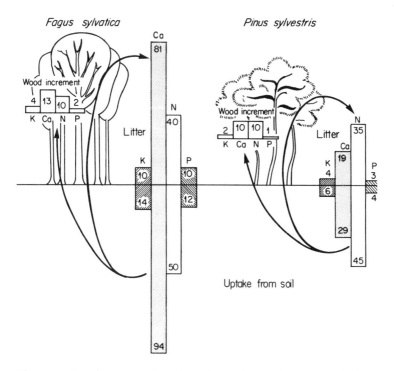

Figure 9.8 The annual cycle of uptake and return of four macronutrient elements in deciduous (hardwood) and coniferous (softwood) forest. The figures are kg/ha; those in the tree crowns represent annual retention in wood increment and those on the soil surface, return in litter. Note that the return represents a large proportion of the uptake. After Duvigneaud and Denaeyer de Smet (1970).

difficult. Levels also vary with geographical situations such as proximity to natural and industrial sources of airborne materials, and with distance from the sea which forms a prolific source of aerosols. For the temperate zones Ovington (1968) suggests annual ranges, in kg/ha, of K, 1–10; Ca, 3–19; Mg, 4–11; N, 0·8–4·9 and P, 0·2–0·6. Additional income may originate from airborne dust which is deposited by impaction on the canopy.

Leaching losses are very difficult to measure and there is little data on this topic. Miller (1963) tentatively suggests, for a *Nothofagus truncata* forest stand in New Zealand, an annual loss in Kg/ha of: K, 15; Ca, 30; Mg, 15; S, 15; N, 2 and P, 0·03. These values may be high by comparison with most temperate forests on eutrophic soils. Crisp (1966), by means of a catchment drainage study, established a nutrient budget for a moorland area of blanket bog, eroding peat and some grassland. He found net annual losses, in kg/ha of K, 8·0; Ca, 50; P, 0·2–0·4 and N, 9·5 but a large proportion of the N and P was contributed from peat erosion and does not represent direct loss from the available plant nutrient pool. The comparison of income and loss provided by these data certainly suggests that a high P/E ratio, combined with nutrient removal by timber cropping, would deplete the resources of the ecosystem. With an originally low supply the depletion may be of considerable significance as elements such as K, Ca and Mg are often limiting under natural conditions.

The idea that nutrient cycling and biomass reach equilibrium under climax conditions must be modified if the long-term balance between income and leaching loss is not compensated by release from parent minerals. The time scale of the sere is thus relevant to the stability of the climax and, given low rates of weathering, coupled with inefficient recycling of nutrients, an apparent climax may fall into decay. External interference may also break the cycle and cause environmental deterioration; for example, the genesis of many N. European heathlands was initiated by human destruction of the climax forest, which enhanced nutrient loss and caused soil deterioration. In northern temperate areas it is not possible to judge the stability of the apparent climax vegetation, as the time period from the early Postglacial until the late Stone Age or Bronze Age, when man began significantly to interfere with vegetation, was probably insufficient for a climax to establish and prove its long-term stability. In tropical areas with very deep, oxic crusts of weathering and a vegetational history of almost geological proportions, the efficiency of nutrient recycling becomes even more important. Went (1970) suggests that only complete recycling of mineral nutrients can ensure the continued growth of luxuriant tropical forest and also implies that mycorrhizal fungi may recycle organic matter directly to the higher plants. In his monograph on mycorrhiza, Harley (1969) provides little support for this view and, more conventionally, suggests that plant photosynthesates nourish the fungi. The topic is important and obviously in need of further research. The slow regeneration of tropical forest after cutting or burning and the degenerate nature of the secondary forest, with its chaotic collection of shrubs and climbers, does, however, suggest that reestablishment of equilibrium is a long process.

Despite the mineral deficiency of the soil under tropical rainforest the efficiency of the system is such that annual production and energy throughput is the highest amongst terrestrial ecosystems. Mineral cycling rate is usually a function of annual production and, as might be expected, the turnover in tropical forest is exceptionally high. Calcium for example, returns in litterfall at 200–300 kg/ha/ann compared with a maximum of about 150 kg/ha/ann in deciduous temperate zone forest. Because of the highly leached nature of the soil and despite its large turnover, calcium forms a lesser proportion of the total annual cycle under tropical conditions compared with temperate forests (Rodin and Bazilevich, 1967).

MINERAL NUTRITION AND ECOLOGY

In the study of living organisms deviant individuals and types are often useful; the very existence of differences may guide the research worker in the formulation of questions. It is probably no coincidence that the ecological study of plant nutrition has been closely related to the development of such concepts as the calcicole–calcifuge problem, the differentiation of tolerance ranges from eutrophic to oligotrophic, halophytism and heavy metal, to mention but a few. In each case the study of these rather abnormal situations has thrown some light on the behaviour of plants, the majority of which inhabit less extreme environments.

Calcicoles and calcifuges

A few moments with a flora will reveal such statements as 'absent from calcareous soils' or 'usually on basic soils' while experimental attempts to grow such plants reveal considerable differences in physiological response or competitive behaviour when they are grown on soils differing in reaction from those which they normally inhabit.

An early example of such experiments was Tansley's (1917) comparison of *Galium saxatile* and *Galium sterneri* syn. (*G. sylvestre*) or (*G. pumilum*). He found that germination and growth of the calcifuge *G. saxatile* was inhibited on calcareous soil and the plants were not able to compete with those of *G. sterneri*. On acidic soil this relationship was reversed, the calcicolous *G. sterneri* being suppressed and giving *G. saxatile* a competitive advantage. No attempt was made to establish the cause, but Tansley commented that it would make 'an interesting investigation in physiological ecology'.

The most obvious chemical differences between calcarous and acidic soils are associated with the high content of calcium carbonate and the near saturation of the cation exchange complex with calcium ions. As might be expected, various workers have found ecologically significant differences in response to calcium supply. Jefferies and Willis (1964) grew the calcifuge *Juncus squarrosus* and the calcicole *Origanum vulgare* in nutrient solutions with differing levels of calcium supply. *J. squarrosus* proved to have a very low calcium requirement and was intolerant of high concentrations. By contrast the growth of *Origanum vulgare* was strongly limited by low calcium supply. Clarkson (1965) found a similar relationship by comparing calcicole, neutrophile and calcifuge species of *Agrostis* grown in solution culture with differing additions of calcium. The growth of the calcicolous and neutrophilous species, *A stolonifera, A. canina* and *A. tenuis* was much improved by increased calcium supply but *A. setacea*, a calcifuge. was little affected (Figure 9.9).

Similar results have been achieved with ecotypic populations of single species derived from contrasting habitats; for example, Snaydon and Bradshaw (1961) found that *Festuca ovina* populations from acidic soils

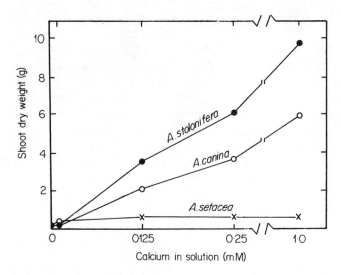

Figure 9.9 The response of four *Agrostis* species to calcium
in solution culture. Data of Clarkson (1965).

have a much lower calcium requirement than those from calcareous soil.
Ramakrishnan (1968; 1970) showed similar differentials in calcium requirement of ecotypes of *Melilotus alba* and, in the second paper, showed that the
differentials of growth response were enhanced when the ecotypes were
grown in competition with each other on acid and on calcareous soils.

A number of nutrient elements and potentially toxic metallic elements
show solubility changes in the pH range encountered in natural soils.
Notable amongst these are iron, manganese and aluminium, all of which
show a greatly increased solubility under acid conditions. Aluminium
differs from the other two metals in being amphoteric and may become
soluble as the aluminate ion at high soil pH values. This may pose
problems when plants are grown on pulverized fuel ash (PFA) in reclamation work (Jones, 1961). Data accumulated during the past 50 years have
shown that the mechanism of tolerance-competition at pH extremes is
generally related to nutrient availability and toxicity effects rather than to
damage from high hydrogen or hydroxyl ion concentrations. The key factors in the physiology of calcicole and calcifuge plants are probably specific
variations in relationship to the availability of calcium, iron, manganese
and some other interacting elements such as phosphorus (see p. 245), and
specific tolerance of aluminium, manganese and possibly iron toxicity.

Aluminium toxicity, as a specifically differential factor, was recognized
very early in agricultural research by Hartwell and Pember (1918) who
showed that rye was little affected by aluminium whereas the growth of
barley was strongly depressed, reflecting the ability of rye to grow on much

more acid soils than barley can. They also noted the interaction of high soil aluminium with phosphorus status; phosphate fertilization suppresses aluminium toxicity, presumably by precipitation of insoluble phosphates. Until the early 1960s aluminium toxicity was not extensively studied as an ecological factor, possibly because its agricultural effects are easily overcome by liming. Jones (1961), however, raised ecological questions concerning the relationship of aluminium to wild plants and suggested that calcifuges tend to be tolerant of high aluminium concentrations compared with the great sensitivity of calcicoles.

Clymo (1962) investigated the growth and distribution of two Carex species, *C. lepidocarpa* and *C. demissa*, which are morphologically similar and both of which inhabit wet soils, thus eliminating the problem of consistent physical differences between acidic and calcareous habitats. By growing plants in nutrient solutions he was able to show that the calcicolous *C. lepidocarpa* was excluded from the habitat of the calcifuge *C. demissa* by Al^{3+} concentrations of above c. 1 p.p.m. and Ca^{2+} concentrations below c. 30 p.p.m. This conclusion supports Jones' suggestion, as also do Rorison's findings (1960) for *Scabiosa columbaria*, a calcicole which showed growth inhibition, in acid soil extracts, caused by high concentrations of Al^{3+}. Hackett (1965) investigated the mineral nutrition of the strongly calcifuge grass, *Deschampsia flexuosa*, and found it to be insensitive to a wide range of nutritional conditions but tolerant of high levels of Al^{3+}. It has an experimental pH optimum of 5·5–6·0 which is above the range of its field occurrence, suggesting that the natural distribution does not reflect its habitat 'preference'. It is more likely that tolerance of high aluminium concentrations permits it to use acid habitats as a 'refuge' from the competition of faster growing species which are more aggressive in their demands on mineral nutrient supply.

Clarkson (1969), in reviewing the aluminium toxicity problem, defined three attributes of the calcifuge plant which may be summarized as: (i) ability to cope with a low calcium supply and its impaired absorption and translodation; (ii) ability to withstand a low phosphorus supply, and fixation of phosphorus by aluminium at the root surface and (iii) tolerance of high aluminium concentrations conferred either by cytoplasmic sites at which the element may be harmlessly accumulated, specific methods for chelating aluminium or a mechanism for precipitating it at the cell wall, thus preventing entry into the cytoplasm.

Recent work has divulged the possibility that the mechanism for coping with aluminium toxicity may also explain the need for high soil iron concentrations shown by most calcifuges. It has been known for many years that calcifuge plants when grown in a calcareous soil develop a 'lime-induced chlorosis' and cease to grow actively. The chlorosis may be cured by foliar application of ferrous iron salts or chelated iron compounds suggesting that it is due to an induced iron deficiency (Hewitt, 1963). By contrast, calcicolous plants do not show such chlorosis and it was assumed that they either have a lower iron demand or they are more efficient at extracting soil iron than calcifuges. However, Grime and Hodgson (1969) have suggested that the mechanism responsible for immobilizing aluminium and rendering it harmless in calcifuges may be non-specific to the extent that it could also immobilize iron, possibly by a chelation process. As a result, in calcareous soil, when aluminium ions are absent, the chelating system absorbs the small amount of iron in the incoming soil solution and induces iron deficiency. In acid soil with a high aluminium concentration the chelating system shows a higher affinity for this than for iron and no deficiency arises.

The agency of induced chlorosis may be more complex than the foregoing would suggest, as other workers have noted chloroses consequent on high calcium or magnesium carbonate in soil, high concentrations of the bicarbonate ion, high soil phosphate status and various heavy metal toxicities, notably of manganese, zinc, copper, cobalt and nickel. The ultimate cause of the chlorosis in all cases appears to be induced iron deficiency and in the case of bicarbonate this may be related to its action in increasing the solubility of phosphate in calcareous soil, followed by the precipitation of insoluble ferric phosphates in both soil and root (Brown, 1960). Woolhouse (1966) provided evidence of ecological differences in the influence of the bicarbonate ion on iron uptake by calcifuge and non-calcifuge grass species. Calcifuges such as *Deschampsia flexuosa* showed inhibition of uptake and subsequent translocation, while *Holcus mollis, Arrhenatherum elatius* and *Koeleria cristata* were insensitive at similar bicarbonate concentrations. Wallace (1961) notes that susceptibility to lime-induced chlorosis is specifically very variable in agricultural plants but, generally, fruit trees and shrubs are more susceptible than cereals and herbaceous crops.

Ecological aspects of the calcicole–calcifuge problem

The distribution patterns of exclusive calcicoles often show extreme fidelity to exposures of limestone and/or calcareous drift (Figure 9.10) but calcifuges rarely show such marked associations. The reason for this is that rendzina soils may leach, if the P/E ratio is well above one, resulting in a superficial acidification and formation of 'limestone heath' in which the sharp differentiation between the plant types is lost. This has been known for many years and Tansley and Rankin (1911) explained the intermixture of calcicole and calcifuge species as a reflection of a root stratification in which the calcifuges exploit the surface acidified layer. Grubb, Green and Merrifield (1969) noted that this is unsatisfactory as it does not account for the seed regeneration of the calcicoles and it is not supported by the excavation of the root systems.

Grubb and his colleagues suggested that pH 5 is a critical lower limit for seedling growth in most calcicoles (Figure 9.11). This would permit their establishment in a great range of soil types but between pH 5 and 6 many

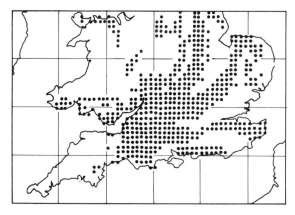

(a) Distribution of limestone outcrops

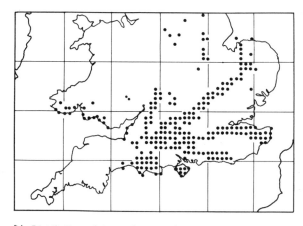

(b) Distribution of *Asperula cynanchica*

Figure 9.10 The distribution of *Asperula cynanchica* in relation to limestone outcrops in southern England. Each dot record represents an occurrence in a 10 km grid square. Redrawn with permission from F. H. Perring and S. M. Walters, *Atlas of the British Flora,* Botanical Soc. of the British Isles, 1962, Map no. 483/2.

large and aggressively competitive grasses exclude the slower growing calcicoles *if* the nutrient and water supply is adequate and *if* grazing pressure is low. Consequently the calcicoles are competitively excluded from many soils, but with heavy grazing, prevalence of nutrient deficiency or drought they form a characteristic limestone association. Superimposed upon this general pattern are other divisions caused by extreme

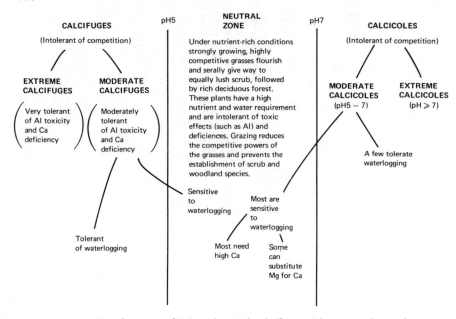

Figure 9.11 A classification of calcicoles and calcifuges. The neutral zone between pH 5–7 may be occupied by the highly demanding but strongly competitive species which exclude the moderate calcicoles and calcifuges. Under heavy grazing, nutrient deficiency or water shortage, the moderate calcicoles and calcifuges may cohabit in this pH range. pH 5 seems to be the upper boundary of Al toxicity effects while pH 7 is probably near the lower boundary of lime-induced iron deficiency. With high rainfall, transient waterlogging and anaerobiosis of soil aggregate centres may extend the pH range by calcifuges by making iron temporarily available in the reduced, ferrous form.

calcicoly, waterlogging relationships and the potential substitution of Mg for Ca in some calicole species. Extreme calcicoly may simply reflect a greater than normal sensitivity to competition, the very alkaline soil serving as a refugium, or it may be a function of excessive physiological calcium demand.

Just as pH is critical for the calcicole so it may also be for the calcifuge, representing an upper limit for seedling survival. After establishment in the surface acidified layer they may be able to tolerate a higher pH, between 5 and 6, provided that grazing, nutrient or water deficiency eliminates competition. The problem for these calcifuges is iron supply and it is under these circumstances that lime-induced chlorosis is often encountered in the wild. Localized or short periods of waterlogging may make iron available in the ferrous form, which could explain why some of the more extreme calcifuges are commoner in limestone heath in the higher rainfall areas of Britain.

Grime (1963a) noted that a number of calcifuges could occur on limestone rendzinas even without superficial acidification, but these observations were all made in N. England or W. Ireland under high P/E regimes, suggesting that the temporary formation of ferrous iron might be significant in these soils. This could be a partial explanation of the observed distribution but, also, many of the species cited are not particularly strict calcifuges, and those which are, were not noted above pH 5–5.5. Grime (1963b) also described the invasion of a rendzina soil by the calcifuge grass *Deschampsia flexuosa* at a surface pH of c. 5. At this pH it was able to compete with the existing calcicole flora, which it replaced, but burning or drought caused reversion to normal limestone rendzina grassland and Grime suggested that the calcifuge cover and accumulation of acid mor humus might be considered a transient phenomenon caused by a sequence of abnormally humid seasons.

From the available data it seems that calcicoles and calcifuges can coexist in a narrow pH range near pH 5 providing that the interaction of nutritional, moisture and grazing factors gives each type an equal competitive status and excludes the vigorously growing neutrophilic species which normally flourish in this pH range. The problem of aluminium toxicity does not appear to arise as concentrations in the soil solution are fairly low at this pH (Grime and Hodgson, 1969). The relationship is further elucidated by Salisbury's observation (1920) that several species which are strict calcicoles in N. Europe and Britain are much more widespread in the drier parts of S. Europe, suggesting that in the northern limestone areas their ecological niche is generated by the complex of factors already discussed but, in S. Europe, the single factor of water shortage protects them from taller, vigorously growing species. As might be expected, the NPK fertilization of calcareous grassland leads to the massive growth of a number of nutrient-demanding grasses which eliminate the calcicoles (Willis, 1963; Bradshaw, 1969). Similarly, the removal of grazing pressure causes an increase in the rankly growing palatable grasses, followed often by invasion of woody plants. Thomas (1960) describes such changes, caused in British chalk grassland in the late 1950s by the myxomatosis killing of the rabbit population.

Nutrient availability

Natural habitats show an extreme range in the availability of the mineral nutrient elements though low availability (oligotrophic) ecosystems are much more common than highly eutrophic types. The extreme of oligotrophy is probably found in rain-fed blanket bog (ombrogenous peat moor) in which the accumulated peat layer may be many metres deep and the upper surface is out of capillary contact with the ground water so that the system is entirely dependent on precipitation and dust income for its mineral nutrients. Such extreme habitats are relatively common but there are far fewer cases of extreme eutrophy. These occur occasionally, for example on phosphate-rich, impure limestones with a high population of nitrogen fixers or, more locally, by enrichment of soil with urine and faeces from herd-forming herbivorous animals. There is, however, a great range of intermediate habitats between which there are considerable nutritional differences while nutrient gradients may occur within associations forming a mosaic with a scale ranging from many metres down to the millimetre level.

The ecological amplitude of a species depends, instantaneously, on its phenotypic plasticity and, over periods of time, on its genetic flexibility. Phenotypic plasticity is important as the plant is anchored to one spot and cannot escape circumstances which it may have encountered for fortuitous seed dispersal or by being excluded from other habitats by competition. Ability to adjust metabolic processes to circumstances, for example by inducible enzyme systems, may be of considerable survival value (Bradshaw, 1969). Differences in such ability may partially explain the varying ecological amplitude of some species, but much recent experimentation suggests that genetic flexibility may be equally, or even more important, as it provides the potential for selection of ecotypes tolerant of a wide range of conditions. It may be suspected that a species which has a wide ecological range incorporates sufficient genetic variability for differentiation of ecotypes adapted to many different habitats, whereas species which are highly habitat specific probably have a much lesser pool of genetic variability.

The differing phenotypic requirement may be seen, subjectively in the diverse habitat requirements of different species, and in experimental work such as Rorison's (1969) comparison of various species in relation to phosphorus supply. *Deschampsia flexuosa* was, for example, rather insensitive to increase in P concentrations between 10^{-7} and 10^{-3} M whereas *Urtica dioica* showed a very marked increase of growth over the same concentration range. The latter finding further confirms the conclusions of Piggot and Taylor (1964) who compared the two species *U. dioica* and *Mercurialis perennis* which have, potentially, the same calcareous woodland floor niche and showed that the significant difference between the two is sensitivity to soil P. *U. dioica* competes very successfully with *M. perennis* at high P levels, coexists at moderate levels and cannot compete under deficiency conditions.

The investigations of various grasses by Bradshaw, Lodge, Jowett and Chadwick (1958; 1960), Bradshaw, Chadwick, Jowett, Lodge and Snaydon (1960) and Bradshaw, Chadwick, Jowett and Snaydon (1964) showed very clearly the strong, specific differences in response to varying calcium, phosphorus and nitrogen supply which, in most cases, correlated strongly with ecological behaviour. Plants of low nutrient potential habitats, such as *Nardus stricta*, were insensitive to increasing nutrient supply, at low levels, and at high levels growth began to be inhibited. By contrast, plants of fertile soils such as *Lolium perenne* and *Agrostis stolonifera* showed much improved growth with increasing nutrient supply, particularly at the lower levels (Figure 9.12).

Genetic flexibility permits the selection, over a period of time, of local ecotypic populations which are suited to specific habitat conditions. Such selection may be imposed by a wide range of environmental variables including nutrients. The work of Snaydon and Bradshaw (1961), previously cited, with *Festuca ovina* ecotypes illustrates the point. Upland populations had a much lower calcium requirement than lowland populations, suggesting that sites of low nutrient potential are likely to support individuals with low nutrient demand. Goodman (1969) studied a collection

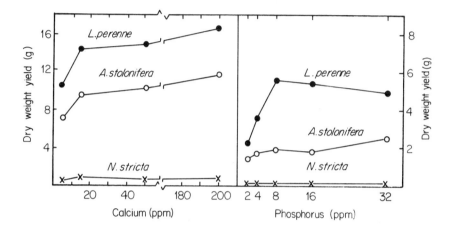

Figure 9.12 The response of three grass species to different levels of calcium (a) and phosphorus (b) in nutrient solution. Data of A. D. Bradshaw, *et al.,J. Ecol.*, **46**, 749–57, part of Figure 1 (1958) and A. D. Bradshaw, *et al., J. Ecol.*, **48**, 631–37, part of Figure 1 (1960).

of *Lolium perenne* populations and concluded that virtually every combination of growth response to N, P and K could be found. The selection of populations with generally low nutrient demand is probably also favoured by heavy grazing pressure which promotes the survival of smaller, slow growing individuals.

Specific differentiation of nutrient demand is one of the factors which permit the development of recognizable ecological associations. Descriptions of various vegetation types indicate that extreme oligotrophy often exceeds the physiological tolerance range for many species, often because of associated factors such as low pH and aluminium toxicity and, thus, oligotrophic habitats have a low species diversity. The species which do survive such conditions often appear to use them as a refugium from competition. At the opposite end of the nutritional scale, eutrophy also encourages low species diversity by producing massive growth of a few nutrient-demanding species which then exclude all others by competition for light. This situation is reflected in the accidental eutrophication of ecosystems by nutrients derived from sewage when single species 'blooms' of green, or blue-green, algae may be produced, and often eliminate most other plant and animal species from the environment.

The intermediate, mesotrophic habitat, is, however, productive of species diversity because it provides an environmental range suitable for a large number of different organisms and is open to niche differentiation. In particular, the balancing of primary production against a population of consumers, both herbivorous and carnivorous, improves the homeostasis of the system by preventing any single species from becoming too competitively aggressive. This last point has a profound significance for the conservationist as the planning of ecosystem management must revolve around the maintenance of this relationship.

One of the most striking and long-standing examples of the changes following altered nutrient balance is the Park Grass experiment established in 1856 at Rothamsted Experimental Station (England). Unfertilized plots still carry the original diverse pasture vegetation with about 60 species. Addition of P and K has caused a large increase in N-fixing Leguminoseae but little reduction in diversity. A full NPK fertilization plus lime, however, has given very heavy hay crops of grasses such as *Alopecurus pratensis, Arrhenatherum elatius* and *Dactylis glomerata* but excludes most of the indigenous species. Many other nutritional modifications have occurred in this experiment, for example acid conditions produced by ammonium sulphate fertilization without liming have caused peat accumulation and dominance by *Agrostis tenuis*. Even within the small area occupied by this experiment, marked ecotype selection has occurred; several of the *Lolium perenne* populations described by Goodman (1969) were collected from Park Grass and Snaydon (1963) has detected marked nutritional ecotyping in *Anthoxanthum odoratum* collected from this site.

The effect of nutrient status is particularly prominent in heavily grazed ecosystems which are in plagioclimax condition. Succession to forest climax often masks some of the lesser differences between habitats but, even so, species constitution and diversity remain as a general function of nutrient availability.

Heavy metal toxicity

Discussion of nutrient status in relation to ecotyping and speciation suggests that, within wide physiological tolerance limits, gene pools exist, amongst the higher plants, which permit the colonization of almost all earthly habitats. Some of the most inhospitable environments apart from desert and polar regions are found on rocks and other materials which contain toxic concentrations of one or more heavy metals. These include some 38 metallic elements, with a density greater than five, which are potently toxic to most living organisms, usually by damaging protein molecules and blocking enzymic processes. The commonest, of ecological importance, are lead, zinc, copper, nickel, manganese, chromium, mercury, silver, cadmium, cobalt, molybdenum, tin and iron. The presence of any one of these elements in soil at greater than the normal trace concentration, due either to pedological factors, geological anomaly or pollution, will expose plants to a strong selective pressure related only to the single toxic factor. The background of environmental conditions may be perfectly normal except for the high level of this one factor.

The earliest interest in this subject centred around species which occur in association with ore-bodies and other rocks with a high heavy metal content and rarely, if at all, in other habitats. Antonovics, Bradshaw and Turner (1971) have reviewed the topic and note that higher plants characteristic of metal-contaminated soils have been recognized since the sixteenth century at least. Some have even been used as indicators in prospecting for heavy metals and owe their specific epithet to the association, for example *Viola calaminaria* has been used to locate zinc deposits in various parts of Europe.

The earliest descriptions of metal-tolerant plants associated with toxic soils were of a subjective nature and, subsequently, various phytosociological techniques have been used to describe the assemblages of plants found on toxic soils. Such correlative work does not critically prove the existence of tolerant species or ecotypes and it is only since the 1950s that good experimental evidence has been produced. Comparative experiments in which 'normal', and supposedly tolerant, populations are grown in nutrient media containing different levels of the toxic metal have shown that the association of certain plant species with metal toxic soils is due to their ability to evolve metal-tolerant races. Thus Kruckberg (1954) found the serpentine endemic, *Streptanthus glandulosus,* to be a serpentine-tolerant race. Populations from the few non-serpentine localities were much less tolerant. Plants with a more widespread occurrence, such as *Agrostis tenuis*, are also frequent colonizers of toxic mining spoil and selection for tolerant ecotypes has again proved to be the cause.

Antonovics, Bradshaw and Turner (1971) suggest that the appearance of metal-tolerant ecotypes is one of the best documented examples of evolution in action. *A. tenuis* populations on toxic spoil have a greater metal tolerance and a lesser variability of tolerance than that of the seed which they produce. This suggests that natural selection is acting to preserve the high tolerance of the adult populations and further investigation shows that this is indeed the case: if non-tolerant seed is sown on contaminated soil, most germinates, but the seedlings fail to root and die. Studies of mine spoil and pasture populations of *A. tenuis* have revealed a higher proportion of tolerant individuals in the mine seed though the level of tolerance is less than that of the adult population because of dilution by non-tolerant pollen from outside the toxic area.

The evolution of tolerant races is extremely rapid as evidenced by their occurrence on spoil heaps between 50 and 100 years old while Snaydon (cited in Bradshaw, McNielly and Gregory, 1965) has shown zinc tolerance in *Festuca ovina* and *A. tenuis* under galvanized fences less than 30 years old. Tolerant individuals may also be selected in one generation from the seed of normal populations (Antonovics, Bradshaw and Turner, 1971). Thus the initial colonization of a toxic soil relies upon selection of tolerant seedlings from the surrounding normal populations followed by a continuous selection for the tolerance characteristic in the face of the diluting effect of gene flow from surrounding populations and a high rate of population turnover. Species which are described as endemic to, or frequently associated with, toxic soil are those in which the normal gene pool permits the occasional appearance of tolerant recombinants. Tolerant individuals are only present at low frequency in normal habitats as they have less competitive vigour than their normal counterparts (Antonovics, Bradshaw and Turner, 1971). There is some difficulty in interpreting this point because mine spoil populations may also have been selected for their ability to survive a physically harsher and chemically more oligotrophic environment than pasture populations. Where tolerant ecotypes are significantly less competitive than normal plants, the toxic habitat may be interpreted as a refugium in the same way that extreme calcifuges and calcicoles escape the competitive pressures of the more normal mesotrophic environment (see p. 288).

The tolerance mechanism is almost always metal-specific; tolerance to one metal does not automatically confer tolerance to another (Gregory and Bradshaw, 1965). The genetic control of the mechanism has been little studied but various workers have shown that it is not an all-or-nothing effect but is a continuous variable in natural populations (Antonovics, Bradshaw and Turner, 1971). As the level of the toxin is reduced, so the number of survivors increases, suggesting that tolerance is a threshold characteristic determined by many genes.

Antonovics, Bradshaw and Turner (1971) conclude their extensive discussion of mechanisms of tolerance by suggesting that higher plants generally keep metal ions away from active metabolic sites by chelation at the cell wall. The nature of the chelation process has been little investigated but it would need to be highly metal-specific to explain the specificity of tolerance. The detoxicating function of the cell wall may well be widespread even under more normal circumstances, for example the immobilization of aluminium and iron by a binding system in calcifuge plants was discussed on p. 266, and it seems likely that marsh plants need to immobilize ferrous iron to prevent excessive uptake from reducing soils.

Metal tolerance in higher plants has some practical application in biogeochemical prospecting and also in the revegetation of toxic mining and smelting wastes which cause industrial dereliction in many areas of the world. Plants may be used as indicators of metal deposits either by studying the distribution of tolerant species and assemblages of species, or by analysis of plant organs in which metals may have become concentrated. Rune (1953) presents a summary of plants which are known to be associated with specific minerals: some examples follow. *Viola calaminare* and *Thlaspi calaminare* occur on the calamine (zinc carbonate and silicate) soils of Belgium, Poland, Germany and Austria: Rune considers them to be, respectively, chemomorphoses (ecotypes) of *Viola lutea* and *Thlaspi alpestre*. A whole assemblage of copper-indicating plants has been described for the Congo, amongst them *Buchnera cupricola* and *Guttenbergia cupricola*. Howard-Williams (1970) has investigated the ecology of *Becium homblei* in Central Africa where it is believed to be a copper indicator and though he showed that it does occur on other soils, it is tolerant of up to 15,000 p.p.m. of soil copper and is generally competitively restricted to contaminated soils. The topic of geobotanical prospecting has been well reviewed by Cannon (1960) and Malyuga (1964).

The revegetation of mine or smelter waste has always posed problems which may be overcome by the expedient of burying the material under a thick layer of non-toxic spoil, but this is usually expensive and not always practicable. Attempts to reduce toxicity by heavy liming and application of organic matter such as sewage sludge have produced promising results but these are greatly improved by the use of seedling material from resistant populations (Smith and Bradshaw, 1970). It is likely that this technique will be used in the future, though there may be some problem where the grass cover is grazed as toxic concentrations of heavy metals may be ingested.

Some further comment is required on the rather widespread, naturally toxic habitat of serpentine. This is a magnesium-iron silicate rock formed by metamorphosis of peridotite which, in many cases, contains chromite ($FeCr_2O_4$) and garnierite ($(Ni, Mg)SiO_3.nH_2O$) (Walker, 1954). Most of the world's serpentine areas are 'barrens' with a very sparse vegetation of unusual species composition, the cause of which may be general infertility due to shortage of major nutrients, nickel and chromium toxicity or imbalance of calcium: magnesium ratio. Kruckberg (1954) established that species associated with serpentine were able to produce races which were tolerant of the soil conditions and Proctor (1971) showed, with *Avena sativa* as an indicator plant, that toxicity may be due to the high ratio of magnesium to calcium in the soil. He also cites previous evidence, derived from indicator plants, that nickel and chromium toxicity occur and that general infertility may be attributed to major nutrient deficiency.

Halophytism

Another long-recognized ecological group of plants is that associated with salt-soils such as the solonchaks of the arid zones and the salt-marshes of the sea coats. The universal attribute of these plants is selective absorption and tolerance of sodium chloride, and the ability to grow in soils so saline that they are either physiologically toxic to normal plants or impose so serious an osmotic water stress that water absorption is impeded.

Sodium chloride is not essential for the healthy growth of most halophytes though, in non-saline soils, they usually cannot compete with normal plants. A few, however, show a strong growth response to additions of sodium chloride well in excess of the trace required by most plants. This response is particularly characteristic of the Chenopodiaceae (Piggot, 1969) many of which contribute to the world's halophilous flora.

The concentration of solutes in sea water is about 33–38 g/l giving an osmotic potential of c. 24 bar (Piggot, 1969). Freshly flooded salt-marsh soil will have this osmotic potential but the upper marsh may have lower values after dilution by rain, or higher values following water loss in dry weather. Salt concentrations in arid zone soils may be very much higher, often reaching saturation, with production of crystalline efflorescences of salt at the soil surface.

Plant cells which are bathed in a medium of high osmotic potential must maintain a higher vacuole osmotic potential or else lose water. This has caused the evolution of a transport mechanism which permits sodium and chloride accumulation, in halophytes, which contrasts with the exclusion of these ions by normal plants. The internal osmotic environment is thus balanced against the external one and, contrary to past opinion, the plant does not suffer 'physiological drought' by osmotic loss of water to the soil medium.

The halophyte has to cope with other consequences of this accumulation mechanism. Firstly, it must have an internal tolerance of sodium and chloride levels which would be directly toxic to normal plants and secondly, it must concentrate essential ions such as potassium from a low environmental concentration in the presence of high concentrations of potentially competitive ions. The transport mechanism must possess sufficient specificity for the required ion to prevent its competitive exclusion by a high concentration of a chemically related ion. More detailed discussion of sodium and potassium uptake in halophytes is given in Chapter 8, p. 241).

CONCLUSION

The differentiation of ecological groupings of plants, in relation to nutritional and other soil chemical conditions, is one of the strongest factors promoting localized variability in vegetation, and it reflects both the physiological status of the individual plant species and their competitive relationship with each other. The acquisition of adaptive characteristics suiting species or ecotypes to specialized soil environments provides a very suitable experimental opportunity for the investigation of mineral nutrition at both the physiological and the genetic control level. This opportunity is now beginning to be exploited and may lead to rapid advances in this particular field of eco-physiology.

Chapter 10

Competition

Almost 200 years ago Malthus (1798) wrote: 'Population, when unchecked, increases in a geometrical ratio. . . . This implies a strong and constantly operating check on population from the difficulty of subsistence.' The full implication of Malthus' words passed unheeded for almost a century as the realization of his predictions was delayed by the Industrial Revolution which permitted the agricultural exploitation of new land areas and the establishment of an international food market. Today the human population again faces the dilemma of numbers versus subsistence without any prospect of relief; any increase in agricultural efficiency deriving from plant breeding and cropping advances now being met by immediate increases in population level.

Population crisis has been an ever-present biological fact since, without restraint, a population will always expand to fill and overfill available space. The subject of competition has attracted the attention of plant ecologists for over 150 years. Clements, Weaver and Hanson (1929) reviewed early work and cited De Candolle (1820) as the first to characterize plant competition: ' . . . all the plants of a given place are in a state of war with respect to each other'. In 1907, Clements, who pioneered so much ecological thought, defined competition in terms which have since been improved only in detail: 'When the immediate supply of a single necessary factor falls below the combined demands of the plants then competition begins.'

Plants may interact by competition for the supply of some particular factor but may also influence each other directly, for example by secretion of toxic metabolic products into the environment. De Candolle's concept of plants at war is, perhaps, a more fitting analogy for the latter type of interaction. The concept of warfare implies a direct and purposive struggle which hardly fits the situation in which a nutrient is removed from the environment before it can be reached by a competitor. As the word 'competition' carries overtones derived from human activities, Harper (1961) suggested its replacement by 'interference' but there is no sign that this has been generally adopted as an alternative.

Another possible approach to competition may be seen in the writings of Margalef (1968), who placed the concept in a cybernetic framework by describing the competition between two species in terms of their respective feedback relationships with a jointly used resource. Each species has its own, stabilizing (negative) feedback relationship with the resource A (Figure 10.1a). If A becomes overtaxed then the feedback loop leads to a reduction in demand and the steady state is maintained. However, if a second species (C) is introduced which utilizes the same resources it is possible for a resultant positive feedback pathway to be established between species B and C (Figure 10.1b). If, for example, B overdraws the resource A then C must suffer a deficiency; some of the C population may consequently die and return part of resource A to the environment. If B immediately takes up the returned portion of A there is nothing to prevent a continuing A deficiency from killing the whole of the population of C. In this case the establishment of the positive feedback pathway between B and C leads to the catastrophic collapse of the C population and a 100% success of B as a competitor. The potential establishment of such feedback loops between all

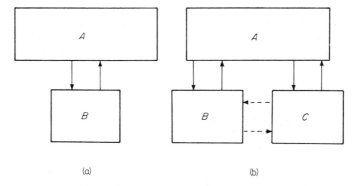

(a)   (b)

Figure 10.1 Diagram (a) represents the negative feedback relationship between a resource A and a consuming species B. This will be a stable relationship, the magnitude of the resource pools and the species pools being determined by relative rates of supply, demand and return. Diagram (b) shows the introduction of a second, competing species C. Each species now has comparable negative feedback relationships with the resource A but the resultant is a possible positive feedback relationship between the two species leading to instability and potential extinction of one species. Reproduced with permission from R. Margalef, Perspectives in Ecological Theory, University of Chicago Press, Illinois, 1968, Figure 2, p. 8. Copyright 1968 by The University of Chicago. All rights reserved.

species and individuals of a complex ecosystem draws attention to the immense intricacy of such networks and to the threat that a minor unbalancing of environmental or species relationships could initiate self-reinforcing population crashes, perhaps leading to species extinction.

This model raises a problem which is discussed by Harper (1967) and has recurred frequently as a difficulty for other workers. If competition is so important as a regulator of ecosystems why should species co-occur? Why is it that competition does not lead to mutual exclusion as Margalef's model would suggest? It is difficult to give a direct answer to these questions for plant ecosystems but experiments have revealed some self-stabilizing properties. For example, Lieth (1960) showed for *Trifolium repens* and *Lolium perenne* that the two species form a mobile mosaic in which low clover density areas are invaded by grass and vice versa. As specialization for difference in niche occupancy has been suggested as a means of avoiding annihilating competition (Harper, 1967) it is tempting to suggest, in cases such as this clover: grass mosaic, that the plants themselves create transient micro-niches in the soil environment. Snaydon (1962) has shown very steep gradients of pH, calcium, phosphorus and potassium in clover associations while the work of Goodman and Perkins (1959) with *Eriophorum angustifolium* and Zinke (1962) with various tree species shows that the plants are capable of establishing considerable micro-environmental gradients of nutrients. As Harper (1967) has suggested, intense niche differentiation would prevent an exclusive struggle by focusing the intense battles within, rather than between, species (see also p. 307).

THE STRUCTURE OF COMPETITION

One approach to plant competition is to analyse it in terms of the factors for which plants compete or which are related to competition between plants. These may be tabulated as follows:

A. Factors for which plants compete	(a)	Space	Above and below ground
	(b)	Light	Above ground
	(c)	Carbon dioxide	
	(d)	Nutrients	Below ground
	(e)	Water	

| B. Plant characteristics which cause competition | (a) | Passive root interactions such as the normal production of respiratory CO_2 |
| | (b) | Direct interaction due to the secretion of specific toxins into the environment (Allelopathy) |

C. Interactions with external factors which influence or cause competition	(a)	Competition for pollinators
	(b)	Competition for agents of dispersal
	(c)	Selective pressure or disturbance of ecological equilibria by animals and man
	(d)	Disturbance of the environment which provides bare soil or seedling niches
	(e)	Influence of temperature, humidity, exposure, wind etc. on other competitive factors
	(f)	Unusual soil conditions such as toxic solute content, heavy metals, excess calcium carbonate etc.

FACTORS FOR WHICH PLANTS COMPETE

Space

Donald (1963) noted that plant competition for space, in the sense of physical interaction, rarely arises. He cites the example suggested by Clements, Weaver and Hanson (1929) of competition amongst close-sown tuberous crops where individual tubers may become polygonal or even be lifted from the soil by their neighbours. Further examples may be seen when clusters of seeds have germinated in the same spot and, in woody plants, when neighbouring branches actually clash, leading to damage, deformity and sometimes infection. Seed establishment is probably the commonest example, most soils presenting only a limited number of niches in which the water supply requirements for germination are satisfied. Harper's (1961) experiments with *Bromus* spp (Figure 10.2) illustrate this point: when sowed on to a compacted soil surface the number of emergent seedlings was limited in relation to seeding-rate by the small number of surface crack sites available for germination. By contrast in the rough-surfaced soil no limitation occurred and the seedling number was proportional to the seeding rate. These thus represent the fairly limited

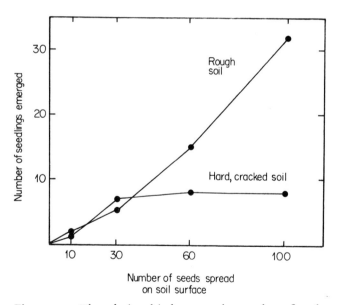

Figure 10.2 The relationship between the number of seeds shown and the number of seedlings emerging for a mixed *Bromus* spp population. Reproduced with permission from J. L. Harper, Approaches to the study of plant competition, *SEB Symp. XV*, 1961, 1–39, Figure 12.

number of examples of competition for space. In the light of the discussion of soil niche differentiation above it may, however, be necessary to reassess general views on competition for space in the soil matrix.

Light

Competition for light is unique in that there is no storage: incoming light must be used or lost. Leaves behave toward light as individual units: when they remain for long periods below compensation point they are not supported by export of assimilates from other parts of the plant and quickly die. For this reason competition for light is between individual leaves rather than between plants. The canopy architecture and its relationship to leaf display and insolation is, thus, a most important factor in determining peak photosynthetic rates and light competitive abilities of plants.

Light intensity variations due to competitive shading effects are further complicated by the regular seasonal and diurnal fluctuations and by transient cloud shadow patterns. In terms of competitive status the photosynthetic surface should receive an integrated daily energy input which supports a positive balance of carbon dioxide assimilation. If the balance becomes negative for more than a few days then individual leaves may begin to die. In the case of whole-plant shading the situation may be rather different: Hutchinson (1967) has shown that some plants, particularly shade-plants, may tolerate fairly long periods of darkness. This is most marked if their growth-rates are limited by nutrient deficiency or soil toxicity. In many cases the final cause of death in over-shaded leaves is attack by pathogenic fungi.

Response to low light intensities is crucially important in the basal part of foliage canopies; in particular of herbaceous associations. Monsi and Saeki (1953) measured light intensities ranging from 28% to 5% of the external value on forest floors and from 4.5% to 0.7% in herbaceous cover. The influence of this light intensity gradient on photosynthesis was shown by Leach and Watson (1968), who experimented with phytometers exposed at different levels in crop canopies. Their phytometers consisted of small plants grown in containers or nutrient solution. In kale the photosynthesis at the canopy base was reduced to about 4% of the rate at the surface of the 60 cm canopy. The absolute rates suggest that leaves at this level had very low photosynthetic rates but remained above compensation point. The leaves of the crop plants are also likely to be better adapted to the low intensities than the phytometer leaves which were only left in position for seven days. The photosynthetic rates were lower than measured light intensities would have suggested, possibly because some of the light was derived from leaf transmission and was in the green part of the spectrum. Salisbury's (1916) studies of woodlands showed that the under-canopy vegetation was complementary to the tree canopy in maximal exploitation of available light. It seems likely that this must also be the case in many other stratified ecosystems.

Light remains important as a competitive factor even in ecosystems showing other types of limitation. Jennings and Aquino (1968) showed that strongly competitive rice varieties receive more light than weak competitors and cite evidence to suggest that competition for other factors such as water and nutrients also operates through light relationships by reducing the amount of light-intercepting foliage.

To permit full exploitation of environmental light gradients there must exist a considerable range of light requirements amongst different types of plant. These requirements must be based upon modifications which maximize the net assimilation rate or relative growth rate for a given energy input (whatever its value) during the life of the photosynthetic organ. They may be associated with (i) minimizing respiratory loss; (ii) maximizing photosynthetic rate for any level of energy input and (iii) alteration of leaf area ratio: an increase will give a larger relative growth rate for the same net assimilation rate.

All three variants are encountered. Grime (1965b) has shown many shade plants to have respiration rates which are lower than those of sun plants (Figure 10.3), leading to conservation of assimilate resources. This fits Hutchinson's (1967) observation, that shade plants tolerated darkness for longer periods than sun plants. A possible example in the field is the long survival of *Fraxinus excelsior* seedlings below a tree canopy. Minimal growth occurs unless a gap is produced by senescence of an adult tree (Wardle, 1959). Ingress of more light promotes rapid growth of the 'dormant' seedlings below the gap and after a short period of intense competition one survives to close the canopy.

The second type of response is also found in shade plants and was discussed by Decker (1955). He noted that the photosynthetic system of shade plants became light saturated at much lower intensities than that of sun plants (Figure 10.4). In consequence the shade plant reaches maximum photosynthesis for longer periods during each diurnal cycle but, as Grime (1965b) points out, individuals which have been selected for this type of adaptation have lost the physiological potential of competing at higher light intensities and, hence, are excluded from open habitats.

At the opposite end of the light intensity scale, plants with the Hatch–Slack CO_2 incorporation pathway and no photorespiration are capable of making better use of high light intensities than normal Calvin cycle plants. Thus *Zea* and *Saccharum* are very high-yielding crop plants (see Chapter 8). Efficient use of high light intensities does not, however, give plants a strong competitive advantage and Jennings and Aquino (1968) noted that high-yielding rice varieties were not always the most successful in intraspecific competition. This is probably, again, a case of niche specialization: high light adaptation is a characteristic of plants of open habitats where light competitive ability is not at a premium while low light adaptation permits plants to colonize sub-canopy situations where competition for light is intense.

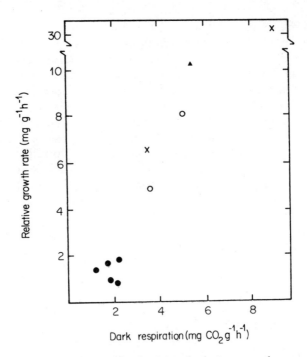

Figure 10.3 The relationship of relative growth rate to dark respiration of leaf discs of a number of plant species of different habitats. The relative growth rates were measured over a period of 15 days of sunny summer weather. ●—shade tolerant species; ○ — shade intolerant species; + — crop plants, ▲ — arable weed: high light plants. Data of Grime (1965b)

The third possible modification, change in leaf area ratio, results from variation in leaf anatomy and in extreme cases is manifested in the development of 'sun' or 'shade' leaves. Haberlandt (1884) commented: 'A sun-leaf may be thrice as thick as a shade-leaf on account of the more abundant development of its palisade tissue'. Shields (1950), in his study of xeromorphy, discussed the influence of light as well as water deficit in modifying leaf anatomy. High light intensity tends to induce the formation of thicker, more sclerified leaves and truly xeromorphic leaves are nearly always characteristics of open, high light intensity habitats.

The thin shade leaf with a single palisade mesophyll layer has a larger area to weight ratio than the sun leaf, carries a lesser bulk of respiring mechanical tissue and shows saturation of its photosynthetic system at comparatively low light intensities. The sun leaf, however, has a much higher light saturation photosynthetic rate as the deeper layers of palisade

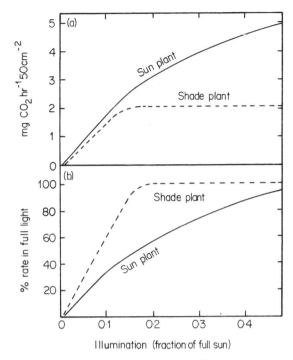

Figure 10.4 Photosynthetic rates of sun and shade plants at different light intensities. (a) Plotted in absolute units. (b) Plotted as percentage of maximum value. Reproduced with permission from J. P. Decker, The uncommon denominator in photosynthesis, *Forest Sci.*, **1**, 88–89 (1955).

cells approach saturation sequentially in the same way that leaf layers in a canopy respond to increasing light intensity.

Whole-plant characteristics which increase the efficiency of light use are: (i) maintainance of optimum leaf area index by sequential loss of lower leaves or 'self pruning'; (ii) selection for optimum leaf display and (iii) avoidance of competition. Loss of lower leaves is common in high-density stands: an example is seen in the exposure of yellowed lower leaves when an overgrown grass sward is cut. Self-pruning is very marked in close-planted conifers where the lower branches rapidly fall below compensation point and die. This type of self-adjustment causes the canopy to reach a ceiling leaf area index which cannot be exceeded as unfolding of new leaves is balanced by the death of old. In consequence, any photosynthesate produced at this time must be incorporated in structural or storage organs and the net assimilation rate will steadily decline to zero as the plant ages. At this stage the plant will have reached ceiling dry weight yield. Went (1957) cites the example of strawberry in which non-photosynthetic tissues had built up to such an extent that the plants barely reached compensation point at 1500 foot candles. Ovington's (1957) study of *Pinus sylvestris* woodland showed that peak annual increments of forest biomass were reached after about 25 years and then declined. Donald (1961) suggested that in dynamic equilibrium there will be some old stand members which make a negative contribution to dry matter accumulation while young members make a positive contribution. In consequence the net weight increase will be zero and the ecosystem could be considered to have achieved a climax equilibrium. As noted in Chapter 8, this situation has not often been observed in the temperate zone and the concept deserves further investigation as it throws doubt on the reality of the climax.

Leaf display as a factor in interspecific competition is most obviously seen in the effects of plant height. Watt's (1925) work on cyclic regeneration in *Fagus sylvatica* woodland showed that mature *Fraxinus excelsior* trees could be overtopped and shaded out of the woodland system by the slower growing but inherently taller and leafier *F. sylvatica*. During the early stages the *Fraxinus* canopy is sufficiently open to permit *Fagus* to grow below it. Similarly, *Calluna vulgaris* moorland may be invaded by *Pteridium aquilinum* rhizomes which enter below the soil level and then throw up leaves which ultimately kill the *Calluna* by shading (Farrow, 1917).

Avoidance of competition for light is most often found in the temperate zones where well-marked seasonal variations permit the development of 'aspect societies'. One of the best examples is the associes of herbs occupying the floor of summer deciduous woodland in which leaves are produced before the tree canopy expands sufficiently to cast significant shade. Blackman and Rutter (1959) made an extensive study of *Endymion non-scripta* and showed that it is not by any means a shade plant and is excluded from habitats in which the March to June light intensity falls below 5–7% of full daylight. For this reason it is typical of deciduous oak and ash woodlands and increases to very high densities in coppice-managed woodlands in which the trees are cut every decade or so, maintaining a rather open canopy. Furthermore, it is excluded from ungrazed grassland by competition for light but survives in some lightly grazed grasslands. It is also characteristic of the *Pteridium aquilinum: Holcus mollis* association which occupies so many degenerate habitats in Britain on light, sandy soils. Again it avoids competition with the *Pteridium* canopy which only completes the unfolding of its leaves in the early summer while the *Endymion* has been photosynthesizing from February or March onward.

In summary, it may be said that competition for light has resulted in the evolution of complicated canopy structure and species relationships which, by means of physiological, anatomical or whole-plant morphological characteristics permit maximum use of the incident light at all levels. Furthermore, as the competition is essentially between leaves, the individual leaf strata of a single species may show different light utilization characteristics which derive from modifications based on both genotype and phenotypic plasticity.

Carbon dioxide

Competition for carbon dioxide is not often discussed and it is usually implied that it is not significant (Milthorpe, 1961). However, it is widely known to occur in glasshouse crop culture where adjustment of CO_2 levels may be required (Bowman, 1968). Crops may reach several times the normal yield with such additions while experiments with CO_2 enrichment in the laboratory show that light saturation values can be raised considerably. It seems that the current atmospheric CO_2 concentration of 300–350 v.p.m. is much below optimum for photosynthesis, in which case competition for CO_2 must take place in the upper parts of leaf canopies at high light intensities. It is difficult, however, to suggest how one leaf could gain a competitive advantage over another unless the CO_2 diffusion gradient to the mesophyll cell surface could be shortened or steepened. Shortening the pathway has the concomitant effect of increasing transpiration but steepening the gradient by loss of photorespiration in Hatch–Slack pathway plants may be significant.

Nutrients

Bradshaw (1969) indicates the range of habitats in which nutrient deficiency, and consequently competition, must occur by suggesting that simple NPKCa fertilization experiments will cause extensive changes in most plant associations. This is exemplified by Willis' (1963) work with sand dune calcareous grassland in which an NPK addition caused a surge of growth in the dominant but normally stunted grasses *Festuca rubra, Agrostis stolonifera* and *Poa pratensis*. The sward depth was much increased and the whole range of rosette and creeping perennials so characteristic of calcareous dune and limestone grassland was quickly ousted by competition for light. The conclusion may be drawn that the physiognomy of such associations is sculpted by the limiting effects of competition for the three major nutrients.

The greatest problem in studying nutritional competition is the complexity of the interaction for so many major, minor and trace elements coupled with the range of competing species, some of which may also show strong differentiation of nutritional ecotypes. The latter point imposes rigorous requirements on the choice of experimental material for work in this field.

Donald (1963) notes that a plant's success in gaining a greater share of the limiting nutrient may cause such an increase in growth that a competing species may be suppressed secondarily by shading. He described several experimental devices by which above and below ground competition effects could be separated in culture. Partitions were used to separate root or shoot systems, permitting factorial investigation of soil and aerial effects. Table 10.1 shows the results of such an experiment using the grasses *Phalaris tuberosa* and *Lolium perenne*. The depression of *P. tuberosa* yield caused by competition for nitrogen was strongly reinforced when the plants were also allowed to compete for light and supports the suggestion by Jennings and Aquino (1968), that competition for many factors may ultimately operate through modified light relationships.

Table 10.1 The interaction of competition for nitrogen and for light in *Lolium perenne* and *Phalaris tuberosa*. (Data of Donald, 1958)

	Yield dry wt (g per treatment)			
	No competition	Competition for light	Competition for nitrogen	Competition for both
L. perenne	4·71	4·19	4·31	4·72
P. tuberosa	4·67	3·19	1·17	0·32

A great deal of the competition literature concerns agricultural plants which are grown either as single species or as mixtures of a few grassland species. The majority have been selected for their high vegetative or reproductive yields and consequently tend to be demanding nutritionally. The situation in natural ecosystems is very different, the habitats forming large- and small-scale mosaics of varying nutrient status which interact with the distribution of species and ecotypes having a very wide range of nutritional requirements.

Tansley (1917), over 50 years ago, grew the calcifuge, *Galium saxatile*, in competition with the calcicole, *Galium pumilum* (*G. sylvestre*), on soils of different pH. Each was capable of growing alone on either an acid peat or a calcareous soil but in competition *G. saxatile* was handicapped on the calcareous soil and *G. pumilum* on the peat. It may be inferred that both species have a wider range of physiological tolerance than is manifested ecologically, competition limiting the potentially wide distribution.

Rorison (1969) illustrates this for several species (Figure 10.5). The physiological response curves of most plants show that they grow best at intermediate pH values and in fertile soils; but their ecological responses may be very different. The work of Hackett (1967) and others shows that *Deschampsia flexuosa*, for example, has a rather low growth rate, a low phosphorus requirement and tolerance of high aluminium levels. When grown alone it performs best in fertile soils but in competition is swamped by other, faster growing, species. Hackett suggested that 'acid soils confer no specific nutritional benefits on *D. flexuosa* but are a refuge for it from competition from more vigorous but less acid tolerant species'.

Another striking example is the competitive relationship between *Mercurialis perennis* and *Urtica dioica*, both rhizomatous plants of the field layer

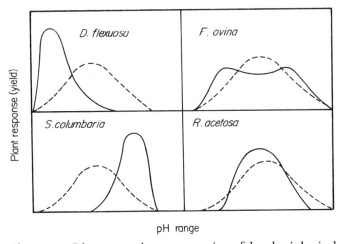

Figure 10.5 Diagrammatic representation of the physiological response curve (– – –) and the ecological response curve (——) to soil pH of *Deschampsia flexuosa*; *Festuca ovina*; *Scabiosa columbaria* and *Rumex acetosa*. Reproduced with permission from I. H. Rorison, *Ecological Aspects of the Mineral Nutrition of Plants*, Blackwell, Oxford, 1969, p. 159, Figure 2.

in some types of British woodland. In culture *M. perennis* grows well in soils of such low phosphate status that *U. dioica* cannot grow beyond the first leaf pair, being limited by P deficiency (Piggot and Taylor, 1964). The authors suggest that *U. dioica* is excluded from many otherwise suitable habitats by low P availability and showed that P fertilization permitted the establishment of *U. dioica* seedlings which could then outcompete *M. perennis* because of their greater height. On a natural gradient of soil phosphorus one would expect to find: first *Mercurialis* alone; then an *Urtica/Mercurialis* mixture in which *Urtica* could not become overcompetitive because of P deficiency size limitation; finally, with high soil P, *Urtica* alone, excluding *Mercurialis* by competition for light.

The intermediate situation in which the plant size is limited by nutrient deficiency introduces another concept: that of nutrient uptake in relation to standing crop biomass. Rennie (1955), Miller (1963) and others have shown that nutrient uptake by plants may remove an appreciable proportion of the whole ecosystem supply from circulation. In oligotrophic ecosystems, though the actual uptake may be quantitatively small it can represent a large proportion of the available nutrient (Fig. 10.6). Under these circumstances the large woody plant which, in consequence of its standing biomass, requires a large amount of nutrient to complete its life cycle, may not be able to compete with a smaller shrub or herbaceous plant. Olsen (1961) suggested this explanation for the existence of grass/shrub clearings in Danish *Fagus sylvatica* woodlands. The glades oc-

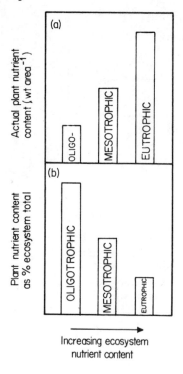

Figure 10.6 Nutrient content of the vegetation in temperate zone ecosystems of differing nutrient status. (a) In a nutrient-deficient ecosystem the plant cover may accumulate a very large proportion of the total habitat supply of nutrient but (b), this may represent only a small amount of nutrient per unit area compared with uptake in eutrophic habitats.

cur on soils of unusually high calcium carbonate content which Olsen believed to limit iron availability. *F. sylvatica* seedlings frequently appeared in the glades but after a few years became chlorotic from iron deficiency and could no longer compete with the grasses and shrubs which could complete their growth on a lesser iron capital (Figure 10.7). Generally speaking, plants which make more dry weight and leaf area for a given nutrient uptake are better fitted to compete when that nutrient is limited.

The limitation of competitive ability by nutrient deficiency is also seen in Willis' (1963) work described previously in this section. The dune grassland showed a remarkable increase in the growth of a few dominant grass species on addition of an NPK fertilizer and the species number of the association was considerably reduced. This would suggest that dry weight production of the dominant grasses is normally nutrient-limited, thus permitting the survival of the characteristic rosette and creeping plants. Generally a single nutrient deficiency in the addition was sufficient to negate the effect so that responses to NP, NK and KP were negligible but for one exception: *Carex flacca* growth was increased by the NK addition. Willis suggested that *C. flacca* must have a low P demand; in other words, it is capable of making a considerable dry weight gain on limited P capital. This could well explain its abundance in the turf of heavily grazed chalk and limestone grasslands which are notoriously P-deficient but usually

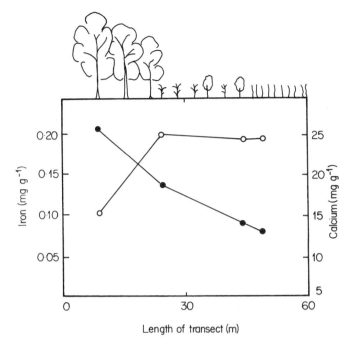

Figure 10.7 Iron and calcium content of *Fagus sylvatica* leaves from plants growing in woodland, and *Cornus* (*Swida*) shrubbery. ○– calcium, ●– iron. Data of Olsen (1961).

have many N-fixing leguminous species and are not highly K-deficient. Under these circumstances *C. flacca* can codominate with grasses which would out-compete it under conditions of better P status.

The concept of niche differentiation in the soil volume demands consideration. Each root has an absorptive region which is surrounded by a 'rhizosphere' of approximately cylindrical form which is related to the root by diffusional gradients of mineral nutrients and root metabolites, mass-flow gradients of water potential and a microbiological association which differs from that of the bulk soil. If the root systems are sufficiently extensive it seems that there may be competition for rhizosphere space and the situation is further complicated by the fact that the root apices will continue to exploit fresh soil, if this is available.

The rhizosphere receives diffusible metabolites from the root and consequently supports a microflora of different nature to that of the unmodified soil; it also serves as an ionic source for the root. Both of these relationships are likely to differ according to the nature and species of the root and, furthermore, Barber (1969) has shown that the presence of a microflora at the root surface grossly alters its nutrient-absorbing characteristics. For these reasons the interlocking of numerous different species' rhizospheres must be considered in terms of microniche differentiation.

The soil also lacks homogeneity, being aggregated into structural units as a result of the previous activities of plants, animals and microorganisms in the soil. Aggregates range in size from a fraction of a millimetre to a centimetre or more and they are not in complete liquid phase contact with each other, being separated by gas-filled pores. As a result they may not be diffusionally uniform, the centres having different gas, liquid and ionic concentrations from the peripheral layers as has been shown, for example, in studies of oxygen concentration (Greenwood and Goodman, 1967).

Finally, plants themselves show considerable variation of rooting depth within the soil profile and are capable of removing and translocating nutrients to impose long-term, persistent patterns which provide a third possible source of nutritionally differing microniches within the soil fabric. The reviews of Newbould (1969) and Barber (1969) suggest that these concepts of variable soil exploitation are now receiving attention though, experimentally, the rooting habitat is the most difficult part of the ecosystem to investigate. Furthermore, the root system is not a static entity but shows continuous extension during the growing season. In relation to slowly mobilized nutrients such as nitrogen and phosphorus, in which conversion of organic to inorganic forms interposes a rate-limiting process, the rate of root extension and exploitation of soil volume can be critical in competition for these nutrients. This also applies to water availability and will be discussed in the next section. Competitive exploitation of soil may also be influenced by the rate of nutrient absorption at the root surface. Woolhouse (1969) has, for example, suggested that acid phosphatase activity at root surfaces influences the rate of P uptake and that this root property is under genetic control varying with species and ecotype.

Water

Competition for water can be compared with competition for nutrients in that plants which produce more dry weight and leaf area for the same uptake are usually better competitors. The further analogy with tolerance of nutrient deficits is the plant which is inherently drought-tolerant.

Because water is so often physiologically limiting, the rate and extent of exploitation of the soil space is important. An early example comes from the work of Karper (1929) with Sorghum varieties (Table 10.2). Var. Kafir is not free-tillering and in wet years has to be close planted to make optimal use of soil water, but in dry years needs to be widely spaced to avoid intraspecific competition for water. By contrast var. Milo tillers freely and the same wide planting gives optimal yield in both wet and dry years as it is able to self-adjust its density to the available water supply.

Utilization of different parts of the soil by the root systems suggests that the mechanism involved may be avoidance of competition: this concept is supported by the conclusions of Goode's (1956) work with *Lolium perenne* S23 and *Poa annua*. When grown in mixed swards *P. annua* is shallow rooted but inherently fairly drought-tolerant while S23 is deep rooting and so avoids the consequences of drought. The sward thus remains healthy during the dry summer months but if it is overgrazed the rooting depth of S23 is reduced: both species come into competition for water and the sward loses its drought resistance.

Avoidance of competition by stratification of root systems was observed as long ago as 1906 when Woodhead described the floor vegetation of woodlands on acid Coal Measures' soils in Northern England. The grass *Holcus mollis* roots in the superficial soil, *Pteridium aquilinum* in the middle

Table 10.2 The yield of tillering and non-tillering sorghum varieties in relation to wet and dry years (after Karper, 1929). The data were collected during a ten-year period.

	var. Kafir		var. Milo	
	2 driest years	2 wettest years	2 driest years	2 wettest years
Row Spacing for maximum yield (cm)	91	7·5	61 and 69	54 and 69
Characteristics of the variety	Kafir does not produce many tillers so that final density is closely related to planting density. Optimum density thus rises with increasing wetness		Milo tillers freely and 'adjusts' its density to suit circumstances. Low-density planting is thus optimum for all conditions	

range and *Scilla (endymion) non-scripta* in the deepest soil. More recently, McWilliams and Kramer (1968) have described an experimental study of the perennial grass *Phalaris tuberosa* and the annual *P. minor* in relation to drought resistance. The shallow rooted *P. minor* is sensitive to drought but *P. tuberosa,* which roots to seven feet or more, appears drought-tolerant. Experiments with tritiated water showed that *P. tuberosa* was able to take up water from a depth of at least four feet.

The rate of exploitation of soil volume is also important in competition for water. Harris and Wilson (1970) found that the undesirable range grass, *Bromus tectorum*, continued root elongation at temperatures down to 3°C and, by elongation during the winter and early spring months, gained a competitive advantage over *Agropyron spicatum*, the roots of which ceased growth at 8–10°C.

FACTORS CAUSING COMPETITION

Passive root interaction

Under normal conditions the soil atmosphere contains less oxygen and more CO_2 than the air above it: a consequence of root and microorganism respiration. There may be a passive interaction between the roots of different species if one is more sensitive than the other to high soil CO_2 or low O_2 concentrations. This interaction is most likely to arise or contribute to competitive effects in wet soils where there is a shortage of air-filled pores and a consequently reduced oxygen diffusion rate. Sheikh (1970), for example, found that *Molinia caerulea* was less sensitive to high CO_2 levels than *Erica tetralix* and, as a result, on soils of moderate nutrient content under waterlogged conditions, *E. tetralix* is outcompeted by *M. caerulea*. The root respiration of the two species must make a passive contribution to this interaction in the sense that the normal respiratory production of CO_2 has become involved in the competitive balance.

Allelopathy

De Candolle (1832) suspected that plants released toxic materials into soils and that these lasted long enough to necessitate the rotation of crops. However, at the time, there was little real evidence for this, or knowledge of other soil effects such as nutrient deficiency and pathogen accumulation.

More recent evidence suggests that such interactions do occur but critical proof is difficult and the ecological importance of allelopathy is still ill-defined. One problem is that *in vitro* effects often disappear if experiments are carried out in soil. Grummer (1961) cites the inhibition of poppy seed germination by bitter almonds in Petri dishes. Bitter almonds are known to form HCN in small amounts: sweet almonds which do not produce HCN have no inhibitory effect. If the same experiment is carried out in soil the inhibition disappears as the cyanide is adsorbed by the soil and rendered harmless. There are many such examples: another arises from Bonner and Galston's (1944) investigation of *Parthenium argentatum* in gravel-culture. They showed that an inhibitor, *trans*-cinnamic acid, accumulated in the culture medium and interfered with the growth of the plants. Bonner (1946), however, was unable to show any inhibition when the plants were grown in soil.

Another difficulty is the problem of establishing causal relationships: it is easy to extract metabolic products which inhibit growth and it is easy to show that one plant depresses the growth of another. It is much less easy to prove that the growth effect is related to the secretion of the inhibitory metabolite into the soil, and not to other competitive interactions. Bedford and Pickering (1919) made an early attempt to overcome this problem while investigating the depression of fruit tree growth and yield by grass cover. They grew grass in perforated trays which could be laid on the surface of the soil in which potted apple seedlings were growing. Frequent lifting of the trays prevented the grass roots from penetrating into the soil below. Water was added either directly to the pot soil or indirectly by feeding it through the grass cover. In the latter case there was a considerable inhibition of the apple growth which appears most likely to have been an allelopathic effect and possibly related to the depression of tree growth observed in grass-covered orchards. The authors noted, however, that irrigation and N fertilization considerably reduced this field effect and that the experimental results cannot be considered as conclusive proof that allelopathy occurs under orchard conditions. Similar effects have been noted by a number of workers with *Juglans nigra* and *J. regia* (Grummer, 1961). Davies (1928) suspected that the phenolic compound juglone might be responsible for the inhibition of herbs below the *Juglans* canopy. Considerable quantities of juglone are washed from the leaves by rainfall and reach the soil below the trees.

Another example in which a toxin has been definitely identified but results still remain controversial arose from the work of Gray and Bonner (1948) with the two desert shrubs, *Encelia farinosa* which supports no underflora of herbs and *Franseria dumosa,* which does. Gray and Bonner showed that watering tomatoes through a mulch of *E. farinosa* leaves caused a reduction in growth and they isolated a toxic component of the leachate which was identified as 3-acetyl-6-methoxy benzaldehyde. It seemed reasonable to suggest that this compound prevented herb growth beneath the *E. farinosa* bushes but Muller (1953), investigating *F. dumosa* more fully, isolated a similar toxin from its leaves. He made the alternative suggestion that the sprawling growth of *F. dumosa* caused the trapping of a layer of dead leaves under each bush, thus providing a favourable habitat for germination of herb seeds. *E. farinosa* is raised above the surface by a short stem and is unlikely to trap leaves in this way. It must be assumed that any toxic activity of the leaf leachates is overcome by the contact with soil; toxins may, for example, be microbiologically degraded and it is possible that the higher soil organic content under *F. dumosa* could influence this process.

A large number of metabolic products can be shown to accumulate in soil (McLaren and Peterson, 1967) and many of these are identifiable as specific phytotoxins (Winter, 1961). It is unlikely, in the presence of this complex of biochemicals, that plants would not show specific responses to the metabolic products of other species. Winter states: 'although it is quite true that the higher plant does not have to absorb organic substances from the soil for its normal development, it cannot be overlooked that some organic substances may be and, in fact, are taken up and from this a number of important consequences follow'.

More recently, a number of reports of allelopathic effects have appeared which seem less ambiguous than much earlier work. Muller (1966) described the role of volatile phytotoxic terpenes which, he suggested, were responsible for the marked patterning of chaparral vegetation of *Salvia leucophylla* and *Artemesia californica*. On the ground and, more obviously, in aerial photographs the zones of contact between these shrub thickets and the interspersed grassland were delimited by bare areas of inhibition. A seedling assay for volatile inhibitors showed that both species were very active and gas chromatography identified a number of volatile terpenes in the assay-chamber atmosphere. The adsorptive properties of the soil caused it to act as a reservoir for the terpenes so that the effect is both soil- and air-borne. Muller suggested that the terpenes find entry by their solubility in the cuticular lipids of the epidermis, mesophyll and root epidermis. Another chaparral scrub plant, *Adenostema fasciculata,* showed a similar inhibitory effect which was caused by a water-soluble toxin. After burning to remove the scrub or, more significantly, after removing the top 5 cm of soil, herbaceous plants were able to establish themselves.

In arid habitats this type of allelopathy, especially if air-borne, is likely to liberate plants from competition for water by imposing a minimum spacing. Muller suggested that if allelopathy is more widespread than has been suspected then traditional theories of competition and ecosystem function will require considerable reevaluation.

Overland (1966) investigated the use of barley (*Hordeum vulgare*) as a 'smother' crop for weed suppression and showed its effectiveness even in the absence of normal competition and, furthermore, it was specific in its effects. Aqueous leachates of the roots were shown to be toxic and to have a similarly specific variation in toxicity. There seems then to be some scientific validity in the traditional choice of different 'smother' crops for different weeds. Brown (1967) also investigated the relationship of a single species (*Pinus banksiana*) to a number of associated herbaceous species. *P. banksiana* seeds were germinated in the presence of water extracts of 56 herb species of which nine were found, consistently, to inhibit germination. Field tests showed that many of these nine plants were active in limiting the establishment of the pine and would consequently protect themselves from subsequent shading.

Currently, it may be said that allelopathy is virtually proven to occur in certain habitats and between some species of plants. In particular it seems well marked when one species is at a much higher density or cover than its competitors (in monoculture and among natural dominants) and also in arid zone ecosystems where a spacing mechanism is important in avoiding competition for water. Additional reviews of the subject may be found in Tukey (1969) and Risser (1969). The most recent developments appear to be the investigations of the biochemical mechanisms of allelopathy. Muller, Lorber, Haley and Johnson (1969) showed, for example, that the terpenes liberated by *Salvia leucophylla* actively inhibited the oxygen uptake of mitochondrial suspensions.

INTERACTION WITH EXTERNAL FACTORS

Competition for pollinators and agents of seed dispersal is generally related to the 'attractiveness' of the flower or fruit to another organism, for example Free (1968) showed that *Taraxacum officinale* in apple orchards seriously reduced the number of honey-bees visiting the apple blossom. Similarly, plants which rely upon the palatability of their fruit for seed dispersal must encounter agent preference as a selective pressure. Posing considerable sampling problems, these aspects of competition have not been extensively worked.

Disturbance of ecosystems and selective pressure by man and animals cause extensive variation in competitive status of plants in natural ecosystems. Such pressures may be visualized as deforming or displacing niche hyperspaces, thus permitting some species to increase in numbers and others to regress. Niche space which is not efficiently occupied may also be taken over by entirely new species which have been introduced by man or animals. The latter change is eptiomized by the recent behaviour of the hybrid of *Spartina maritima* and *S. alterniflora* in Britain subsequent to its origin in Southampton Water shortly before 1870 (Goodman *et al.*, 1969). The hybrid originated as a result of the importation of *S. alterniflora* from N. America in the early nineteenth century. Present day populations consist of a sterile F$_1$ hybrid (*S.* × *townsendii*) and a fertile amphiploid (*S. anglica*) which is believed to have arisen in c. 1890, the accounting for the very sudden spread of the plant after that date.

S. anglica is a colonizer of bare, tidal mud at the lower end of the salt-marsh association, entering the niche which was formerly occupied by various annual *Salicornia* spp. It is a most successful competitor as it spreads vegetatively with great vigour from tussocks established from seed or vegetative fragments. Its perennial habit permits rapid invasion of the *Salicornia* zone and subsequent entry into other parts of the marsh. In the century since its appearance it has spread to the majority of British salt-marshes, with a consequent alteration in their physiognomy and reduction in species diversity.

Similar consequences have followed the introduction of many species to habitats outside their normal geographical range, sometimes with disastrous effects on land-use. *Opuntia* spp., for example, spread unchecked over many square miles of Australian grazing land where they were free from serious competition or predation and were brought under control only when a predator, the moth *Cactoblastis cactorum* was introduced as a biological controller. Many examples are cited in great detail by Salisbury (1961).

Plants of specialized habitats which require bare soil for their establishment have been allowed to enter other ecosystems through man's agricultural activities since the late Stone Age or early Bronze Age, the pollen record showing a sudden increase of 'weed' species associated with the appearance of cereals. A good example is *Poa annua*, which is very common in cultivated and trampled ground in Britain but rather rare in other habitats.

A rather different situation arises from the interaction of competition and grazing of grassland ecosystems. The persistence of many rosette and creeping annuals depends on grazing pressure to limit the height and competitive ability of the more vigorous species. The plagioclimax short-grassland of rendzines in Britain is an excellent example in which the grazing pressure derives from rabbits and sometimes sheep. In the mid-1950s the rabbit populations were greatly reduced by myxomatosis, from which time there has been a change to tall-grass or limestone-scrub cover in many areas with the consequent loss of many low growing species.

The question of the interaction of predators, pathogens and vector organisms with the ecology of plant distribution requires reexamination, in terms of the whole ecosystem, as population levels in natural associations must be extremely sensitive to these effects. A great deal of information has come from the accidental introduction of pathogens and predators to new habitats and also from the introduction of plant species to new geographical ranges in which they are free from natural pathogens and predators. In the one case, explosive outbreaks of damage or disease have occurred, massively reducing existing populations and in the other, waves of uncontrolled insurgence have led to plants occupying great areas of new territory. The fact that biological control may be used to reverse such invasions is further evidence of the importance of disease and predation in delineating the natural distribution and population levels of plant species.

We have to ask the question: is a climatic correlation with plant distribution mediated by a direct physiological response or is it perhaps that a predator, disease organism or pollinator is the limiting factor? The complication may be even greater in cases where a disease vector is involved: neither host nor disease organism may be limited by an environmental factor but the absence of the vector will impose a limitation. Disease-causing organisms and phytophagous insects are most likely to influence such distributional interactions as they are often highly specific—sometimes even to a racial level—and therefore exert intense selective pressure. Almost all plants in natural ecosystems are affected in this way as the diversity of insects, fungi, bacteria and viruses is such that very few plants are immune to some form of attack. Pollinators are not quite as selective but even so must exert some preferential pressure.

The interaction of competitive effects with other environmental variables may be seen as reflections of topographic moisture gradients, altitudinal and exposure effects, etc. Many of these relationships centre around the concept of the 'ecological refuge' which has been previously mentioned. Plant species which cannot withstand competition from others under favourable environmental conditions are often able to survive when the more demanding species are excluded by an environmental rigour. An example is seen in the oscillating boundary of *Pteridium aquilinum*: grassland associations according to the severity of the previous winter's frost. After a hard winter the grass is able to reinvade the edge of the *P. aquilinum* colony from which it was previously excluded by shading. The deep rhizomes of *P. aquilinum* rapidly permit the recouping of the loss unless further severe frosts are encountered in the subsequent year.

In its extreme form, this type of competition habitat interaction has permitted the evolution of highly specialized populations on unusual soils. The most widespread examples are the halophyte flora of salt-marshes and deserts and the calcicole flora of limestone rendzina soils. In the first case, the soil conditions make it so difficult for normal plants to grow satisfactorily that the halophytes are virtually spared competition except from plants of their own type. The calcicole flora of rendzinas may be interpreted similarly though the soil conditions here are not so exclusive except to obligate calcifuges. Cole (1969) has written: 'And genuinely difficult habitats . . . produce almost incredible concentrations of one or a very few species.'

Other examples are seen in the evolution of local populations of ecotypes and species tolerant of heavy-metal contaminated soils now common as artificial habitats surrounding mining and smelting works. Evolution of tolerant genotypes has been shown for a number of species, for example *Agrostis tenuis* forms distinct ecotypes tolerant of lead, nickel, copper and zinc (Gregory and Bradshaw, 1965) while *Festuca rubra* has a lead-tolerant ecotype. These tolerances permit the plants to colonize habitats which are otherwise devoid of plant cover, so that they meet very little competition. In grass swards on normal soils the metal-tolerant ecotypes are incapable of competing and are reduced to a small proportion of the population (see Chapter 9).

THE BEHAVIOUR OF COMPETING PLANTS: INTRASPECIFIC COMPETITION

The foregoing sections have reviewed the relationship of plant competition to environmental factors. The pressures of competition are such that the plant growing in an ecosystem behaves differently from an isolated plant in terms of growth morphogenesis and physiology.

Clements, Weaver and Hanson's (1929) early experiments with *Helianthus annuus* planted at different densities provided dramatic proof of the individual species' plastic response to competitive stress. The plants were grown in plots with spacing of 2, 4, 8, 16, 32 and 64 inches. Comparison of their photographs of the plants after two months shows remarkable differences: the 64-inch plants were large and robust with well-developed leaves inserted at all of the nodes while the poorly developed 2-inch plants were etiolated, with spindly stems and the majority of the lower leaves lost by senescence (Figure 10.8).

Plastic responses of this type have attracted the agronomists' attention for many years as the interaction of such responses with density is reflected

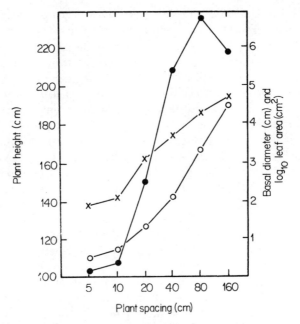

Figure 10.8 The response of *Helianthus annuus* to intraspecific competition at plant spacings between 5 and 160 cm. The plants were harvested at 10 weeks shortly prior to flowering. Note that leaf area is plotted on a log scale which covers the range 10 to 100,000 sq. cm. Data of Clements, Weaver and Hanson (1929).

in crop yield. Agriculture is generally concerned with yield per unit area and for this reason the usually high sowing densities result in intense intraspecific competition and strong plastic modification of individuals. Only in cases where the quality of the individual product is crucial, such as fruit production, are planting densities less extreme.

The data shown in Figure 10.9 were extracted from the work of Puckridge (1962) with wheat and show the contrast between individual plant behaviour and the yield characteristics of the whole crop. Both plant dry weight and seed number per plant were strongly depressed by increasing the density above the minimum of 1·4 per square metre but the grain yield per unit area was increased by densities up to 35 per square metre, hence the optimum agricultural yield was associated with a density which placed individual plants in a considerably suboptimal environment.

There have been few studies of individual plants under competitive circumstances as the agronomic requirement has been for data on area yield and *mean* plant response. In cases where individual plants have been

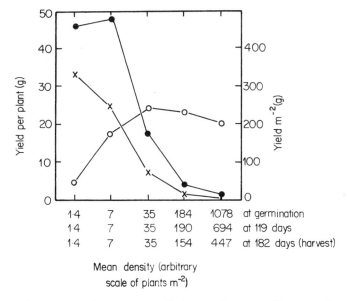

Figure 10.9 The response of wheat to intraspecific competition with a planting density ranging from 1·4 to 1078 plants per square metre at germination. Note the decline in plant numbers which occurred at higher densities during the course of the experiment. Note also the slight reduction in plant weight at the lowest density, an effect which is also seen in the plant height data of Figure 10·9. This may be a consequence of an 'oasis' effect in which individual plants are exposed to a greater water stress than more closely spaced plants. Data of Puckridge (1962).

followed there appears to be an ecological example of the Biblical maxim: 'Unto everyone that hath shall be given.' Larger plants tend to become larger and smaller plants relatively smaller until some are eliminated. Such trends may often be traced back to the influence of seed weight, individuals with larger than average food resources giving seedlings which are most likely to become the dominants of the population (Black, 1958). The consequence is that weight variation, which at the seedling stage is normally distributed, becomes steadily more skewed until the population is dominated by a few large plants and contains large numbers of small plants some of which are destined to be eliminated.

This process of increasing dominance of large individuals is also deducible from the work of Hozumi, Koyama and Kira (1955) who grew *Zea mais* in single rows at two different spacings. Autocorrelation analysis between the weights of the nth plant in a row and the $(n + 1)$th to $(n + 5)$th showed the markedly alternating effect seen in Figure 10.10. The first $(n + n + 1)$ correlation was always negative and the subsequent values then fluctuated alternately between positive and negative, suggesting that a large plant tended to depress the weight of its neighbours while a small plant caused enhanced neighbour weight. It should be mentioned that the authors also detected another effect in these experiments which would tend to cancel the consequences of this alternation. This was the negative correlation between shoot length and elongation rate which, of course, would tend to equalize plant size. The behaviour of the crop as a whole must be governed by the balance between these two effects.

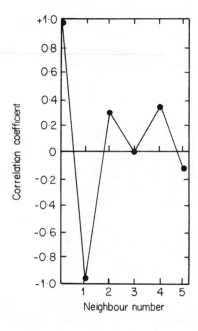

Figure 10.10 The correlation between shoot weights of a *Zea mais* plant and its first to fifth neighbours. After K. Hozumi *et al.*, *J. Inst. Polytech. Osaka City Univ. Series D*, **6**, 121–30 (1955).

Harper (1967) summarizes the direct consequences of density stress as (i) elicitation of a plastic response from individuals as they adjust to share limiting resources; (ii) increasing of mortality; (iii) exaggeration of differentials within the population and encouragement of a hierarchy of exploitation. The second point, the increase of mortality, is a density-dependent, self-regulating process and leads to examples such as the seed/seedling number curves shown in Figure 10.2. As seeding density increases the relationship is at first linear but limiting factors such as germination niche saturation and seedling competition are encountered, causing a declining response to increased seed rate and a maximum achievable density.

Patterns of population growth and balance are, to a great extent, governed by the influence of competition mediated through density-dependent mortality and interference with reproductive capacity. The increase of plant populations is often assumed to be logistic:

$$\frac{\mathrm{d}N}{\mathrm{d}t} = rN \left(\frac{K - N}{K} \right)$$

where $\mathrm{d}N/\mathrm{d}t$ = change of number (N) with time (t), r = reproductive rate and K is a limiting constant.

$$\begin{array}{ccc} \text{Rate of population} \\ \text{growth} \end{array} = \begin{array}{c} \text{Intrinsic rate of} \\ \text{increase} \end{array} \times \begin{array}{c} \text{Actual realization of} \\ \text{potential increase} \end{array}$$

In stable populations N is likely to remain nearly equal to K for long periods, density-dependent mortality and limited reproduction dominating the regulatory process. Harper (1967) points out that many plant populations spend a large proportion of their time in recovering from environmental catastrophes. These populations will generally be increasing at something near to their intrinsic rates and population size will be a function of the magnitude of the last catastrophe and the time available for regrowth.

The value of r is twofold, being represented in local vegetative multiplication and also in the further flung effects of seed production. The first may be looked upon as a large capital investment producing a small but secure return in increasing population size while the second represents a small capital investment with a potentially high return in increased numbers but a much greater risk of total loss. Harper discussed the strategy of the plant life-cycle in governing its potential competitive relationships. In terms of seed production the rate of population increase is critically governed by precocity of reproduction, for example an individual producing two offspring in the first year, then dying, has the same potential rate of increase as an individual producing one offspring per year for ever. Harper contrasted the biennial *Digitalis purpurea* producing c. 100,000 seeds every two years with an annual counterpart which would achieve the same population growth rate with only 330 seeds p.a. However, high risk of seed mortality must affect the economy of these two types of plant; the annual risks total seed loss but this is much less likely with the biennial.

INTERSPECIFIC COMPETITION

The competition of two or more species is a source of complication which has limited most such work to agricultural investigations of

weed/crop interactions and studies of pasture mixtures. Notable exceptions to this background are to be found in the work of Clements, Weaver
and Hanson (1929) who experimented with the interspecific competition
of various constituent plants of natural ecosystems in addition to comparing agricultural plants.

Agronomic experiments have had two different aims, the first being the
investigation of yield depression by weed species and the second the investigation of the advantages arising from the mixture of two or more
species in forage or fodder crops.

In a population of a single species slight genetic variations or even random variations deriving from seed weight may influence the behaviour of
the competing individuals. Similarly, in mixed populations interspecific
differences may be exploited and exaggerated by density stress. One of the
most satisfactory approaches to these problems derives from the
experimental models of de Wit (1960). Two species are sown or planted
together in different ratios but the overall plant density is maintained constant. If there is no competition and growth rates are equal then there is no
change in ratio during the experimental period (Figure 10.11a) but if one
grows more rapidly than the other, the relationship becomes distorted as
in Figure 10.11b. There is still, however, no competitive effect of one upon
the other. If competition occurs then the successful plant increases its yield
at the expense of the unsuccessful competitor (Figure 10.11c).

When data from such experiments are expressed as species ratios at
sowing, versus the ratios at harvest (input ratio v. output ratio), they may be
used to propound four different behavioural situations (Figure 10.12a and
b). When there is no drift of species ratio from input to output the association will remain stable and the *relative reproductive rate* may be expressed as
unity:

$$RRR = \frac{A_{harvest}/A_{sown}}{B_{harvest}/B_{sown}} = 1 \cdot 0$$

The slope of the input/output relationship will thus be 45° (Figure
10.12a). If, however, one species has a reproductive rate which is higher
than the other, irrespective of the sowing ratio, then the slope remains at
45° but it is displaced either above or below the unity line and, with the
course of time, the species with the lower reproductive rate will become
extinct. If the relative reproductive rate of the two species is sensitive to
sowing ratio, the slope of the line will be altered. If the slope is greater than
45° one, or the other, species will drift toward extinction according to the
initial seed ratio. If the slope is less than 45° the two species will converge
toward an intermediate value lying on the unity line and the association
will become stable.

The situations which promote stability are, therefore, those in which
$RRR = 1$ and there is no competitive interaction, or those in which the in-

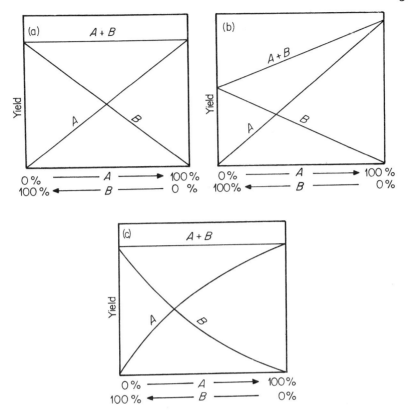

Figure 10.11 The relationship of vegetative or propagule yield of two species, *A* and *B*, to proportions of each in the sowing or planting mixture. (a) No competitive interaction and identical growth or seed production rates; (b) No competitive interaction but different growth or seed production rates; (c) Competitive interaction in which the proportion of *B* in the harvest is suppressed and that of *A* enhanced. Redrawn with permission from C. T. de Wit, *Verslag. Landbouwk. Onderz. Ned.*, **66**, 8 (1960).

creased ratio of one species in a seed mixture is reflected as a decrease of that species in the harvest mix or *vice versa*. Such a response may be interpreted as 'self-competitive limitation' at high initial ratios and 'space exploitation' at low ratios. In the unstable situation with a slope of more than 45° the species which is at a high level in the initial seed mix gains a long-term competitive advantage which compounds with time.

Examination of these four potential deviations from equilibrium suggests that most circumstances are likely to eliminate one species of a competing pair and yet it is notable that natural ecosystems support a great diversity of coexisting species. In fact, it is only the unusually harsh en-

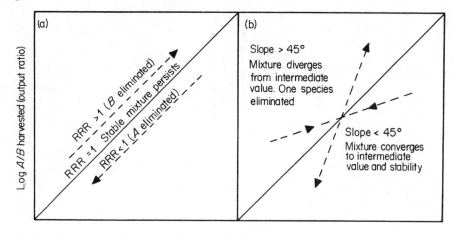

Figure 10.12 The relationships which may exist between input (sown) and output (harvested) ratios of two species grown in competition. (a) The continuous line with a slope of 45° represents an input-output ratio of unity giving a stable situation in which both *A* and *B* will persist. Lying above or below this line, but still with unit slope, are the two pecked lines representing a constant reduction in the proportion of one species at each harvest. Both of these situations are unstable and lead to the ultimate extinction of one species. (b) In this case the two pecked lines represent a situation in which the input-output ratios are density dependent so that the slope is >45° or <45°. With a slope >45° the situation is unstable; above the intersection with the continuous line, *B* will gradually disappear, and below the intersection *A* will gradually disappear. With a slope of <45° the mixture converges to an intermediate value and *A* and *B* persist together. Redrawn from Donald (1963) after de Wit, *Verslag. Landbouwk, Onderz. Ned.*, **66**, 8 (1960).

vironment, sometimes a grossly polluted or damaged environment, which is characterized by very low species diversity. The problem of coexistence has exercised many ecologists and a number of explanations have been put forward. One, for example, is Hutchinson's (1965) modification of the usual logistic equation for population growth:

$$\frac{\mathrm{d}N_1}{\mathrm{d}t} = r_1 N_1 \frac{(K_1 - N_1 - aN_2)}{K_1}$$

$$\frac{\mathrm{d}N_2}{\mathrm{d}t} = r_2 N_2 \frac{(K_2 - N_2 - \beta N_1)}{K_2}$$

where N_1 and N_2 are the respective numbers of species 1 and 2, r_1 and r_2 are their relative growth rates and K_1 and K_2 are constants for each species. The condition for continued cosurvival is that $\alpha < K_1/K_2$ and $\beta < K_2/K_1$. In words: the population growth of one species will inhibit its own further growth more than that of the other species, thus preventing competition from culminating in an exclusive struggle.

Returning to the Margalef (1968) feedback concept discussed at the beginning of the chapter, it was postulated that a positive feedback could be established between a pair of species competing for a single limiting resource. This feedback could result in one species 'mopping up' the resource to such an extent that the second species would be driven to extinction. Hutchinson's model, however, averts this catastrophe, as a density-dependent negative feedback within each species prevents the population explosion which would be necessary to remove all of the resource from the environment. The equilibrium proportions of the two species in the population will depend on the rate of this internal feedback mechanism and its critical density dependence.

Such density-dependent population control of the 'exploding' species might, for example, be caused by a soil niche specialization in which differential exploitation of rooting volume prevented the otherwise successful species (A) from obtaining all of a limited supply of a nutrient. In the early phase of its growth, species A would not be nutrient-limited and might compete strongly with species B for light, but as A reached nutrient limitation, its competitive function would become increasingly intraspecific and some niche space would remain in which species B might persist at low density. The negative feedback between species A and the environment is, thus, a nutrient limitation to which species B is not sensitive.

Density and population regulation have been more widely studied by zoologists than botanists, perhaps because the more obvious characteristics of density-dependent physiological and behavioural response are involved. Wynne-Edwards (1969) states that: '. . . it can be accepted that more or less effective methods of regulating their own numbers have been evolved by most types of animal'. Man appears to be unique as the only animal for which this is patently untrue but it has only ceased to be true since the Stone Age when the territorial limitations of a hunting life were replaced by the apparently unlimited resources made available by agriculture.

Amongst the animals self-regulation appears to be mediated through density-dependent physiological/behavioural changes and through establishment of territorial rights. Amongst plants the latter situation, the establishment of territorial boundaries, may occur through niche specialization and would serve to ensure the persistence of species diversity. Harper (1967) suggested that competition in a mixture, passing through an initial oscillatory phase with strong selective pressure, leads to the avoidance of intergroup struggles by the acquisition of such niche specialization characteristics. He further suggested that, if the evolution of a balanced ecosystem in which none of the competitive pressures causes extinctions leads to diversity and effective niche utilization, then such a stable ecosystem is likely to be highly efficient in its use of natural resources. There is no evidence, however, that mixed populations can out-yield pure stands and Harper suggests that this may be due to lack of in-herent combining ability in agriculturally bred plants. May it not, however, be that a part of the 'exploitation efficiency' is the very stability which comes from full niche utilization? In a sense the price of diversity may be that some low-yielding plants are supported in certain niches and by their presence reduce the total production of the ecosystem though adding to its stability. If this is so one would not expect an *ad hoc* mixture of agricultural plants to behave as a stable, diverse ecosystem and indeed it does not.

Bibliography

Alberda, Th. (1962). Actual and potential production of crops. *Neth. J. Agric. Sci.*, **10**, 325–33.

Alexander, M. (1961). *Introduction to soil microbiology.* John Wiley, New York.

Allam, A-W, I. (1971). Soluble sulphides in rice fields and their *in vitro* effects on rice seedlings. Ph.D. Dissertation. Louisiana State Univ.

Allison, F. E. (1965). Organic carbon. In *Methods of Soil Analysis*, pp. 1367–78, Am. Soc. Agron., Wisconsin.

Allison, F. E. (1968). Soil aggregation—some facts and fallacies as seen by a microbiologist. *Soil Sci.*, **106**, 136–43.

Altmann, P. S. and Dittmer, D. S. (1966). *Environmental Biology,* Fed. Am. Soc. Exp. Biol., Bethesda.

Antonovics, J., A. D. Bradshaw and R. G. Turner (1971). Heavy metal tolerance in plants. *Adv. ecol. Res.*, **7**, 1–85.

Arber, A. (1920). *Water plants.* Cambridge University Press, Cambridge.

Arikado, H. and Y. Aduchi (1965). Anatomical and ecological responses of barley and some forage crops to the flooding treatment. *Bull. Fac. Agric. Mie Univ.*, **11**, 1–29.

Arikado, H. (1959). Comparative studies on the development of the ventilating system between lowland and upland rice plants growing under flooded and upland soil conditions. *Bull. Fac. Agric. Mie Univ.*, **19**, 1–10.

Aristovskaya, T. V. and G. A. Zavarzin (1971). Biochemistry of iron in soil. In *Soil Biochemistry*, Vol. 2, pp. 385–408, Dekker, New York.

Armstrong, W. (1964). Oxygen diffusion from the roots of some British bog plants. *Nature, Lond.*, **204**, 801–2.

Armstrong, W. (1967a). The relationship between oxidation-reduction potentials and oxygen diffusion levels in some waterlogged organic soils. *J. Soil Sci.*, **18**, 27–34.

Armstrong, W. (1967b). The use of polarography in the assay of oxygen diffusing from roots in anaerobic media. *Physiol. Pl.*, **20**, 540–53.

Armstrong, W. (1967c). The oxidizing activity of roots in waterlogged soils. *Physiol. Pl.*, **20**, 920–6.

Armstrong, W. (1968). Oxygen diffusion from the roots of woody species. *Physiol. Pl.*, **21**, 539–43.

Armstrong, W. (1970). Rhizosphere oxidation in rice and other species: a mathematical model based on the oxygen flux component. *Physiol. Pl.*, **23**, 623–30.

Armstrong, W. (1971a). Oxygen diffusion from the roots of rice grown under non-waterlogged conditions. *Physiol. Pl.*, **24**, 242–7.

Armstrong, W. (1971b). Radial oxygen losses from intact rice roots as affected by distance from the apex, respiration and waterlogging. *Physiol. Pl.*, **25**, 192–7.

Armstrong, W. (1972). A re-examination of the functional significance of aerenchyma. *Physiol. Pl.*, **27**, 173–77.

Armstrong, W. and Boatman, D. J. (1967). Some field observations relating the growth of bog plants to conditions of soil aeration. *J. Ecol.*, **55**, 101–10.

Armstrong, W. and D. J. Read (1972). Some observations on oxygen transport in conifer seedlings. *New Phytol.*, **71**, 55–62.

Army, T. J. and F. A. Green (1967). Photosynthesis and crop production systems. In *Harvesting the Sun*, pp. 321–32, Academic Press, New York.

Arnon, D. L. and C. M. Johnson (1942). Influence of hydrogen ion concentration on the growth of higher plants under controlled conditions. *Pl. Physiol., Lancaster*, **17**, 525–39.

Ashton, T. (1948). *Techniques of breeding for drought resistance in crops.* Comm. Bur. Pl. Breed and Gen. Tech. Comm. 14. Commonwealth Agricultural Bureau.

Ashton, T. (1956). Effects of a series of cycles of alternating low and high soil water content on the rate of apparent photosynthesis in sugarcane. *Pl. Physiol., Lancaster*, **31**, 266–74.

Atkins, J. G. (1958). *Rice diseases.* U.S. Dept. Agric. Farmers Bull. 2120.

Baba, I., K. Inada and K. Tajima (1964). Mineral nutrition and the occurrence of physiological diseases. In *The Mineral Nutrition of the Rice Plant*, pp. 173–95, John Hopkins, Baltimore.

Baba, I., I. Iwata and K. Tajima (1963). Physiological injury. In *Theory and practice of growing rice*, pp. 149–72, Fuji, Tokyo.

Babalola, O., L. Boersma and C. T. Youngberg (1968). Photosynthesis and transpiration of Monterey pine seedlings as a function of soil water suction and soil temperature. *Pl. Physiol., Lancaster*, **43**, 515–21.

Bange, G. G. J. (1953). On the quantitative explanation of stomatal transpiration. *Acta. bot. neerl.*, **2**, 255–97.

Bannister, P. (1964a,b,c). The water relations of certain heath plants with reference to ecological amplitude. I. Introduction. Germination and establishment. II. Field studies. III. Experimental studies and general conclusions. *J. Ecol.*, **52**, 423–32, 481–97, 499–509.

Barber, D. A. (1969). The influence of the microflora on the accumulation of ions by plants. In *Ecological Aspects of the Mineral Nutrition of Plants*, pp. 191–200, Blackwell, Oxford.

Barber, D. A., M. Ebert and N. T. S. Evans (1962). The movement of ^{15}O through barley and rice plants. *J. exp. Bot.*, **13**, 397–403.

Barber, S. A., J. M. Walker and E. H. Vasey (1963). Mechanisms for the movement of plant nutrients from the soil and fertilizer to the plant root. *J. agric. Fd. Chem.*, **11**, 204–7.

Barley, K. P. (1961). The abundance of earthworms in agricultural soils and their possible significance in agriculture. *Agronomy*, **13**, 249–68.

Barley, K. P. (1963). Influence of soil strength on the growth of roots. *Soil Sci.*, 96, 175–86.

Barrs, H. D. (1968). Determination of water deficits in plant tissues. In *Water Deficits and Plant Growth*, pp. 236–68, Academic Press, New York.

Bartlett, R. J. (1961). Iron oxidation proximate to plant roots. *Soil Sci.*, 92, 372–79.

Bear, F. E. (1964). *Chemistry of the Soil*, Reinhold, New York.

Bedford, D. of and S. U. Pickering (1919). *Science and Fruit Growing*, Macmillan, London.

Begg, J. E., J. F. Bierhuizen, E. R. Lemon, D. K. Misra, R. O. Slatyer and W. R. Sten (1964). Diurnal energy and water exchanges in bullrush millet in an area of high solar radiation. *Agric. Meteorol.*, 1, 294–312.

Bjorkman, O. (1966). The effect of oxygen concentration on photosynthesis in higher plants. *Physiol. Pl.*, 19, 618–33.

Bjorkman, O. (1968). Carboxydismutase activity in shade-adapted and sun-adapted species of higher plants. *Physiol. Pl.*, 21, 1–10.

Black, C. A. (1965). *Methods of Soil Analysis*, Vols. I and II, Am. Soc. Agron., Wisconsin.

Black, C. A. (1968). *Soil–Plant Relationships* (2nd. edn.), Wiley, New York.

Black, J. D. F. and D. W. West (1969). Solid state reduction at a platinum microelectrode in relation to measurement of oxygen flux in soil. *Aust. J. Soil Res.*, 7, 67–72.

Black, J. N. (1958). Competition between plants of different initial seed size in swards of subterranean clover (*Trifolium subterraneum*) with particular reference to leaf area and microclimate. *Aust. J. agric. Res.*, 9, 299–318.

Black, J. N. (1964). An analysis of the potential production of swards of subterranean clover (*Trifolium subterraneum* L.) at Adelaide, South Australia. *J. app. Ecol.*, 1, 3–18.

Blackman, F. F. (1905). Optima and limiting factors. *Ann. Bot.*, 19, 281–95.

Blackman, G. E. and A. J. Rutter (1959). Physiological and ecological studies in the analysis of plant environment. V. An assessment of the factors controlling the distribution of the bluebell (*Scilla nonscripta*) in different communities. *Ann. Bot.* N.S., 14, 487.

Blackman, V. H. (1919). The compound interest law and plant growth. *Ann. Bot.*, 33, 353–60.

Bloomfield, C. (1953a, b). A study of podsolization. I. The mobilization of iron and aluminium by Scots Pine needles. II. The mobilization of iron and aluminium by the leaves and bark of *Agathis australis* (Kauri). *J. Soil Sci.*, 4, 5–16, 17–23.

Bloomfield, C. (1954a, b, c). A study of podsolization. III. The mobilization of iron and aluminium by Rima (*Dicradium cupressinum*). IV. The mobilization of iron and aluminium by picked and fallen larch needles. V. The mobilization of iron and aluminium by aspen and ash leaves. *J. Soil. Sci.*, 5, 39–45, 46–49, 50–59.

Bloomfield, C. (1965). Some processes of podsolization. In *Experimental Pedology*, pp. 257–66, Butterworth, London.

Boatman, D. J. and W. Armstrong (1968). A bog type in N. West Sutherland. *J. Ecol.*, 56, 129–41.

Boeke, J. E. (1940). On the origin of the intercellulary channels and cavities in the rice root. *Ann. Jardin Bot. Buitenzorg*, 50, 199–208.

Boggie, R. (1972). Effect of water table height on root development of *Pinus contorta* on deep peat in Scotland. *Oikos*, 23, 304–12.

Bond, G. (1952). Some features of root growth in nodulated plants of *Myrica gale* L. *Ann. Bot.* N.S., **16**, 467–75.

Bonner, J. (1946). Further investigation of toxic substances which arise from guayule plants: relationship of toxic substances to the growth of guayule in soil. *Bot. Gaz.,* **107,** 343–51.

Bonner, J. and A. W. Galston (1944). Toxic substances from the culture medium of Guayule which may inhibit growth. *Bot. Graz.,* **106,** 185–98.

Bould, C. (1963). Mineral nutrition of plants in soils and in culture media. Part I Mineral nutrition of plants in soils. In *Plant Physiology,* Vol. III, pp. 15–96, Academic Press, New York.

Boulter, D., D. A. Coult and G. G. Henshaw (1963). Some effects of gas concentrations on metabolism of the rhizome of *Irish pseudacorus* L. *Physiol. Pl.,* **16,** 541–48.

Bowman, G. E. (1968). The measurement of carbon dioxide in the atmosphere. In *The Measurement of Environmental Factors in Terrestrial Ecology,* pp. 131–9, Blackwell, Oxford.

Bradley, E. F. and O. T. Denmead (1967). *The Collection and Processing of Field Data,* Interscience, New York.

Bradshaw, A. D. (1969). An ecologists viewpoint. In *Ecological aspects of the Mineral Nutrition of Plants,* pp. 415–27, Blackwell, Oxford.

Bradshaw, A. D., M. J. Chadwick, D. Jowett, R. W. Lodge and R. W. Snaydon (1960). Experimental investigations into the mineral nutrition of several grass species. III. Phosphate level. *J. Ecol.,* **48,** 631–7.

Bradshaw, A. D., M. J. Chadwick, D. Jowett and R. W. Snaydon (1964). Experimental investigation into the mineral nutrition of several grass species. IV. Nitrogen level. *J. Ecol.,* **52,** 665–76.

Bradshaw, A. D., R. W. Lodge, D. Jowett and M. J. Chadwick (1958). Experimental investigations into the mineral nutrition of several grass species. I. Calcium level. *J. Ecol.,* **46,** 749–57.

Bradshaw, A. D., R. W. Lodge, D. Jowett and M. J. Chadwick (1960). Experimental investigations into the mineral nutrition of several grass species. II. Calcium and pH. *J. Ecol.,* **48,** 143–50.

Bradshaw, A. D., T. S. McNielly and R. P. G. Gregory (1965). Industrialization, evolution and the development of heavy metal tolerance in plants. In *Ecology and the Industrial Society,* pp. 327–43, Blackwell, Oxford.

Braun-Blanquet, J. (1932), *Plant Sociology: the Study of Plant Communities.* McGraw Hill, New York (translation).

Bray, R. J. and J. T. Curtis (1957). An ordination of the upland forest communities of southern Wisconsin. *Ecol. Monogr.,* **27,** 325–49.

Bremner, J. M. and K. Shaw (1958). Denitrification in soil. *J. agric. Sci.,* **51,** 22–52.

Bremner, J. M. (1967). Nitrogenous compounds. In *Soil Biochemistry,* Vol. 1, pp. 19–66, Dekker, New York.

Briggs, L. J. (1897). The mechanics of soil moisture. *U.S. Dept. Agr. Soils Bull.,* **10.**

Brix, H. (1962). The effect of water stress on the rates of photosynthesis and respiration in Tomato plants and Loblolly Pine. *Physiol. Pl.,* **15,** 10–20.

Brouwer, R. (1965). Water movement across the root. In *Water in living organisms,* pp. 131–49, Cambridge University Press, Cambridge.

Brown, J. C. (1960). An evaluation of bicarbonate induced iron chlorosis. *Soil Sci.*, **89**, 246–7.

Brown, J. M. A., H. A. Outred and C. F. Hill (1969). Respiratory metabolism in mangrove seedlings. *Pl. Physiol., Lancaster*, **44**, 287–94.

Brown, R. T. (1967). Influence of naturally occurring compounds on germination and growth of Jack Pine. *Ecology*, **48**, 542–6.

Brown, W. V. and G. A. Pratt (1965). Stomatal inactivity in grasses. *Southwestern Naturalist*, **10**, 48–50.

Bryan, R. B. (1971). The efficacy of aggregation indices in the comparison of some English and Canadian soils. *J. Soil Sci.*, **22**, 166–78.

Buckingham, E. (1907). Studies on the movement of soil moisture. *U.S. Dept. Agr. Bur. Soils Bull.*, **38**.

Bunting, B. T. (1965). *The Geography of the Soil*, Hutchinson, London.

Burg, S. P. and E. A. Burg (1965). Ethylene action and the ripening of fruits. *Science*, **148**, 1190–6.

Burges, A. (1967). The decomposition of organic matter in the soil. In *Soil Biology*, pp. 479–92, Academic Press, London.

Burges, A. and F. Raw (1967). *Soil Biology*, Academic Press, London.

Burnett, J. H. (1964). *The Vegetation of Scotland*, Oliver & Boyd, Edinburgh.

Butzer, K. W. (1965). Climatic change in arid regions since the Pliocene. In *A History of Land Use in Arid Regions*, pp. 31–56, UNESCO, Paris.

Campbell, N. E. R. and H. Lees (1967). The nitrogen cycle. In *Soil Biochemistry*, Vol. 1, pp. 194–215, Dekker, New York.

Cannon, H. L. (1960). Botanical prospecting for ore deposits. *Science*, **132**, 591–8.

Cannon, W. A. and E. E. Free (1925). Physiological features of roots with especial reference to the relation of roots to the aeration of the soil. *Carnegie Inst. Wash. Publ.*, **368**, 1–168.

Carr, D. J. and D. F. Gaff (1961). The role of cell-wall water in the water relations of leaves. In *Plant Water Relationships in Arid and Semi-arid Conditions*, pp. 117–25, UNESCO, Paris.

Carrol, J. C. (1943). Effects of drought, temperature and nitrogen on turf grasses. *Pl. Physiol., Lancaster*, **18**, 19–36.

Chapman, H. D. and P. F. Pratt (1961). *Methods of Analysis for Soils, Plants and Water.* University of California.

Chew, R. M. and A. E. Chew (1965). The primary productivity of a desert shrub (*Larrea tridentata*) community. *Ecol. Monogr.*, **35**, 355–75.

Chirkova, T. V. (1968). Oxygen supply to roots of certain woody plants kept under anaerobic conditions. *Soviet Plant Physiol.*, **15**, 475–8.

Clapham, A. R. (1956). Autecological studies and the 'Biological Flora' of the British Isles. *J. Ecol.*, **44**, 1–11.

Clapham, A. R. (1969). Introduction. In *Ecological Aspects of the Mineral Nutrition of Plants*, Blackwell, Oxford.

Clark, J. (1961). Photosynthesis and respiration in White Spruce and Balsam Fir. *Syracuse University State Univ. Coll. For. Tech. Pub.*, **85**.

Clarkson, D. T. (1965). Calcium uptake by calcicole and calcifuge species in the genus *Agrostis. J. Ecol.*, **53**, 427–35.

Clarkson, D. T. (1969). Metabolic aspects of aluminium toxicity and some possible mechanisms for resistance. In *Ecological Aspects of the Mineral Nutrition of Plant*, pp. 381–97, Blackwell, Oxford.

314

Clausen, J., D. D. Keck and W. M. Heisey (1948). Experimental studies on the nature of species. I Environmental responses of climatic races of *Achillea*. *Carnegie Inst. Wash. Publ.*, **581.**

Clements, F. E. (1905). *Research Methods in Ecology,* Nebraska Univ. Publ. Co. Lincoln.

Clements, F. E. (1907). *Plant Physiology and Ecology,* Holt, New York.

Clements, F. E. (1916). *Plant Succession,* Carnegie Inst., Washington.

Clements, F. E. (1920). *Plant Indicators,* Carnegie Inst., Washington.

Clements, F. E. (1928). *Plant Succession and Indicators,* Wilson, N. York.

Clements, F. E., J. E. Weaver and H. C. Hanson (1929). *Plant Competition,* Carnegie Inst., Washington.

Clymo, R. S. (1962). An experimental approach to part of the calcicole problem. *J. Ecol.*, **50,** 701–31.

Coffey, G. N. (1912). A Study of the Soils of the United States. *U.S. Dept. Agr. Bur. Soils Bull.*, **85.**

Cole, L. C. (1969). The impending emergence of ecological thought. *BioScience,* **14,** 30–2.

Coleman, E. A. (1946). A laboratory study of lysimeter drainage under controlled conditions of moisture tension. *Soil Sci.*, **62,** 365–82.

Colwell, W. E. and R. W. Cummings (1944). Chemical and biological studies on aqueous solutions of boric acid and of calcium, sodium and potassium metaborate. *Soil Sci.*, **57,** 37–49.

Connell, W. E. and W. H. Patrick (1968). Sulphate reduction in soil: effects of redox potential and pH. *Science,* **159,** 86–7.

Cook, C. W. (1943). A study of the roots of *Bromus inermis* in relation to drought resistance, *Ecology,* **24,** 169–82.

Cooper, C. F. (1969). Ecosystem models in watershed management. In *The Ecosystem Concept in Natural Resource Management,* pp. 309–24, Academic Press, New York.

Cosgrove, D. J. (1967). Metabolism of organic phosphates in soil. In *Soil Biochemistry,* Vol. 1, pp. 216–28, Dekker, New York.

Cowan, I. R. and F. L. Milthorpe (1968). Plant factors influencing the water status of plant tissue. In *Water Deficits and Plant Growth,* Vol. 1, pp. 137–93, Academic Press, New York.

Cowles, H. C. (1899). The ecological relationships of the vegetation on sand dunes of Lake Michigan. *Bot. Gaz.*, **27,** 95–117; 167–202; 281–308; 361–91.

Cowles, H. C. (1911). The cause of vegetation cycles. *Bot. Gaz.*, **51,** 161–83.

Crafts, A. S. (1968). Water deficits and physiological processes. In *Water Deficits and Plant Growth,* Vol. 2, pp. 85–133. Academic Press, New York.

Crampton, C. B. (1963). The development and morphology of iron pan podsols in Mid- and South Wales. *J. Soil Sci.*, **14,** 282–302.

Crampton, C. B. (1968–70). The evolution of soils in Morgannwg. *Proc. Cardiff Nat. Soc.*, **95,** 41–52.

Crawford, R. M. M. (1966). The control of anaerobic respiration as a determining factor in the distribution of the genus *Senecio. J. Ecol.*, **54,** 403–13.

Crawford, R. M. M. (1969). The physiological basis of flooding tolerance. *Ber. dt. bot. Ges.*, **82,** 111–14.

Crawford, R. M. M. (1971). Some metabolic aspects of ecology. *Trans. Bot. Soc. Edinb.*, **41,** 309–22.

Crawford, R. M. M. and M. McManmon (1968). Inductive responses of alcohol and malic dehydrogenases in relation to flooding tolerance in roots. *J. Exp. Bot.*, **19**, 435–41.

Crawford, R. M. M. and P. D. Tyler (1969). Organic acid metabolism in relation to flooding tolerance in roots. *J. Ecol.*, **57**, 237–46.

Crisp, D. T. (1966). Input and output of minerals for an area of Pennine moorland: the importance of precipitation, drainage, peat erosion and animals. *J. app. Ecol.*, **3**, 327–48.

Crocker, R. L. and B. A. Dickinson (1957). Soil development on the recessional moraines of the Herbert and Mendenhall glaciers, south eastern Alaska. *J. Ecol.*, **45**, 169–85.

Crompton, E. (1956). The environmental and pedological relationships of peaty gleyed soils. *Trans. 6th Internat. Congr. Soil Sci.*, **6**, 155–61.

Currie, J. A. (1962). The importance of aeration in providing the right conditions for plant growth. *J. Sci. Food. Agric.*, **13**, 380–5.

Curtis, J. T. and R. P. McIntosh (1951). An upland forest continuum in the prairie-forest border region of Wisconsin. *Ecology*, **32**, 476–96.

Dainty, J. (1962). Ion transport and electrical potentials in plant cells. *A. Rev. Pl. Physiol.*, **13**, 379–402.

Danserau, P. (1957). *Biogeography: an Ecological Perspective*, Ronald, New York.

Darwin, C. (1881). *The Formation of Vegetable Mould Through the Action of Worms*, Murray, London.

Davies, C. H. (1940). Absorption of water by maize roots. *Bot. Gaz.*, **101**, 791–805.

Davies, C. H. (1942). Response of *Cyperus rotundus* (L) to five moisture levels. *Plant Physiol., Lancaster*, **17**, 311–16.

Davies, E. F. (1928). The toxic principle of *Juglans nigra* as identified with synthetic juglone, and its toxic effects on tomato and alfalfa plants. *Am. J. Bot.*, **15**, 620–9.

Davies, M. S. and R. W. Snaydon (1973). Physiological differences among populations of *Anthoxanthum oderatum* L. collected from the Park Grass Experiment, Rothamsted. I. Response to calcium. *J. app. Ecol.*, **10**, 33–45.

Day, P. R. (1965). Particle fractionation and particle-size analysis. In *Methods of Soil Analysis*, pp. 545–67, Am. Soc. Agron., Madison.

Deb, B. C. (1949). The movement or precipitation of iron oxide in podzol soils. *J. Soil Sci.*, **1**, 112–22.

De Candolle, A. P. (1820). *Essai elementaire de geographie botanique.* Cited in *Plant Competition*, Carnegie Inst., Washington.

De Candolle, A. P. (1832). *Physiologie Vegetale.* Cited in *Plant Competition*, Carnegie Inst., Washington.

Decker, J. P. (1955). The uncommon denominator in photosynthesis as related to tolerance. *For. Sci.*, **1**, 88–9.

Dickson, R. E. and T. C. Broyer (1972). Effects of aeration, water supply and nitrogen source on growth and development of Tupelo Gum and Bald Cypress. *Ecology*, **53**, 626–34.

Dijkshoorn, W. (1969). The relation of growth to the chief ionic constituents of the plant. In *Ecological Aspects of the Mineral Nutrition of Plants*, pp. 201–13, Blackwell, Oxford.

Dimblebey, G. W. (1961). Soil pollen analysis. *J. Soil Sci.*, **12**, 1–11.

Dimblebey, G. W. (1962). *The Development of the British Heathlands and their Soils*, Clarendon, Oxford.

316

Dixon, A. H. (1914). *Transpiration and Ascent of Sap in Plants,* Macmillan, London.

Dixon, H. and J. Joly (1894). On the ascent of sap. *Ann. Bot.,* **8,** 468–70.

Doi, Y. (1952). Studies on the oxidizing power of roots of crop plants. I. The difference with species of crop plant and wild grass. *Proc. Crop. Sci. Soc. Japan,* **21,** 12–13.

Dokuchayev (1900). *Pektsie o Pohovedenie.* In *Collected Works,* Vol. 7, pp. 257–96, Moscow 1955.

Donald, C. M. (1958). The interaction of competition for light and for nutrients. *Aust. J. agr. Res.,* **9,** 421–35.

Donald, C. M. (1961). Competition for light in crops and pastures. In *Mechanisms in Biological Competition,* pp. 282–313, Cambridge University Press, Cambridge.

Donald, C. M. (1963). Competition among crop and pasture plants. *Adv. Agron.,* **15,** 1–117.

Downton, W. J. S. and E. B. Tregunna (1968). Carbon dioxide compensation and its relation to photosynthetic carboxylation reaction, systematics of the Gramineae and leaf anatomy. *Can. J. Bot.,* **46,** 207–15.

Drift, J. van der (1965). The effect of animal activity in the litter layer. In *Experimental Pedology,* pp. 277–35, Butterworths, Oxford.

Duncan, W. G. (1967). Model building in photosynthesis. In *Harvesting the Sun,* pp. 309–14, Academic Press, New York.

Duvigneaud, P. and S. Denaeyer-de-Smet (1970). Biological cycling of minerals in temperate deciduous forests. In *Analysis of Temperate Forest Ecosystems,* pp. 199–225, Chapman and Hall, London.

Eavis, B. W. and D. Payne (1969). Soil physical conditions and root growth. In *Root Growth,* Butterworths, Oxford.

Eckardt, F. D. (1965). *Methodology of Plant Eco-physiology,* UNESCO, Paris.

Eckardt, F. D. (1968). *Functioning of terrestrial ecosystems at the primary production level,* UNESCO, Paris.

Edlefsen, N. E. (1941). Some thermodynamic aspects of the use of soil moisture by plants. *Trans Am. geophys Un.,* **22,** 917–40.

Edwards, C. A., D. E. Reichle and D. A. Crosby (1970). The role of soil invertebrates in the turnover of soil organic matter. In *Analysis of Temperate Forest Ecosystems,* pp. 147–72, Chapman and Hall, London.

Ehlig, C. F. and W. R. Gardner (1964). Relationship between the transpiration and internal water balance of plants. *Agron. J.,* **56,** 127–30.

Ehrlich, H. L. (1971). Biochemistry of the minor elements in soil. In *Soil Biochemistry,* Vol. 2, pp. 361–84, Dekker, New York.

Epstein, E. (1969). Mineral metabolism of halophytes. In *Ecological Aspects of the Mineral Nutrition of Plants,* pp. 345–55, Blackwell, Oxford.

Epstein, E. and R. L. Jefferies (1964). The genetic basis of selective ion transport in plants. *A. Rev. Pl. Physiol.,* **15,** 169–84.

Etherington, J. R. (1962). The growth of *Alopecurus pratensis* L and *Agrostis tenuis* Sibth in relation to soil moisture conditions. Ph.D. Thesis, University of London.

Etherington, J. R. and A. J. Rutter (1964). Soil water and the growth of grasses. I. The interaction of watertable depth and irrigation amount on the growth of *Agrostis tenuis* and *Alopecurus pratensis. J. Ecol.,* **52,** 677–89.

Etherington, J. R. (1967). Soil water and the growth of grasses. II. Effects of soil water potential on growth and photosynthesis of *Alopecurus pratensis. J. Ecol.,* **55,** 373–80.

Evans, H. J. and G. J. Sorger (1966). Role of mineral elements with emphasis on univalent cations. *A. Rev. Pl. Physiol.*, **17**, 47–76.

Evans, L. T. (1963). Extrapolation from controlled environments to the field. In *Environmental Control of Plant Growth*, pp. 421–37, Academic Press, New York.

Evans, N. T. S. and M. Ebert (1960). Radioactive oxygen in the study of gas transport down the root of *Vicia faba. J. exp. Bot.*, **11**, 246–57.

Farrow, E. P. (1917). On the ecology of the Breckland III. General effects of rabbits on the vegetation. *J. Ecol.*, **5**, 1–18.

Felbeck, G. T. (1971). Chemical and biological characterization of humic material. In *Soil Biochemistry*, Vol. 2, pp. 36–59, Dekker, New York.

Felgan, R. S. and C. H. Low (1967). Clinal variation in the surface-volume relationship of the columnar cactus *Lophocereus schotii* in northwestern Mexico. *Ecology*, **48**, 530–6.

Finkle, B. J. (1965). Soil humic acid as a hydroxy-polystyrene: a biochemical hypothesis. *Nature, Lond.*, **207**, 604–5.

Fuscus, E. L. and P. J. Kramer (1970). Radial movement of oxygen in plant roots. *Pl. Physiol., Lancaster*, **45**, 667–69.

Fisher, R. A. (1925). *Statistical methods for research workers*, Oliver and Boyd, Edinburgh.

Fisher, R. A. (1935–1966). *The design of experiments*, Oliver and Boyd, Edinburgh.

Fitzpatrick, E. A. (1956). An indurated soil horizon formed by permafrost. *J. Soil Sci.*, **7**, 248–54.

Floyd, R. A. and A. J. Ohlorogge (1971). Gel formation on nodal root surfaces of *Zea mays*. Some observations relevant to understanding its action at the root-soil interface. *Plant and Soil*, **34**, 595–606.

Fortescue, J. A. C. and G. C. Martin (1970). Micronutrients: forest ecology and systems analysis. In *Analysis of Temperate Forest Ecosystems*, pp. 173–98, Chapman and Hall, London.

Franco, C. M. and A. C. Magelhaes (1965). Techniques for the measurement of transpiration of individual plants. In *Methodology of Plant Ecophysiology*, pp. 211–24, UNESCO, Paris.

Free, J. B. (1968). Dandelion as a competitor to fruit trees for bee visitors. *J. app. Ecol.*, **5**, 169–78.

Freny, J. R. (1967). Sulfur containing organics. In *Soil Biochemistry*, Vol. 1, pp. 229–59, Dekker, New York.

Fried, M. and H. Broeshart (1967). *The Soil-Plant System*, Academic Press, New York.

Fripiat, J. J. (1965). Surface chemistry and soil science. In *Experimental Pedology*, pp. 3–13, Butterworth, London.

Fukui, J. (1953). Studies on the adaptability of green manure and forage crops to paddy field conditions 1. *Proc. Crop Sci. Soc. Japan*, **22**, 110–12.

Gardner, W. R. (1960). Dynamic aspects of water availability to plants. *Soil Sci.*, **89**, 63–73.

Gardner, W. R. and R. H. Nieman (1964). Lower limit of water availability to plants. *Science*, **143**, 1460–2.

Gates, C. T. (1955a, b). The response of the young tomato plant to a brief period of water shortage I. The whole plant and its principal parts. II. Individual leaves. *Aust. J. biol. Sci.*, **8**, 196–214, 215–30.

318

Gates, C. T. (1957). The response of the young tomato plant to a brief period of water stress. III. Drifts in nitrogen and phosphorus. *Aust. J. biol. Sci.,* **10,** 125–46.

Gates, C. T. (1968). Water deficits and growth of herbaceous plants. In *Water Deficits and Plant Growth,* Vol. 2, pp. 135–90, Academic Press, New York.

Gates, C. T. and J. Bonner (1959). The response of the young tomato plant to a brief period of water shortage. IV. Effects of water stress on nucleic acid metabolism of tomato leaves. *Pl. Physiol., Lancaster,* **34,** 49–55.

Gates, D. M. (1962). *Energy Exchange in the Biosphere,* Harper and Row, New York.

Gates, D. M. (1965). Energy, plants and ecology. *Ecology,* **46,** 1–13.

Gates, D. M. (1968). Transpiration and leaf temperature. *A. Rev. Pl. Physiol.,* **19,** 211–38.

Gates, D. M. and La V. E. Papian (1971). *Atlas of energy budgets of plant leaves,* Academic Press, New York.

Geiger, R. (1965). *The Climate Near the Ground* translated from the German 4th Edition (1961), Harvard U.P., Cambridge, Mass.

Gilbert, O. L. (1970). Biological flora of the British Isles. *Dryopteris villarii* (Bellandi) Woynar. *J. Ecol.,* **58,** 301–13.

Gilbert, O. L. (1970a, b). Further studies on the effect of sulphur dioxide on lichens and bryophyte. A biological scale for the estimation of sulphur dioxide pollution. *New Phytol.,* **69,** 605–27, 629–34.

Gleason, H. A. (1926). The individualistic concept of the plant association, *Bull. Torrey bot. Club,* **53,** 7 : 26.

Gleason, H. A. and A. Cronquist (1964). *The natural geography of plants,* Columbia U.P., New York.

Goodall, D. W. (1954). Objective methods for the classification of vegetation. III. An essay on the use of factor analysis. *Aust. J. Bot.,* **2,** 304–24.

Goode, J. E. (1956). Soil moisture deficits under swards of different grasses species in an orchard. *Ann. rept. East Malling Res. Sta.,* 69–72.

Goodman, G. T. and D. F. Perkins (1959). Minteral uptake and retention in Cotton Grass (*Eriophorum vaginatum*) *Nature, Lond.,* **184,** 467–8.

Goodman, P. J. (1969). Intraspecific variation in mineral nutrition of plants from different habitats. In *Ecological Aspects of the Mineral Nutrition of Plants,* pp. 237–53, Blackwell, Oxford.

Goodman, P. J., E. M. Braybrook, J. M. Lambert and C. J. Marchant (1969). Biological flora of the British Isles. *Spartina Schreb. J. Ecol.,* **57,** 285-3–3.

Goodman, P. J. and W. T. Williams (1961). Investigations into 'die-back' in *Spartina townsendi* agg. III. Physiological correlates of 'die-back'. *J. Ecol.,* **49,** 391–8.

Goto, Y. and K. Tai (1957). On differences of oxidizing powers of paddy rice seedling roots among some varieties. *Soil and Plant Food,* **2,** 198–200.

Grable, A. R. (1966). Soil aeration and plant growth. *Adv. Agron.,* **18,** 57–106.

Graecen, E. L., K. P. Barley and D. A. Farrel (1969). The Mechanics of root growth in soils with particular reference to the implications for root distribution. In *Root Growth,* pp. 256–9, Butterworth, Oxford.

Gray, R. and J. Bonner (1948). An inhibitor of plant growth from the leaves of *Encelia farinosa. Am. J. Bot.,* **35,** 52–7.

Greene, H. (1963). Perspectives in soil science. *J. Soil Sci.,* **14,** 1–11.

Greenwood, D. J. (1961). The effect of oxygen concentration on the decomposition of organic materials in soils. *Plant and Soil,* **14,** 360–76.

Greenwood, D. J. (1962). Nitrification and nitrate dissimilation in soil. II. Effect of oxygen concentration. *Plant and Soil,* **17,** 378–91.

Greenwood, D. J. (1967a). Studies on the transport of oxygen through the stems and roots of vegetable seedlings. *New Phytol.,* **66,** 337–47.

Greenwood, D. J. (1967b). Studies in oxygen transport through mustard seedlings (*Sinapsis alba* L.). New Phytol., **66, 597–606.**

Greenwood, D. J. (1969). Effect of oxygen distribution in the soil on plant growth. In *Root Growth,* pp. 202–23, Butterworths, Oxford.

Greenwood, D. J. (1970). The distribution of carbon dioxide in the aqueous phase of aerobic soils. *J. Soil Sci.,* **21,** 314–29.

Greenwood, D. J. (1971). Studies in the distribution of oxygen around the roots of mustard seedlings (*Sinapsis alba* L.). *New Phytol.,* **70, 97–101.**

Greenwood, D. J. and D. Goodman (1967). Direct measurement of the distribution of oxygen in soil aggregates and in columns of fine soil crumbs. *J. Soil. Sci.,* **18,** 182–96.

Gregory, R. P. G. and A. D. Bradshaw (1965). Heavy metal tolerance in populations of *Agrostis tenuis* Sibth and other grasses. *New Phytol.,* **64,** 131–43.

Greig-Smith, P. (1964). *Quantitative Plant Ecology,* Butterworths, London.

Grim, R. E. (1953). *Clay Mineralogy,* McGraw-Hill, London.

Grime, J. P. (1963a). Factors determining the occurrence of calcifuge species on shallow soils over calcareous substrata. *J. Ecol.,* **51,** 375–90.

Grime, J. P. (1963b). An ecological investigation of a junction between two plant communities in Coombsdale on the Derbyshire limestone. *J. Ecol.,* **51,** 391–402.

Grime, J. P. (1965a) Comparative experiments as a key to the ecology of flowering plants. *Ecology,* **46,** 513–15.

Grime, J. P. (1965b). Shade tolerance in flowering plants. *Nature, London,* **208,** 161–2.

Grime, J. P. and J. G. Hodgson (1969). An investigation of the ecological significance of lime chlorosis by means of large scale comparative experiments. In *Ecological Aspects of the Mineral Nutrition of Plants,* pp. 67–99, Blackwell, Oxford.

Grubb, P. J., H. E. Green and R. C. J. Merrifield (1969). The ecology of chalk heath: its relevance to the calcicole–calcifuge and acidification problem. *J. Ecol.,* **57,** 175–210.

Grummer, G. H. (1961). The role of toxic substances in the interrelationships between higher plants. In *Mechanisms in Biological Competition,* pp. 219–28, Cambridge University Press, Cambridge.

Haberlandt, G. (1884). *Physiologische Pflanzenanatomie,* Engelmann, Leipzig.

Hackett, C. (1965). Ecological aspects of the nutrition of *Deschampsia Flexuosa* (L.). Trin II. The effects of Al, Ca, Fe, K, Mn, P and pH on the growth of seedlings and established plants, *J. Ecol.,* **53,** 315–33.

Hackett, C. (1967). Ecological aspects of the mineral nutrition of *Deschampsia flexuosa* (C.) Trin III. Investigation of phosphorus requirement and response to aluminium in water culture and a study of growth in soil. *J. Ecol.,* **55,** 831–40.

Haines, W. B. (1930). The hysteresis effect in capillary properties and the modes of moisture distribution associated therewith. *J. agric. Sci.,* **20,** 97–116.

Hall, A. D. (1903). *The Soil: an Introduction to the Scientific Study of the Growth of Crops,* Murray, London.

Hardy, R. F. W., R. D. Holster, E. K. Jackson and R. C. Burns (1968). The acetylene–ethylene assay for N_2 fixation: laboratory and field evaluation. *Pl. Physiol., Lancaster,* **43,** 1185–1207.

Harley, J. L. (1969). A physiologist's viewpoint. In *Ecological Aspects of the Mineral Nutrition of Plants,* pp. 437–47, Blackwell, Oxford.

Harper, J. L. (1961). Approaches to the study of plant competition. In *Mechanisms in Biological Competition,* pp. 1–39, Cambridge University Press, Cambridge.

Harper, J. L. (1967). A Darwinian approach to plant ecology. *J. Ecol.,* **55,** 247–70.

Harper, J. L. and R. A. Benton (1966). The behaviour of seeds in soil. II. The germination of seeds on the surface of water supplying substrate. *J. Ecol.,* **54,** 151–66.

Harper, J. L. and G. A. Sagar (1953). Some aspects of the ecology of buttercups in permanent grassland. *Proc. Brit. Weed Cont. Conf.,* 256–65.

Harris, G. A. and A. M. Wilson (1970). Competition for moisture among seedlings of annual and perennial grasses as influenced by root elongation at low temperature. *Ecology,* **51,** 530–4.

Hartwell, B. L. and F. R. Pember (1918). The presence of aluminium as a reason for the difference in effect of so-called acid soil on barley and rye. *Soil Sci.,* **6,** 259–77.

Harvey, H. W. (1950). On the production of organic matter in the sea off Plymouth. *J. mar. biol. Ass. U.K.,* NS **20,** 97–137.

Hassouna, M. G. and P. F. Wareing (1964). Possible role of rhizosphere bacteria in the nitrogen nutrition of *Ammophila areneria. Nature, Lond.,* **202,** 467–9.

Hatch, M. D. and C. R. Slack (1966). Photosynthesis by sugar-cane leaves. A new carboxylation reaction and the pathway of sugar formation. *Biochem. J.,* **101,** 103–11.

Head, W. S. and G. C. Rogers (1969). Factors affecting the distribution and growth of roots of perennial woody species. In *Root Growth,* pp. 280–95, Butterworths, London.

Healy, M. T. and W. Armstrong (1972). The effectiveness of internal oxygen transport in a mesophyte (*Pisum sativum* L.). *Planta,* **103,** 302–309.

Heath, O. V. S. (1967). Resistance to water transport in plants. *Nature, Lond.,* **213,** 741.

Heide, H. van der, M. H. van Raalte and B. M. de Boer-Bolt (1963). The effect of low oxygen content of the medium on the roots of barley seedlings. *Acta Bot. Neerl.,* **12,** 131–47.

Hewitt, E. J. (1952). A biological approach to the problem of soil acidity. *Trans. Int. Soc. Soil Sci. Dublin,* **1,** 107–18.

Hewitt, E. J. (1963). The essential nutrient elements: requirements and interactions. In *Plant Physiology,* Vol. 3, pp. 137–360, Academic Press, New York.

Hewitt, L. F. (1948). *Oxidation reduction potentials in bacteriology and biochemistry* (5th Edn.), London County Council.

Hibbert, A. R. (1967). Forest treatment effects on water yield. In *Forest Hydrology,* pp. 527–43, Pergamon, Oxford.

Hill, R. (1956). Oxidation-reduction potentials. In *Modern Methods of Plant Analysis,* Vol. 1, pp. 393–414, Springer-Verlag, Berlin.

Hislop, J. and I. J. Cooke (1968). Anion exchange resin as a means of assessing soil phosphate status: a laboratory technique. *Soil Sci.,* **105,** 8–11.

Hodges, J. D. (1967). Patterns of photosynthesis under natural environmental conditions. *Ecology*, **48**, 234–42.

Hogg, W. H. (1967). *Atlas of long-term irrigation needs for England and Wales*, Ministry of Agriculture Fisheries and Food, London.

Hollis, J. P. (1967). Toxicant diseases of rice. *La. Agr. Exp. Sta. Bull.*, 614.

Holmgren, P., P. G. Jarvis and M. S. Jarvis (1965). Resistance to carbon dioxide and water vapour transfer in leaves of different species. *Physiol. Pl.*, **18**, 557–73.

Hook, D. D. and C. L. Brown (1972). Permeability of the cambium to air in trees adapted to wet habitats. *Bot. Gaz.*, **133**, 304–10.

Hook, D. D., C. L. Brown and R. H. Wetmore (1972). Aeration in trees. *Bot. Gaz.*, **133**, 443–54.

Hook, D. D., C. L. Brown and P. P. Kormanik (1971). Inductive flood tolerance in Swamp Tupelo (*Nyssa sylvatica var. biflora* (Walt.) Sarg.). *J. exp. Bot.*, **22**, 78–89.

Howard, P. J. (1969). The classification of humus types in relation to soil ecosystems. In *The Soil Ecosystem*, pp. 41–54, Systematics Assn. Publ. No. 8.

Howard-Williams, C. (1970). The ecology of *Becium homblei* in Central Africa with special reference to metaliferrous soils. *J. Ecol.*, **58**, 745–63.

Hozumi, K., H. Koyama and T. Kira (1955). Intraspecific competition among higher plants. IV. A preliminary account of the interaction between adjacent individuals. *J. Inst. Polytech. Osaka City Univ. Series D*, **6**, 121–30.

Hudson, J. P. (1965). Gauges for the study of evapotranspiration rates. In *Methodology of Plant Ecophysiology*, pp. 443–51, UNESCO, Paris.

Huikari, O. (1954). Experiments on the effect of anaerobic media upon birch, pine and spruce seedlings. *Commun. Inst. Forest. Fenn.*, **42**, 1–13.

Hurst, H. M. and A. Burges (1967). Lignin and humic acids. In *Soil Biochemistry*, Vol. 1, pp. 260–86, Dekker, New York.

Hutchinson, G. E. (1957). *A treatise on limnology*, Vol. 1, Wiley, New York.

Hutchinson, G. E. (1965). *The Ecological Theatre and the Evolutionary Play*. Yale U.P., Newhaven.

Hutchinson, T. E. (1967). Comparative studies of the ability of species to withstand prolonged periods of darkness. *J. Ecol.*, **55**, 291–9.

Hutchinson, T. C. (1968). A physiological study of *Teucrium scorodonia* ecotypes which differ in their susceptibility to lime induced chlorosis and iron deficiency chlorosis. *Plant and Soil*, **28**, 81–105.

Isherwood, F. A. (1965). Biosynthesis of lignin. In *Biosynthetic Pathways in Higher Plants*, pp. 133–46, Academic Press, London.

Ivanov, L. (1928). Zur methodik der transpiration bestimmung am standart. *Ber. dtsch. bot. Ges.*, **46**, 306–10.

Jacks, J. V. (1965). The role of organisms in the early stages of soil formation. In *Experimental Pedology*, pp. 219–26, Butterworth, Oxford.

Jackson, M. L. (1958). *Soil Chemical Analysis*. Prentice Hall, New Jersey.

Jarvis, M. S. (1963). A comparison between the water relations of species with contrasting types of geographical distribution in the British Isles. In *The Water Relations of Plants*, pp. 289–312, Blackwell, Oxford.

Jarvis, P. G. and M. S. Jarvis (1963). Effects of several osmotic substrates on the growth of *Lupinus albus* seedlings. *Physiol. Pl.*, **16**, 485–500.

Jarvis, P. G. and M. S. Jarvis (1963a, b, c, d). The water relations of tree seedlings. I. Growth and water use in relation to soil water potential. II. Transpiration in relation to soil water potential. III. Transpiration in relation to the osmotic potential of the root media. IV. Some aspects of tissue water relations and drought resistance. *Physiol. Pl.,* **16,** 215–35, 236–53, 269–75 and 501–76.

Jarvis, P. G. and M. S. Jarvis (1965). The water relations of tree seedlings. V. Growth and root respiration in relation to the osmotic potential of the medium. In *Water Stress in Plants,* pp. 167–82, Czech. Acad. Sci., Prague.

Jefferies, R. L. and A. J. Willis (1964). Studies on the calcicole–calcifuge habit II. The influence of calcium on the growth and establishment of four species in soil and sand culture. *J. Ecol.,* **52,** 691–707.

Jeffery, D. W. (1964). The formation of polyphosphate in *Banksia ornata,* an Australian heath plant. *Aust. J. biol. Sci.,* **17,** 845–54.

Jennings, D. H. (1967). Electrical potential measurements, ion pumps and root exudation—a comment and a model explaining cation selectivity by the root. *New Phytol.,* **66,** 357–69.

Jennings, D. H. (1968). Halophytes, succulence and sodium in plants—a unified theory. *New Phytol.,* **67,** 899–911.

Jennings, D. H. (1969). The physiology of the uptake of ions by the growing plant cell. In *Ecological Aspects of the Mineral Nutrition of Plants,* pp. 261–79, Blackwell, Oxford.

Jennings, P. R. and R. C. Aquino (1968). Studies in competition on rice. III. The mechanism of competition among phenotypes. *Evolution, Lancaster, Pa.,* **22,** 529–42.

Jenny, H. and C. D. Leonard (1934). Functional relationships between soil properties and rainfall. *Soil Sci.,* **38,** 363–81.

Jensen, M. E. (1968). Water consumption by agricultural plants. In *Water Deficits and Plant Growth,* Vol. 2, pp. 1–22, Academic Press, New York.

Joffe, J. S. (1936). *Pedology,* Rutgers U.P., New Brunswick.

Jones, H. E. (1971a). Comparative studies of plant growth and distribution in relation to waterlogging. II. An experimental study of the relationship between transpiration and the uptake of iron in *Erica cinerea* L. and *E. tetralix* L. *J. Ecol.,* **59,** 167–78.

Jones, H. E. (1971b). Comparative studies in plant growth and distribution in relation to waterlogging. III. The response of *Erica cinerea* L. to waterlogging in peat soils of differing iron content. *J. Ecol.,* **59,** 583–91.

Jones, H. E. and J. R. Etherington (1970). Comparative studies of plant growth and distribution in relation to waterlogging. I. The survival of *Erica cinerea* L. and *E. tetralix* L. and its apparent relationship to iron and manganese uptake in waterlogged soil. *J. Ecol.,* **58,** 487–96.

Jones, L. H. (1961). Aluminium uptake and toxicity in plants. *Plant and Soil,* 13, 297–310.

Jones, R. (1967). The relationship of dune-slack plants to soil moisture and chemical conditions. Ph.D. Thesis, University of Wales.

Jones, R. (1972a). Comparative studies of plant growth and distribution in relation to waterlogging. V. The uptake of iron and manganese by dune and slack plants. *J. Ecol.,* **60,** 131–40.

Jones, R. (1972b). Comparative studies of plant growth and distribution in relation to waterlogging. VI. The effect of manganese on the growth of dune and slack plants. *J. Ecol.,* **60,** 141–46.

Jones, R. (1973). Comparative studies of plant growth and distribution in relation to waterlogging. VII. The influence of watertable fluctuation on iron and manganese availability in dune slacks. *J. Ecol.*, **61**, 107–116.

Jones, R. and J. R. Etherington (1971). Plant growth and distribution in relation to waterlogging. IV. The growth of dune and dune slack plants. *J. Ecol.*, **59**, 793–801.

Karper, R. E. (1929). The contrast in response of Kafir and Milo to variations in spacing. *J. Am. Soc. Agron.*, **21**, 344–54.

Kemper, W. D. and W. S. Chepil (1965). Size distribution of aggregates. In *Methods of Soil Analysis*, pp. 499–510, Am. Soc. Agron., Wisconsin.

Kershaw, K. A. (1963). Pattern in vegetation and its causality. *Ecology*, **44**, 377–88.

Kershaw, K. A. (1973). *Quantitative and Dynamic Ecology*, 2nd. edn., Arnold, London.

Kilian Ch. and G. Lemee (1956). Les Xerophytes: leur economie d'eau. In *Encyclopedia of Plant Physiology*, Vol. 3, pp. 787–824, Springer, Berlin.

Kimball, B. A. and E. R. Lemon (1971). Air turbulence effects upon soil gas exchange. *Soil Sci. Soc. Amer. Proc.*, **35**, 16–21.

Klepper, B. (1968). Diurnal pattern of water potential in woody plants. *Pl. Physiol. Lancaster*, **43**, 1931–4.

Klikoff, L. G. (1965). Photosynthetic response to temperature and moisture stress of three timberline meadow species. *Ecology*, **46**, 516–17.

Kononova, M. (1961). *Soil Organic Matter*, Pergamon, London.

Kormondy, E. J. (1965). *Readings in ecology*, Prentice Hall, New Jersey.

Kozlowski, T. T. (1964). *Water Metabolism in Plants*, Harper and Row, New York.

Kramer, P. J. (1938). Root resistance as a cause of absorption lag. *Am. J. Bot.*, **25**, 110–13.

Kramer, P. J. (1940). Causes of decreased absorption of water by plants in poorly aerated media. *Am. J. Bot.*, **27**, 216–20.

Kramer, P. J. (1969). *Plant and Soil Water Relationships*. McGraw-Hill, New York.

Kramer, P. J. and W. T. Jackson (1954). Causes of injury to flooded tobacco plants. *Pl. Physiol. Lancaster*, **29**, 241–5.

Kramer, P. J. and T. T. Kozlowski (1960). *Physiology of Trees*, McGraw-Hill, New York.

Kramer, P. J., W. S. Riley and T. T. Bannister (1952). Gas exchange of Cypress knees. *Ecology*, **33**, 117–21.

Kruckberg, A. R. (1954). The ecology of serpentine soils. III. Plant species in relation to serpentine soils. *Ecology*, **35**, 267–87.

Kubiena, W. L. (1953). *Soils of Europe*, Murby, London.

Kuhl, A. (1962). Inorganic phosphorus uptake and metabolism. In *Physiology and Biochemistry of the Algae*, pp. 211–29, Academic Press, New York.

Lambert, J. M. and M. B. Dale (1964). The use of statistics in phytosociology. *Adv. ecol. Res.*, **2**, 59–99.

Lang, A. (1963). Achievements, challenges and limitations of phytotrons. In *Environmental Control of Plant Growth*, pp. 405–19, Academic Press, New York.

Lang, A. (1965). Effects of some internal and external conditions on seed germination. in *Encyclopedia of Plant Physiology*, Vol. 15, pp. 848–93, Springer, Berlin.

Larsen, S. (1967). Soil phosphorus. *Adv. Agron.*, **19**, 151–210.

Lazenby, A. (1955). Germination and establishment of *Juncus effusus* L. II. The interaction of moisture and competition. *J. Ecol.*, **43**, 595–605.

Leach, G. J. and D. J. Watson (1968). Photosynthesis in crop profiles measured by phytometers. *J. app. Ecol.*, **5**, 381–408.

Leblanc, F. and J. de Sloover (1970). Relation between industrialization and the distribution and growth of epiphytic lichens and mosses in Montreal. *Can. J. Bot.*, **48**, 1485–96.

Lemon, E. (1965). Micrometeorology and the physiology of plants in their natural environment. In *Plant Physiology*, Vol. 4A, pp. 203–27, Academic Press, New York.

Lemon, E. (1967). Aerodynamic studies of CO_2 exchange. In *Harvesting the Sun*, pp. 263–90, Academic Press, New York.

Lemon, E. R. and A. E. Erickson (1952). The measurement of oxygen diffusion in the soil with a platinum micro-electrode. *Proc. Soil Sci. Soc. Amer.*, **16**, 160–3.

Lemon, E. R. and A. E. Erickson (1955). Principle of the platinum micro-electrode as a method of characterizing soil aeration. *Soil Sci.*, **79**, 382–92.

Lewis, H. and F. Went (1945). Plant growth under controlled conditions. IV. Response of California annuals to photoperiod and temperature. *Am. J. Bot.*, **32**, 1–12.

Lewis, T. and L. R. Taylor (1967). *Introduction to Experimental Ecology*, Academic Press, London.

Leyton, L. and L. Z. Rousseau (1957). Root growth of tree seedlings in relation to aeration. In *Physiology of Forest Trees*, pp. 467–75, Ronald, New York.

Leyton, L., E. R. C. Reynolds and F. B. Thompson (1968). Interception of rainfall by trees and moorland vegetation. In *The Measurement of Environmental Factors in Terrestrial Ecology*, pp. 97–108, Blackwell, Oxford.

Liebig, J. (1840). *Chemistry and its Application to Agriculture and Physiology*, Taylor and Walton, London.

Lieth, H. (1960). Patterns of change within grassland communities.In *The Biology of Weeds*, pp. 27–39, Blackwell, Oxford.

Lieth, H. (1968a). The measurement of calorific values of biological materials and the determination of ecological efficiency. In *Functioning of Terrestrial Ecosystems at the Primary Production Level*, pp. 233–41, UNESCO, Paris.

Lieth, H. (1968b). The determination of plant dry matter production with special emphasis on the underground parts. In *Functioning of Terrestrial Ecosystems at the Primary Production Level*, pp. 179–86, UNESCO, Paris.

Lieth, H. (1970). Phenology in productivity studies. In *Analysis of Temperate Forest Ecosystems*, pp. 29–46, Chapman and Hall, London.

Lindeman, R. L. (1942). The trophic dynamic aspect of ecology. *Ecology*, **23**, 399–418.

Long, I. F. (1968). Instruments and techniques for measuring the micro-meteorology of crops. In *The Measurement of Environmental Factors in Terrestrial Ecology*, pp. 1–32, Blackwell, Oxford.

Loomis, R. S., W. A. Williams and W. G. Duncan (1967). Community architecture and the productivity of terrestrial plant communities. In *Harvesting the Sun*, pp. 291–308, Academic Press, New York.

Loughman, B. C. (1969). The uptake of phosphorus and its transport within the plant. In *Ecological Aspects of the Mineral Nutrition of Plants*, pp. 309–22, Blackwell, Oxford.

Lounama, J. (1956). Trace elements in plants growing wild on different rocks in Finland. *Ann. Bot. Soc. Vannamo*, **29**, 1–196.

Luxmore, R. J., L. H. Stolzy, and J. Letey (1970). Oxygen diffusion in the soil-plant system. *Agron. J.*, **62**, 317–32.

Luxmore, R. J. and L. H. Stolzy (1972). Oxygen diffusion in the soil-plant system V and VI. *Agron. J.*, **64**, 720–29.

McCloud, D. E. and L. S. Dunavin (1954). Agrohydric balance studies at Gainsville, Florida. In *Publication in Climatology*, pp. 55–68, Seabrook, New Jersey.

McGregor, A. N. and D. E. Johnson (1971). Capacity of desert algal crusts to fix atmospheric nitrogen. *Soil Sci. Soc. Amer. Proc.*, **35**, 843–4.

McIntyre, D. S. (1970). The platinum microelectrode method of soil aeration measurement. *Adv. Agron.*, **22**, 235–83.

Macklon, A. E. S. and P. E. Weatherly (1965). Controlled environment studies of the nature and origins of water deficits in plants. *New Phytol.*, **64**, 414–27.

Mackney, D. (1961). A podzol development sequence in oakwoods and heath in central England. *J. Soil Sci.*, **12**, 23–40.

McKell, C. M., E. R. Perrier and G. L. Stebbins (1960). Responses of two subspecies of Orchard Grass (*Dactylis glomerata* subsp. *lusitanica* and *judaica*) to increasing soil moisture stress. *Ecology*, **41**, 772–8.

McLaren, A. D. and G. H. Peterson (1967). *Soil Biochemistry*, Vol. 1, Dekker, New York.

MacMannon, M. and R. M. M. Crawford (1971). A metabolic theory of flooding tolerance: the significance of enzyme distribution and behaviour. *New Phytol.*, **70**, 299–306.

McRae, I. C. and T. F. Castro (1967). Nitrogen fixation in some tropical rice soils. *Soil Sci.*, **103**, 277–80.

MacRobbie, E. A. C. (1971). Fluxes and compartmentation in plant cells. *A. Rev. Pl. Physiol.*, **22**, 75–96.

McVean, D. N. (1955). Ecology of *Alnus glutinosa* (1.) Gaertn. IV. Root system. *J. Ecol.*, **43**, 219–225.

McVean, D. N. and D. A. Ratcliffe (1962). *Plant Communities of the Scottish Highlands*, HMSO, London.

McWilliams, J. R. and P. J. Kramer (1968). The nature of the perennial response in Mediterranean grasses. I. Water relations and survival in *Phalaris*. *Aust. J. agric. Res.*, **19**, 381–95.

Majmudar, A. M. and J. P. Hudson (1957). The effect of different water regimes on the growth of plants under glass. II. Experiments with Lettuces (*Lactuca sativa* Linn.) *J. hort. Sci.*, **32**, 201–13.

Major, J. (1969). Historical development of the ecosystem concept. In *The Ecosystem Concept and Natural Resource Management*, pp. 9–22, Academic Press, New York.

Malthus, T. R. (1798). *The principles of population*.

Malyuga, D. P. (1964). *Biogeochemical Methods of Prospecting*, Consultants Bureau (translation), New York.

Mandal, L. N. (1961). Transformations of iron and manganese in waterlogged rice soils. *Soil Sci.*, **91**, 121–6.

Margalef, R. (1968). *Perspectives in Ecological Theory*, Univ. Chicago Press, Chicago.

Martin, M. H. (1968). Conditions affecting the distribution of *Mercurialis perennis* L. in certain Cambridgeshire woodlands. *J. Ecol.*, **56**, 777–93.

Mather, J. R. (1954). The measurement of potential evapo-transpiration. Publication in Climatology 225, Seabrook, New Jersey.

Mather, K. (1967). *The Elements of Biometry*, Methuen, London.

Mather, K. (1943–65). *Statistical Analysis in Biology*, Methuen, London.

Maximov, N. A. (1929). *The Plant in Relation to Water,* Allen and Unwin, London.

Mayer, A. M. and E. Gorham (1951). The iron and manganese content of plants present in the natural vegetation of the English Lake District. *Ann. Bot., NS* 15, 247–63.

Mazelis, M. and B. Vennesland (1957). Carbon dioxide fixation in oxaloacetate in higher plants. *Pl. Physiol, Lancaster,* 32, 591–600.

Meidner, H. (1965). Stomatal control of transpirational water loss. In *The State of Water in the Living Organism,* pp. 185–204, Cambridge University Press, Cambridge.

Meider, H. and T. A. Mansfield (1968). *Physiology of Stomata,* McGraw-Hill, London.

Meyer, F. H. and D. Göttsche (1971). Distribution of root tips and tender roots of beech. In *Integrated Experimental Ecology,* Chapman and Hall, London.

Milburn, J. A. and R. P. C. Johnson (1966). The conduction of sap. II. Detection of vibrations produced by sap cavitation in *Ricinus* xylem. *Planta,* 69, 43–52.

Millar, A. A., M. E. Duyser and G. E. Wilkinson (1968). Internal water balance of barley under soil moisture stress. *Pl. Physiol., Lancaster,* 43, 968–72.

Milles, L. P. (1949). Rapid formation of high concentrations of hydrogen sulphide by sulphate reducing bacteria. *Contrib. Boyce Thompson Inst.* 15, 437–65.

Miller, R. B. (1963). Plant nutrients in Hard Beech. III. The cycle of nutrients. *N.Z. J. Soil. Sci.,* 6, 388–413.

Milthorpe, F. L. (1950). Changes in drought resistance of wheat seedlings during germination. *Ann. Bot., NS* 14, 79–89.

Milthorpe, F. L. (1961). The nature and analysis of competition between plants of different species. In *Mechanisms in Biological Competition,* pp. 330–55, Cambridge University Press, Cambridge.

Mitsui, S., S. Aso, K. Kumazawa and T. Ishiwara (1954). The nutrient uptake of rice plants as influenced by hydrogen sulphide and butyric acid abundantly evolving under waterlogged soil conditions. *Trans, 5th Int. Congr. Soil. Sci.,* 2, 364.

Mizrahi, T., A. Blumenfield and A. A. Richmond (1972). The role of abscisic acid and salination in the adaptive response of plants to reduced aeration. *Plant and Cell Physiol.,* 13, 15–21.

Mohr, E. C. J. and F. A. van Bahren (1959). *Tropical Soils,* van Hoeve, The Hague.

Molisch, H. (1888). Uber wurzelausschiedungen und deren einwerkung auf organische substanzen. *Sitzungsber. Akad. Wiss. Wien. Math. Nat. Kl.,* 96, 84.

Monsi, M. and T. Saeki (1953). Uber den lichtfaktor in den pflanzengesellschaften und seine bedeutung fur die stoffproduction. *Jap. J. Bot.,* 14, 22–52.

Monteith, J. L. (1968). Analysis of the photosynthesis and respiration of field crops from vertical fluxes of carbon dioxide. In *Functioning of Terrestrial Ecosystems at the Primary Production Level,* pp. 349–58, UNESCO, Paris.

Mooney, H. A. and W. D. Billings (1961). Comparative physiological ecology of arctic and alpine populations of *Oxyria digyna. Ecol. Monogr.,* 31, 1–29.

Moss, D. N. (1962). The limiting carbon dioxide concentrations for photosynthesis. *Nature, Lond.,* 193, 587.

Muckenhirn, R. J., E. P. Whiteside, E. H. Templin, R. F. Chandler and L. T. Alexander (1949). Soil classification and the genetic factors of soil formation. *Soil Sci.,* 67, 93–105.

Mulder, E. G., T. A. Lie and J. W. Woldendorp (1969). Biology and soil fertility. In *Soil Biology*, pp. 163–208, UNESCO, Paris.

Muller, C. H. (1953). The association of desert annuals with shrubs. *Am. J. Bot.*, **40**, 53–60.

Muller, C. H. (1966). The role of chemical inhibition (allelopathy) in vegetational composition. *Bull. Torrey Bot. Club*, **93**, 332–51.

Muller, W. H., P. Lorber, B. Haley and K. Johnson (1969). Volatile growth inhibitors produced by *Salvia leucophylla*: effect on oxygen uptake by mitochondrial suspensions. *Bull. Torrey Bot. Club*, **96**, 89–95.

Newbould, P. (1969). The absorption of nutrients by plants from different zones in the soil. In *Ecological Aspects of the Mineral Nutrition of Plants*, pp. 177–90, Blackwell, Oxford.

Newbould, P. J. (1967). *Methods for Estimating the Primary Production of Forests*, Blackwell, Oxford.

Newbould, P. J. (1968). Methods of estimating root production. In *Functioning of Terrestrial Ecosystems at the Primary Production Level*, pp. 187–90, UNESCO, Paris.

Newman, E. I. (1967). Response of *Aira praecox* to weather conditions. I. Response to drought in spring. *J. Ecol.*, **55**, 539–56.

Newman, E. I. (1969a, b). Resistance to water flow in soil and plant. I. Soil resistance in relation to amounts of roots: theoretical estimates. II. A review of experimental evidence on rhizosphere resistance, *J. app. Ecol.*, **6**, 1–12, 261–72.

Nichiporovich, A. A. (1969). The role of plants in the bioregenerative system. *A. Rev. Pl. Physiol.*, **20**, 185–208.

Nicholson, M. (1970). *The Environmental Revolution*, Hodder and Stoughton, London.

Nir, I. and A. Poljakoff-Mayber (1967). Effect of water stress on the photochemical activity of chloroplasts. *Nature, Lond.*, **213**, 418–19.

Njoku, E. (1957). The effect of mineral nutrition and temperature on leaf shape in *Ipomoea caerulea*. *New Phytol.*, **56**, 154–71.

Oden, S. (1962). Electrometric methods for oxygen studies in water and soil: IV. Fundamental problems involved with the design and use of oxygen diffusion electrodes. C.S.I.R.O. Aust. translation No. 6480 C from *Grundforbattring*, **3**, 117–210 (Swedish).

Odum, E. P. (1953 and 1959). *Fundamentals of Ecology*, 1st and 2nd Edns, Saunders, Philadelphia.

Odum, E. P. (1971). *Fundamentals of Ecology*, 3rd Edn., Saunders, Philadelphia.

Odum, H. T. (1957). Trophic structure and productivity of Silver Springs, Florida. *Ecol. Monogr.*, **27**, 55–112.

Odum, H. T. (1971). *Environment, Power and Society*, Wiley, New York.

Odum, H. T. and E. P. Odum (1955). Trophic structure and productivity of a windward coral reef community on Eniwetok atoll. *Ecol. Monogr.*, **25**, 291–320.

Odum, H. T. and R. Pigeon (1971). *A Tropical Rain Forest*, AEC Div. Tech. Inf., Oak Ridge.

Okudu, A. and E. Takahashi (1964). The role of silicon. In *The Mineral Nutrition of the Rice Plant*, pp. 123–46, Johns Hopkins, Baltimore.

Olsen, C. (1961). Competition between trees and herbs for nutrient elements in calcareous soils. In *Mechanisms in Biological Competition*, pp. 145–55, Cambridge University Press, Cambridge.

Olsen, R. A. and J. E. Robbins (1971). The cause of the suspension effect in resin-water systems. *Soil Sci. Soc. Amer. Proc.*, **35**, 260–5.

Oosting, H. J. (1956). *The Study of Plant Communities: an Introduction to Ecology*, Freeman, San Francisco.

Oppenheimer, H. R. (1960). Adaptation to drought: xerophytism. In *Plant Water Relationships in Arid and Semi-arid Conditions*, pp. 105–38, UNESCO, Paris.

Oppenheimer, H. R. and K. Mendel (1939). Orange leaf transpiration under orchard conditions. I. Soil moisture high. *Palestine J. Bot.*, **2**, 171–250.

Orloci, L. (1966). Geometric models in ecology. I. The theory and application of some ordination methods. *J. Ecol.*, **54**, 193–215.

Ordin, L. (1958). The effect of water stress on the cell wall metabolism of plant tissue. In *Radio Isotopes in Scientific Research*, Vol. 4, pp. 553–64, Pergamon, London.

Ordin, L. (1960). Effect of water stress on cell wall metabolism of *Avena coleoptile* tissue. *Pl. Physiol. Lancaster*, **35**, 443–50.

Overland, L. (1966). The role of allelopathic substances in the 'smother crop' barley. *Am. J. Bot.*, **53**, 423–32.

Overstreet, R. and H. Jenny (1939). The significance of the suspension effect in the uptake of cations by plants from soil water systems. *Soil Sci. Soc. Amer. Proc.*, **24**, 257–61.

Ovington, J. D. (1957). Dry matter production by plantations of *Pinus sylvestris*. *Ann. Bot.*, NS**21**, 287–314.

Ovington, J. D. (1961). Some aspects of energy flow in plantations of *Pinus sylvestris* L., *Ann. Bot.*, NS**25**, 121–20.

Ovington, J. D. (1968). Some factors affecting nutrient distribution within ecosystems. In *Functioning of the Terrestrial Ecosystem at the Primary Production Level*, pp. 95–105, UNESCO, Paris.

Owen, P. C. (1952). The relationship of the germination of wheat to water potential. *J. Exp. Bot.*, **3**, 276–90.

Park, Y. D. and A. Tanaka (1968). Studies of the rice plant on an 'Akiochi' soil in Korea. *Soil. Sci. Plant. Nutr.*, **14**, 27–34.

Parker, J. (1968). Drought resistance mechanisms. In *Water Deficits and Plant Growth*, pp. 195–234, Academic Press, New York.

Patrick, W. H. and I. C. Mahapatra (1968). Transformation and availability to rice of nitrogen and phosphorus in waterlogged soil. *Adv. Agron.*, **20**, 323–59.

Paul, E. A., C. A. Campbell, D. A. Rennie and K. J. McCallum (1964). Investigations of the dynamics of soil humus utilizing carbon dating techniques. *Trans. 8th. Int. Congr. Soil Sci.*, 201–8.

Pauling, L. (1930). The structure of mica and related minerals. *Proc. Nat. Acad. Sci. U.S.*, **16**, 123–9.

Pearsall, W. H. (1938). The soil complex in relation to plant communities. *J. Ecol.*, **26**, 180–93.

Pearsall, W. H. (1952). The pH of natural soils and its ecological significance. *J. Soil Sci.*, **3**, 41–51.

Pearsall, W. H. (1964). The development of ecology in Britain (1964). *J. Ecol.*, **52**, (supplement); 1–12.

Penman, H. L. (1948). Natural evaporation from open water, bare soil and grass. *Proc. R. Soc.*, A **193**, 120–45.

Penman, H. L. (1956). Evaporation: an introductory survey. *Neth. J. agric. Sci.,* 4, 9–29.

Penman, H. L. (1963). *Vegetation and Hydrology,* Comm. Bureau Soils, Harpenden.

Penman, H. L. and I. F. Long (1960). Weather in wheat: an essay in micrometeorology. *Q. Jl. R. met. Soc.,* 86, 16–50.

Perrier, A. (1971). Leaf temperature measurement. In *Plant Photosynthetic Production: Manual of Methods,* pp. 632–71, Junk, The Hague.

Perrin, R. M. S. (1965). The use of drainage water analysis in soil studies. In *Experimental Pedology,* pp. 73–96, Butterworths, London.

Perring, F. H. and S. M. Waters (1968). *Atlas of the British Flora,* Botanical Soc. of the British Isles, Nelson, London.

Petrie, A. H. K. and J. G. Wood (1938a, b). Studies on the nitrogen metabolism of plants. I. The relation between the content of amino acids, protein and water in leaves. III. On the effect of water content on the relationship between proteins and amino acids. *Ann. Bot.,* NS2, 33–60, 881–98.

Pfeffer, W. (1880). *The Physiology of Plants* (translation), Clarendon, Oxford (1900–6).

Phillip, J. R. (1966). Plant water relationships: some physical aspects. *A Rev. Pl. Physiol.,* 17, 245–68.

Pielou, E. C. (1969). *An Introduction to Mathematical Ecology,* Interscience, New York.

Piggot, C. D. (1969). Influence of mineral nutrition on the zonation of flowering plants in coastal salt marshes. In *Ecological Aspects of the Mineral Nutrition of Plants,* pp. 25–35, Blackwell, Oxford.

Piggot, C. D. and K. Taylor (1964). The distribution of some woodland herbs in relation to the supply of nitrogen and phosphorus in the soil. *J. Ecol.,* 52 (supplement), 175–85.

Piper, C. S. (1944). *Soil and Plant Analysis,* Univ. Adelaide.

Pitts, R. G. (1969). Explorations in the chemistry and microbiology of Louisiana Rice plant-soil relations. *Ph.D. Diss.* La. State Univ. La. USA.

Platt, R. B. and J. F. Griffiths (1964). *Environmental Measurement and Interpretation,* Reinhold, New York.

Polunin, N. (1960). *Introduction to Plant Geography and Some Related Sciences,* Longmans, Oxford.

Pomeroy, L. R. (1970). The strategy of mineral cycling. *Ann. Rev. Ecol. Syst.,* 1, 171–90.

Ponnamperuma, F. N. (1965). Dynamic aspects of flooded soils and the nutrition of the rice plant. In *The Mineral Nutrition of the Rice Plant,* John Hopkins, Baltimore.

Ponnamperuma, F. N. (1972). The chemistry of submerged soils. *Adv. Agron.,* 24, 29–95.

Ponnamperuma, F. N., R. Bradfield and M. Peech (1955). Physiological disease of rice attributable to iron toxicity. *Nature, Lond.,* 175, 265.

Ponnamperuma, F. N., T. A. Loy and F. M. Tianco (1969). Redox equilibria in flooded soils: II. The manganese oxide systems. *Soil Sci.,* 108, 48–57.

Ponnamperuma, F. N., E. Martinez and T. Loy (1966). Influence of redox potential and partial pressure of carbon dioxide on pH value and the suspension effect of flooded soils. *Soil Sci.,* 101, 421–31.

Ponnamperuma, F. N., E. M. Tianco and T. Loy (1967). Redox equilibria in flooded soils: I. The iron hydroxide systems. *Soil Sci.,* 103, 374–82.

Poore, M. E. D. (1955a, b, c). The use of phytosociological methods in ecological investigations. I. The Braun Blanquet system. II. Practical issues involved in an attempt to apply the Braun Blanquet system. III. Practical applications. *J. Ecol.,* **43,** 226–44, 245–69, 606–51.

Poore, M. E. D. (1956). The use of phytosociological methods in ecological investigations. IV. General discussion of phytosociological problems. *J. Ecol.,* **44,** 28–50.

Postgate, G. R. (1959). Sulphate reduction in bacteria. *Ann. Rev. Microbiol.,* **13,** 505–20.

Proctor, J. (1971). The plant ecology of serpentine. II. Plant response to serpentine soils. *J. Ecol.,* **59,** 397–410.

Puckridge, D. W. (1962). Ph.D. thesis. Dept. of Agronomy, Univ. of Adelaide. Cited by Donald (1963).

Quastel, J. H. (1963). Microbial activities of soil as they affect plant nutrition. In *Plant Physiology,* Vol. III, pp. 671–756, Academic Press, New York.

Raalte, M. H. van (1941). On the oxygen supply of rice roots. *Ann. Jard. Bot. Buitzenzorg,* **51,** 43–57.

Raalte, M. H. van (1943–4). On the oxidation of the environment by the roots of rice (*Oryza sativa* L.). *Hort. Bot. Bogoriensis, Java. Syokubutu-Iho,* **1,** 15–34.

Raciborski, M. M. (1905). Utleniajace: redukajace wlaśnosei kómorki żywej. *I, II and III. Bull. Int. de L'Acad. Sciences (Cracovie),* 338, 668, 693.

Ramakrishnan, P. S. (1968). Nutritional requirements of the edaphic ecotypes in *Melilotus alba* Medic. I. pH, calcium and phosphorus. *New Phytol.,* **67,** 145–57.

Ramakrishnan, P. S. (1970). Nutritional requirements of the edaphic ecotypes in *Melilotus alba* Medic. III. Interference between the calcareous and acidic populations in the two soil types. *New Phytol.,* **69,** 81–6.

Ratcliffe, D. A. (1961). Adaptation to habitat in a group of annual plants. *J. Ecol.,* **49,** 187–293.

Raw, F. (1967). Arthropoda (except Acari and Collembola). In *Soil Biology,* pp. 323–62, Academic Press, New York.

Read, D. J. and W. Armstrong (1972). A relationship between oxygen transport and the formation of the ectotrophic mycorrhizal sheath in conifer seedlings. *New Phytol.,* **71,** 49–53.

Redman, F. H. and W. H. Patrick (1965). The effect of submergence on several biological and chemical soil properties. *La. Agric. Expl. Sta. Bull.,* 592.

Reinhart, K. G. (1967). Watershed calibration methods. In *Forest Hydrology,* pp. 715–23, Pergamon, Oxford.

Rennie, P. J. (1955). The uptake of nutrients by mature forest growth. *Plant and Soil,* **7,** 49–95.

Revelle, R. and R. Fairbridge (1957). Carbonates and carbon dioxide. In *Treatise on Marine Ecology and Palaeocology,* Vol. I, pp. 239–96, Geol. Soc. Amer. Mem. 6.

Revelle, R. W. Broeker, H. Craig, C. D. Keeling and J. Smagorinsky (1965). Atmospheric carbon dioxide. In *Restoring the Quality of Our Environment,* pp. 111–33, Env. Poll. Panel, Presidents Science Advisory Comm., Washington.

Reynolds, E. R. C. (1967). Transpiration as related to internal water content. *Nature, London,* **207,** 1001–2.

Richards. F. J. (1941). The diagrammatic representation of the results of physiological and other experiments designed factorially. *Ann. Bot.,* NS5, 249–61.

Richards, P. W. (1952). *The Tropical Rainforest,* Cambridge University Press, Cambridge.

Richards, L. A. (1965). Physical condition of water in soil. In *Methods of Soil Analysis,* pp. 128–52, Am. Soc. Agron., Wisconsin.

Richards, L. A. and W. E. Loomis (1942). Limitations of autoirrigators for controlling soil moisture under growing plants. *Pl. Physiol. Lancaster,* **17,** 223–35.

Richards, L. A. and C. H. Wadleigh (1952). Soil water and plant growth. In *Soil Physical Conditions and Plant Growth,* pp. 73–251, Am. Soc. Agron. Wisconsin.

Risser, P. G. (1969). Competitive relationships among herbaceous grassland plants. *Bot. Rev.,* **35,** 251–84.

Robinson, G. W. (1932). *Soils: Their Origin, Constitution and Classification,* Murby, London.

Rodin, L. E. and N. I. Bazilevich (1967). *Production and Mineral Cycling in Terrestrial Vegetation* (translation), Oliver and Boyd, London.

Roger, R. W., R. T. Lang and D. J. D. Nicholas (1966). Nitrogen fixation by lichens and soil crusts. *Nature, London,* **209,** 96–7.

Rogers, W. S. (1969). The East Malling root-observation laboratories. In *Root Growth,* pp. 361–76, Butterworths, London.

Rogers, W. S. and G. C. Head (1969). Factors affecting the distribution and growth of roots of perennial woody species. In *Root Growth,* pp. 280–95, Butterworths, London.

Roo, H. C. de (1969). Tillage and root growth. In *Root Growth,* pp. 339–58, Butterworths, London.

Rorison, I. H. (1960a). Some experimental aspects of the calcicole–calcifuge problem. I. The effects of competition and mineral nutrition upon seedling growth in the field. *J. Ecol.,* **48,** 585–99.

Rorison, I. H. (1960b). Some experimental aspects of the calcicole–calcifuge problem. II. The effects of mineral nutrition on seedling growth in solution culture. *J. Ecol.,* **48,** 679–88.

Rorison, I. H. (1969). Ecological inferences from laboratory experiments on mineral nutrition. In *Ecological Aspects of the Mineral Nutrition of Plants,* pp. 155–75, Blackwell, Oxford.

Rouschal, E. (1937–8). Eine physiologische studie an *Ceterach officinarum.* Flora (Jena), **132,** 305–18.

Ruhland, W. (1958). *Encyclopedia of Plant Physiology Vol. IV Mineral Nutrition of Plants,* Springer, Berlin.

Rune, O. (1953). Plant life on serpentine and related rocks in the north of Sweden. *Acta. phytogogr. suec.,* **31,** 1–139.

Russel, E. W. (1961). *Soil Conditions and Plant Growth,* 9th Edn. Longmans, London.

Russel, E. W. (1971). Soil structure: its maintenance and improvement. *J. Soil Sci.,* **22,** 137–51.

Rutter, A. J. (1968). Water consumption by forests. In *Forest Hydrology,* pp. 23–84, Academic Press, New York.

Sachs, J. von (1887). *Lectures on the Physiology of Plants* (translation), Clarendon, Oxford.

Salisbury, E. J. (1916). The oak-hornbeam woods of Hertfordshire. Parts I & II. *J. Ecol.,* **4,** 83–117.

Salisbury, E. J. (1920). The significance of the calcicolous habit. *J. Ecol.,* **8,** 202–15.

Salisbury, E. J. (1925). Note on the edaphic succession in some dune soils with special reference to the time factor. *J. Ecol.*, **13**, 322–8.

Salisbury, E. J. (1952). *Downs and Dunes*, Bell, London.

Salisbury, E. J. (1961). *Weeds and Aliens*, Collins, London.

Salter, P. J. and J. P. Williams (1967). The influence of texture on the moisture characteristics of soils. *J. Soil Sci.*, **18**, 174–81.

Satchell, J. E. (1967). Lumbricidae. In *Soil Biology*, pp. 259–322, Academic Press, London.

Scharpenseel, H. W. (1971). Special methods of chromatographic and radiometric analysis. In *Soil Biochemistry*, Vol. 2, pp. 96–128, Dekker, New York.

Schimper, A. F. D. (1898). *Pflanzen geographie auf physiologische grundlage*. Fischer, Jena. (English translation, 1903, Clarendon Press, Oxford.)

Schneider, G. W. and N. F. Childers (1941). Influence of soil moisture on photosynthesis, respiration and transpiration of apple leaves. *Pl. Physiol., Lancaster*, **16**, 565–83.

Schnitzer, M. (1971). Characterization of humic constituents by spectroscopy. In *Soil Biochemistry*, Vol. 2, pp. 60–95, Dekker, New York.

Schofield, R. K. (1935). The pF of water in soil. *Trans. 3rd Int. Congr. Soil Sci.*, **2**, 37.

Scholander, P. F., L. van Dam, and S. I. Scholander (1955). Gas exchange in the roots of mangroves. *Am. J. Bot.*, **42**, 92–8.

Schrenk, H. (1889). Ueber das aerenchym ein dem kork homologes gewebe bei sumpfpflanzen. *Jahrb. Wiss. Bot.*, **20**, 526–74.

Scott, A. D. and D. D. Evans (1955). Dissolved oxygen in saturated soil. *Soil Sci. Soc. Amer. Proc.*, **19**, 7–16.

Scott, E. G. (1960). Effect of supra-optimal boron levels on respiration and carbohydrate metabolism of *Helianthus annuus*. *Pl. Physiol. Lancaster*, **35**, 653–61.

Sculthorpe, C. D. (1967). *The Biology of Aquatic Vascular Plants*, Arnold, London.

Sears, P. B. (1964). Ecology—a subversive science. *BioScience*, **14**, 11.

Sestak, Z., J. Catsky and P. G. Jarvis (1971). *Plant Photosynthetic Production: Manual of Methods*, Junk, The Hague.

Sheikh, K. H. (1970). The responses of *Molinia caerulea* and *Erica tetralix* to soil aeration and related factors. II. Effects of different gas concentrations on growth in solution culture and general conclusions. *J. Ecol.*, **58**, 141–54.

Shields, L. M. (1950). Leaf xeromorphy as related to physiological and structural influences. *Bot. Rev.*, **16**, 399.

Skujins, J. J. (1967). Enzymes in soil. In *Soil Biochemistry*, Vol. 1, pp. 371–414, Dekker, New York.

Slatyer, R. O. (1957). Significance of the permanent wilting percentage in studies of plant and soil water relations. *Bot. Rev.*, **23**, 585–636.

Slatyer, R. O. (1961). Effects of several osmotic substrates on the water relations of tomato plants. *Aust. J. Bio. Sci.*, **14**, 519–40.

Slatyer, R. O. (1961). Methodology of a water balance study concluded on a desert woodland (*Acacia aneura* F. Muell.) community in central Australia. In *Plant-Water Relationships in Arid and Semiarid Conditions*, pp. 15–26, UNESCO, Paris.

Slayer, R. O. (1967). *Plant-Water Relationships*, Academic Press, New York.

Slatyer, R. O. and H. D. Barrs (1965). Modifications to the relative turgidity technique with notes on its significance as an index of the internal water status of leaves. In *Methodology of Plant Ecophysiology*, pp. 331–41, UNESCO, Paris.

Slatyer, R. O. and S. A. Taylor (1960). Terminology in soil-plant water relations. *Nature, Lond.,* **187,** 922.

Smith, K. A. and S. W. F. Restall (1971). The occurrence of ethylene in anaerobic soil. *J. Soil Sci.,* **22,** 430–43.

Smith, K. A. and R. Scott-Russell (1969). Occurrence of ethylene and its significance in anaerobic soil. *Nature, Lond.,* **222,** 769–71.

Smith, R. A. H. and A. D. Bradshaw (1970). The reclamation of toxic metalliferous wastes. *Nature, Lond.,* **227,** 376–7.

Snaydon, R. W. and A. D. Bradshaw (1961). Differential responses to calcium within the species *Festuca ovina* L. *New Phytol.,* **60,** 219–34.

Snaydon, R. W. (1962). Micro-distribution of *Trifolium repens* and its relation to soil factors. *J. Ecol.,* **50,** 133–43.

Snaydon, R. W. (1963). Final report on investigations of physiological adaptations in ecotypes of *Trifolium repens, Festuca ovina* and other herbage species. Cyclostyled rept., Dept. Agric. Bot. Univ. of Reading.

Snedecor, G. W. (1956). *Statistical Methods,* 5th Edn. Iowa State Coll. Press, Ames.

Soil Survey Staff (1960). *Soil Classification: a Comprehensive System. 7th Approximation,* US Dept. Agric.

Southwood, T. R. E. (1966). *Ecological Methods with Particular Reference to the Study of Insects,* Methuen, London.

Specht, R. L. (1957). Dark Island Heath (Ninety Mile Plain, South Australia). IV. Soil moisture patterns produced by rainfall interception and stemflow. *Aust. J. Bot.,* **5,** 137–50.

Spurr, S. H. (1969). The natural resource ecosystem. In *The Ecosystem Concept in Natural Resource Management,* pp. 3–7, Academic Press, New York.

Stace, H. C. T. (1956). Chemical characteristics of terra rossas and rendzinas of southern Australia. *J. Soil Sci.,* **7,** 280–93.

Stanhill, G. (1957). The effect of differences in soil moisture status on plant growth. A review and analysis of soil moisture regime experiments. *Soil Sci.,* **84,** 205–14.

Stanhill, G. (1965). The concept of potential evapo-transpiration in arid zone agriculture. In *Methodology of Plant Ecophysiology,* pp. 109–17, UNESCO, Paris.

Starkey, R. L. (1966). Oxidation and reduction of sulphur compounds in soils. *Soil Sci.,* **101,** 297–306.

Stevenson, F. J. (1967). Organic acids in soil. In *Soil Biochemistry,* vol. 1, pp. 119–46, Dekker, New York.

Steward, F. C. (1963). *Plant Physiology,* Vol. 3, Academic Press, New York.

Stewart, W. D. P. (1966). *Nitrogen Fixation in Plants,* University of London.

Stocker, O. (1929). Eine feldmethode zur bestimmung der momentanen transpirations und evaporationsgroesse. *Ber. dt. bot. Ges.,* **47,** 126–36.

Stocker, O. (1960). Physiological and morphological changes in plants due to water deficiency. In *Plant Water Relationships in Arid and Semiarid Conditions* (Review of research), pp. 63–104, UNESCO, Paris.

Sutton, C. D. and D. Gunary (1969). Phosphate equilibria in soil. In *Ecological Aspects of the Mineral Nutrition of Plants,* pp. 127–34, Blackwell, Oxford.

Szeicz, G. (1968). Measurement of radiant energy. In *The Measurement of Environmental Factors in Terrestrial Ecology,* pp. 109–30, Blackwell, Oxford.

Takahashi, J. (1960a). Review of investigations on physiological diseases of rice. Part 1. *Inter. Rice Commis. N. L.,* **9,** 1–6.

334

Takahashi, J. (1960b). Review of investigations on physiological diseases of rice. *Inter. Rice. Commis. N. L.,* **9,** 17–24.

Takai, Y. and T. Kamura (1966). The mechanism of reduction in waterlogged paddy soil. *Folia Microbiol. (Prague),* **11,** 304–13.

Takijima, T. (1963). Studies on the behaviour of the growth inhibiting substances in paddy soils with special reference to the occurrence of root damage in paddy fields. *Bull. Natl. Inst. agr. Sci. Japan Ser. B.,* **13,** 117–252.

Tanaka, A., R. P. Mulleriyawa and T. Yasu (1968). Possibility of hydrogen sulphide induced toxicity of the rice plant. *Soil Sci. Plant Nutr.,* **14,** 1–6.

Tanner, C. B. (1968). Evaporation of water from soil and plants. In *Water Deficits and Plant Growth,* Vol. 1, pp. 73–106, Academic Press, New York.

Tansley, A. G. (1911). *Types of British Vegetation,* Cambridge University Press, Cambridge.

Tansley, A. G. (1917). On competition between *Galium saxatile* L. (*G. hercynicum* Weig.) and *G. sylvestre* Poll. (*G. asperum* Schreb.) on different types of soil. *J. Ecol.,* **7,** 173–9.

Tansley, A. G. (1935). The use and abuse of vegetational concepts and terms. *Ecology,* **16,** 284–307.

Tansley, A. G. (1939). *The British Isles and their Vegetation,* Cambridge University Press, Cambridge.

Tansley, A. G. and M. M. Rankin (1911). The plant formations of calcareous soils. B. The sub-formation of the chalk. In *Types of British Vegetation,* pp. 161–86, Cambridge University Press, Cambridge.

Taylor, H. M. and L. F. Ratliff (1969). Root elongation rates of cotton and peanuts as a function of soil strength and water content. *Soil Sci.,* **108,** 113–19.

Taylor, S. A. (1968). Terminology in plant and soil water relations. In *Water Deficits and Plant Growth,* Vol. 1, pp. 49–72, Academic Press, New York.

Taylor, S. A. and R. O. Slatyer (1960). Water–soil–plant relations terminology. *Trans 7th Intern. Congr. Soil Sci.,* **1,** 395.

Teal, J. M. and J. W. Kanwisher (1966). Gas transport in the marsh grass *Spartina alterniflora. J. exp. Bot.,* **17,** 355–61.

Thomas, A. S. (1960). Changes in vegetation since the advent of myxomatosis. *J. Ecol.,* **48,** 287–306.

Thornthwaite, C. W. (1948). An approach toward a rational classification of climate. *Geog. Rev.,* **38,** 85–94.

Thornthwaite, C. W. (1954). A re-examination of the concept and measurement of potential transpiration. In *Publication in Climatology,* **7,** Seabrook, N. Jersey.

Thorp, J. and G. D. Smith (1949). Higher categories of soil classification: Order, Suborder and Great Soil Groups. *Soil Sci.,* **67,** 117–26.

Thurston, J. M. (1969). The effect of liming and fertilization on the botanical composition of permanent grassland and on the yield of hay. In *Ecological Aspects of the Mineral Nutrition of Plants,* pp. 3–10, Blackwell, Oxford.

Thut, H. F. and W. E. Loomis (1944). Relation of light to the growth of plants. *Pl. Physiol. Lancaster,* **19,** 117–30.

Todd, G. W. and D. L. Webster (1965). Effects of repeated drought periods on photosynthesis and survival of cereal seedlings. *Agron. J.,* **37,** 399–404.

Tranquillini, W. (1964). The physiology of plants at high altitudes. *Ann. Rev. Pl. Physiol.,* **15,** 345–62.

Tregunna, E. B. and J. Downton (1967). Carbon dioxide compensation in members of the Amaranthaceae and some related families. *Can. J. Bot.,* 45, 2385–7.

Troughton, A. (1972). The effect of aeration in the nutrient solution on the growth of *Lolium perenne. Plant and Soil,* 36, 93–108.

Tukey, H. B. (1969). Implications of allelopathy in agricultural plant science. *Bot. Rev.,* 35, 1–16.

Turner, F. T. and W. H. Patrick (1968). Chemical changes in waterlogged soils as a result of oxygen depletion. *Trans. 9th Int. Congr. Soil Sci.,* 4, 53–65.

Tyler, P. D. and R. M. M. Crawford (1970). The role of shikimic acid in waterlogged roots and rhizomes of *Iris pseudacorus* L. *J. exp. Bot.,* 21, 677–82.

UNESCO (1968). *Agroclimatological methods,* UNESCO, Paris.

Vamos, R. (1959). 'Brusone' disease of rice in Hungary. *Plant and Soil,* 11, 103–9.

Van Bavel, C. H. M. (1965). Composition of the soil atmosphere. In *Methods of Soil Analysis,* pp. 315–18, Am. Soc. Agron., Wisconsin.

Van Bavel, C. H. M. and R. J. Reginato (1965). Precision lysimetry for direct measurement of evaporative flux. In *Methodology of Plant Ecophysiology,* pp. 129–35, UNESCO, Paris.

Van den Honert, T. H. (1948). Water transport in plants as a catenary process. *Faraday Soc. Disc.,* 3, 146–53.

Van Dyne, G. M. (1969). *The Ecosystem Concept in Natural Resource Management,* Academic Press, New York.

Vartapetian, B. B. (1964). Polarographic study of oxygen transport in plants, *Fiziologia Rastenni,* 11, 774.

Veihmeyer, F. J. (1927). Some factors affecting the irrigation requirements of deciduous orchards. *Hilgardia,* 2, 125–288.

Veihmeyer, F. J. and A. H. Hendrickson (1927). Soil moisture conditions in relation to plant growth. *Pl. Physiol., Lancaster,* 2, 71–82.

Veihmeyer, F. J. and A. H. Hendrickson (1950). Soil moisture in relation to plant growth. *Ann. Rev. Pl. Physiol.,* 7, 285–304.

Virtanen, A. I. and J. K. Meitinen (1963). Biological nitrogen fixation. In *Plant Physiology,* Vol. 2, pp. 539–668, Academic Press, New York.

Vomocil, J. A. (1965). Porosity. In *Methods of Soil Analysis,* Vol. 1, pp. 299–34, Am. Soc. Agron., Wisconsin.

Wadleigh, C. H. and H. G. Gauch (1948). Rate of leaf elongation as affected by the intensity of the total soil moisture stress. *Pl. Physiol., Lancaster,* 23, 485–95.

Wadleigh, C. H., H. G. Gauch and O. C. Magistad (1946). Growth and rubber accumulation in guayule as conditioned by salinity and irrigation. regime. *US Dept. Agr. Tech. Bull.,* 925, 1–34.

Wadsworth, R. M. (1968). *The Measurement of Environmental Factors in Terrestrial Ecology,* Blackwell, Oxford.

Waksman, S. A. (1952). *Soil Microbiology,* Wiley, New York.

Walker, R. B. (1954). The ecology of serpentine soils. 11. Factors affecting plant growth on serpentine soils. *Ecology,* 35, 259–66.

Walker, T. W. (1965). The significance of phosphorus in pedogenesis. In *Experimental Pedology,* pp. 295–316, Butterworths, London.

Wallace, T. (1961). *The Diagnosis of Mineral Deficiencies in Plants by Visual Symptoms,* HMSO, London.

Wang, T. S. C., S. Y. Cheng and H. Tung (1967). Dynamics of soil organic acids. *Soil Sci.,* **104,** 138–44.

Wardle, P. (1959). The regeneration of *Fraxinus excelsior* in woods with a field layer of *Mercurialis perennis. J. Ecol.,* **47,** 483–97.

Warcup, J. H. (1951). The ecology of soil fungi. *Trans. Brit. Mycol. Soc.,* 34, 376–99.

Warming, E. (1891). *De psammofile vegetationer in Danmark,* Vid. Medd. Foren.

Warming, E. (1896). *Lerbuch der Oekologischen.*

Wassink, E. C. (1968). Light energy conversion in photosynthesis and growth of plants. In *Functioning of Terrestrial Ecosystems at the Primary Production Level,* pp. 53–66, UNESCO, Paris.

Watson, D. J. (1947). Comparative physiological studies on the growth of field crops. I. Variation in net assimilation rate and leaf area between species and varieties and within and between years. *Ann. Bot.,* NS11, 41–76.

Watt, A. S. (1925). On the ecology of British beechwoods with special reference to their regeneration. II. Sections II and III. The development and structure of beech communities on the Sussex Downs. *J. Ecol.,* **13,** 27–73.

Watt, K. E. F. (1966). *Systems Analysis in Ecology,* Academic Press, New York.

Weatherly, P. E. (1950). Studies in the water relations of the cotton plant. I. The field measurement of water deficits in leaves. *New Phytol.,* 49, 81–7.

Weatherly, P. E. (1951). Studies in the water relations of the cotton plant. II. Diurnal and seasonal fluctuations and environmental factors. *New Phytol.,* 50, 204–16.

Weatherly, P. E. (1963). The pathway of water movement across the root cortex and leaf mesophyll of transpiring plants. In *The Water Relations of Plants,* pp. 85–100, Blackwell, Oxford.

Weatherly, P. E. (1969). Ion movement within the plant and its integration with other physiological processes. In *Ecological Aspects of the Mineral Nutrition of Plants,* pp. 323–40, Blackwell, Oxford.

Weaver, J. E. (1926). *Root Development of Field Crops,* McGraw-Hill, New York.

Weavind, T. E. F. and J. F. Hodgeson (1971). Iron absorption by wheat roots as a function of distance from the root tip. *Plant and Soil,* 34, 697–705.

Webley, D. M., D. J. Eastwood and C. H. Gimingham (1952). Development of a soil microflora in relation to plant succession on sand dunes including the 'rhizosphere' flora associated with colonising species. *J. Ecol.,* **40,** 168–78.

Wellbank, P. J. and E. D. Williams (1968). Root growth of a barley crop estimated by sampling with portable powered soil coring equipment. *J. app. Ecol.,* 5, 477–81.

Went, F. W. (1957). *The Experimental Control of Plant Growth,* Chronica Botanica, Waltham.

Went, F. W. (1970). Plants and the chemical environment. In *Chemical Ecology,* pp. 71–82, Academic Press, New York.

West, C., G. E. Briggs and F. Kidd (1920). Methods and significant relations in a quantitative analysis of plant growth. *New Phytol.,* 19, 200–7.

Whitehead, F. H. (1965). The effect of wind on plant growth and soil moisture relations: a reply to the re-assessment by Humphries and Roberts. *New Phytol.,* 64, 319–22.

Whittaker, R. H. (1962). Classification of natural communities. *Bot. Rev.,* 28, 1–239.

Whittaker, R. H. (1970). *Communities and Ecosystems,* Collier-Macmillan, London.

Whyte, R. O. (1965). *Crop Production and Environment,* Faber, London.

Wiersum, L. K. (1957). The relationship of size and structural rigidity of pores to their penetration by roots. *Plant and Soil,* **9**, 75–85.

Williams, W. T. and D. A. Barber (1961). The functional significance of aerenchyma in plants. In *Mechanisms in Biological Competition,* pp. 132–44, Cambridge University Press, London.

Williams, W. T. and J. M. Lambert (1959). Multivariate methods in plant ecology. I. Association analysis in plant communities. *J. Ecol.,* **47**, 83–101.

Williamson, R. E. (1964). The effect of root aeration on plant growth. *Proc. Soil. Sci. Soc. Am.,* **28**, 86–90.

Willis, A. J. (1963). Braunton Burrows: the effects on the vegetation of the addition of mineral nutrients to the dune soil. *J. Ecol.,* **51**, 353–74.

Willis, A. J. and E. W. Yemm (1961). Braunton Burrows: the mineral nutrient status of the dune soils. *J. Ecol.,* **49**, 377–90.

Willis, A. J., B. F. Folkes, J. F. Hope-Simpson and E. W. Yemm (1959a). Braunton Burrows: the dune system and its vegetation. Part I. *J. Ecol.,* **47**, 1–24.

Willis, A. J., B. F. Folkes, J. F. Hope-Simpson and E. W. Yemm (1959b). Braunton Burrows: the dune system and its vegetation. Part II. *J. Ecol.,* **47**, 249–88.

Winter, A. G. (1961). New physiological and biological aspects in the interrelationships between higher plants. In *Mechanisms in Biological Competition,* pp. 229–44, Cambridge University Press, Cambridge.

Wit, C. T. de (1959). Potential photosynthesis of crop surfaces. *Neth. J. agric. Sci.,* **7**, 141–9.

Wit, C. T. de (1960). On competition. *Versl. landbouwk. Onderz. Ned.,* **66**, 8.

Wood, J. T. and D. J. Greenwood (1971). Distribution of carbon dioxide and oxygen in the gas phase of anaerobic soils. *J. Soil Sci.,* **22**, 281–8.

Woodell, S. R. J., H. A. Mooney and A. J. Hill (1969). The behaviour of *Larrea divaricata* (Creosote Bush) in response to rainfall in California. *J. Ecol.,* **57**, 37–44.

Woodhams, D. H. and T. T. Kozlowski (1954). Effects of soil moisture stress on carbohydrate development and growth in plants. *Am. J. Bot.,* **41**, 316–20.

Woolhouse, H. W. (1966). The effect of bicarbonate on the uptake of iron in four related grasses. *New Phytol.,* **65**, 372–5.

Woolhouse, H. W. (1969). Differences in the properties of the acid phosphatases of plant roots and their significance in the evolution of edaphic ecotypes. In *Ecological Aspects of the Mineral Nutrition of Plants,* pp. 357–80, Blackwell, Oxford.

Wright, T. W. (1956). Profile development in the sand dunes of Culbin Forest, Morayshire. II. Chemical properties. *J. Soil Sci.,* **7**, 33–42.

Wynne-Edwards, V. C. (1969). Self-regulating systems in populations of animals. In *The Subversive Science,* pp. 99–111, Houghton Mifflin, New York.

Yamada, N. and Y. Ota (1958). Study on the respiration of crop plants. 7. Enzymatic oxidation of ferrous iron by root of rice plant. *Proc. Crop. Sci. Soc. Japan,* **26**, 205–10.

Yong, T. W. (1967). Ecotypic variation in *Larrea divaricata. Am. J. Bot.,* **54**, 1041–44.

Youngs, E. G. (1965). Water movement in soils. In *The State and Movement of Water in Living Organisms,* pp. 89–112, Cambridge University Press, Cambridge.

Zahner, R. (1968). Water deficits and the growth of trees. In *Water Deficits and Plant Growth,* Vol. 2, pp. 191–254, Academic Press, New York.

Zahner, R. and J. R. Donelly (1967). Refining correlations of rainfall and radial growth in young red pine. *Ecology,* **48,** 525–30.

Zelitch, I. (1965). Environmental and biochemical control of stomatal movement in leaves. *Biol. Rev.,* **40,** 463–82.

Zelitch, I. (1969). Stomatal control. *Ann. Rev. Pl. Physiol.,* **20,** 329–50.

Zelitch, I. (1971). *Photosynthesis, Photorespiration and Plant Productivity,* Academic Press, New York.

Zinke, P. J. (1962). The pattern of individual trees on soil properties. *Ecology,* 43, 130–3.

Zeobell, C. E. (1958). Ecology of sulphate reducing bacteria. *Producers Monthly,* 22, 12–29.

Zohlkevich, V. N. and T. F. Koretskaya (1959). Metabolism of pumpkin roots during soil drought. *Fiziol. Rast.,* **6,** 690–5.

Index